Ecology and Evolution of Parasitism

Ecology and Evolution of Parasitism

EDITED BY

Frédéric Thomas,
Jean-François Guégan,
and
François Renaud

OXFORD

UNIVERSITY PRESS

OXFORD
UNIVERSITY PRESS

Great Clarendon Street, Oxford OX2 6DP

Oxford University Press is a department of the University of Oxford.
It furthers the University's objective of excellence in research, scholarship,
and education by publishing worldwide in

Oxford New York

Auckland Cape Town Dar es Salaam Hong Kong Karachi
Kuala Lumpur Madrid Melbourne Mexico City Nairobi
New Delhi Shanghai Taipei Toronto

With offices in

Argentina Austria Brazil Chile Czech Republic France Greece
Guatemala Hungary Italy Japan Poland Portugal Singapore
South Korea Switzerland Thailand Turkey Ukraine Vietnam

Oxford is a registered trade mark of Oxford University Press
in the UK and in certain other countries

Published in the United States
by Oxford University Press Inc., New York

English translation © Oxford University Press 2009

Translation of *Écologie et évolution des systèmes parasites* by F. THOMAS,
J.-F. GUÉGAN, F. RENAUD
Originally published in French by De Boeck & Larcier s.a. 2007, Éditions
De Boeck Université, Rue des Minimes 39, B 1000 Bruxelles

First published in English 2009

British Library Cataloguing in Publication Data

Data available

Library of Congress Cataloging in Publication Data

Data available

Typeset by Newgen Imaging Systems (P) Ltd., Chennai, India
Printed in Great Britain
on acid-free paper by
CPI Antony Rowe, Chippenham, Wiltshire

ISBN 978–0–19–953532–3 (Hbk)   978–0–19–953533–0 (Pbk)

10 9 8 7 6 5 4 3 2 1

Foreword

A book on ecology and evolutionary biology focusing on parasites and other pathogens would have been unthinkable only 20 years ago but has now become unavoidable: ecology has fundamentally changed and this is fortunate—because what is more universal than host–parasite systems? The days are over when ecology text-books haughtily ignored the impact of parasites and their evolution on the dynamics of ecological systems. As living organisms, we are all an integral part of the host–parasite systems with which this book is concerned.

Putting an end to the practice that for too long has marginalized parasites and other pathogens, Frédéric Thomas, Jean-François Guégan, and François Renaud offer us here an entirely new—or rather refocused—subject with a broad perspective. By abandoning the traditional approach and treating the ecology and evolution of host–parasite systems, Frédéric Thomas and his colleagues show us the ubiquity and subtleties of parasite biology: no ecologist, biologist, or naturalist can now ignore pathogens or parasites under the pretext of studying only bird behaviour, animal (mammals and birds) population regulation, sexual selection, human immune defences, or the functioning of ecosystems.

This book is thus a truly innovative ecology or general biology text, which should motivate students, scientists, or anyone intrigued by the wonders and mysteries of the living world and its biodiversity. Addressing the biology, ecology, and evolution of living organisms—*all* living organisms—through their host–parasite relationships is an excellent strategy to get to the foundations of the dynamics and evolution of biodiversity. Parasites and pathogens 'are part of the book of life, just as we are', and their study offers a wide-ranging introduction to the biology and evolution of living organisms.

'What are the roles of pathogens in the ecology and evolution of free-living organisms?' ask Thomas, Guégan, and Renaud. This is the thread that links the different chapters of the book and provides its consistency and educational purpose. I will give three examples, arbitrarily selected from the many possibilities provided by the original approach of this group of authors, concerning first epidemiology, second immunology, and third ethology.

Taking the perspective of parasites, even microparasites such as viruses or bacteria, does not by any means narrow the ecologist's line of vision: Jean-François Guégan, Guillaume Constantin de Magny, Patrick Durand, and François Renaud declare that it broadens the scope of epidemiology and parasitology, opening up the 'window of observation' in order to set the foundations of a truly ecological and evolutionary epidemiology. The point actually is to develop more comprehensive systematic analyses that take into account the set of forces, actors, and parameters concerned in the emergence and spread of infectious and parasitic agents. This is explained with the example of cholera and the effects of the El Niño phenomenon, and Lyme disease and the fluctuations that touch ecosystems, landscapes, and their inhabitants—rodents, ticks, deer, etc.

The evolutionary ecology of immune responses, addressed by Gabriele Sorci, Thierry Boulinier, Michel Gauthier-Clerc, and Bruno Faivre, reminds us that the immune system is critical for the survival and fitness of individuals. The authors clearly explain the emergence of immuno-ecology—a new discipline that combines the study of immunological mechanisms with the selection pressures acting on hosts and their parasites.

Finally Frank Cézilly and Marie-Jeanne Perrot-Minnot help us discover up to what point parasites can affect host behaviour—in its sexual, feeding, or social aspects since they are potent factors in selection (sexual selection or otherwise).

In conclusion, these authors offer us a text on the ecology and evolution of host–parasite relationships that attests to the tremendous development of evolutionary ecology, the variety of approaches, and possible applications, all in a very consistent manner that still allows for a diversity of points of view—an essential addition to 'traditional' ecology manuals.

Robert Barbault

Contents

7 Parasitism and biological control **107**

Éric Wajnberg and Nicolas Ris

8 Health ecology: a new tool, the macroscope **129**

*Guillaume Constantin de Magny, François Renaud, Patrick Durand, and
Jean-François Guégan*

List of Contributors

Jacques Blondel, Directeur of Research Emeritus at CNRS, CEFE-CNRS, 34293 Montpellier cedex 5, France.

Thierry Boulinier, CEFE, CNRS UMR 5175, 1919 Route de Mende, 34293 Montpellier cedex 5, France.

Frank Cézilly, Équipe Écologie Évolutive, UMR CNRS 5561 Biogéosciences, 6 Boulevard Gabriel, 21000 Dijon, France.

Patrice David, CNRS, 1919 Route de Mende, 34000 Montpellier, France.

Julie Deter, Université Montpellier 2, CNRS, ISEM, 34095 Montpellier cedex 05, France.

Patrick Durand, Génétique et Evolution des Maladies Infectieuses, UMR IRD-CNRS 2724, Montpellier, France.

Bruno Faivre, BioGéoSciences, CNRS UMR 5561, Université de Bourgogne, 6 Boulevard Gabriel 21000 Dijon, France.

Michel Gauthier-Clerc, Station Biologique de la Tour du Valat, Le Sambuc, 13200 Arles, France.

Jean-François Guégan, Génétique et Evolution des Maladies Infectieuses, UMR IRD-CNRS 2724, Montpellier, France and The French School of Public Health, The French Center on Globalization and Health, Hôtel-Dieu Central Hospital, Paris, France.

Philipp Heeb, UMR 5174, 118 Route de Narbonne, 31062 Toulouse cedex 9, France.

Pierre Joly, UMR 5023 Ecology of Fluvial Hydrosystems, Université Lyon I, 69622 Villeurbanne, France.

Camille Lebarbenchon, GEMI, 911 Avenue Agropolis, BP 64501, 34394 Montpellier cedex 5, France and Centre de Recherche de la Tour du Valat, Le Sambuc, 13200 Arles, France.

Thierry Lefèvre, GEMI, 911 Avenue Agropolis, BP 64501, 34394 Montpellier cedex 5, France.

Guillaume Constantin de Magny, Génétique et Evolution des Maladies Infectieuses, UMR IRD-CNRS 2724, Montpellier, France. Present address: University of Maryland Institute for Advanced Computer Studies, College Park, Maryland, USA.

Yannis Michalakis, GEMI, IRD, 911 Avenue Agropolis, 34 000, Montpellier, France.

Guillaume Mitta, UMR 5555 CNRS Université de Perpignan, 66860 Perpignan, France.

Serge Morand, Université Montpellier 2, CNRS, ISEM 34095 Montpellier cedex 05, France.

Catherine Moulia, UMR CNRS-UMII, Place Eugène Bataillon, Université Montpellier 2, 34095 Montpellier, France.

Marie-Jeanne Perrot-Minnot, Équipe Écologie Évolutive, UMR CNRS 5561 Biogéosciences, 6 Boulevard Gabriel, 21000 Dijon, France.

Robert Poulin, Department of Zoology, University of Otago, PO Box 56, Dunedin, New Zealand.

François Renaud, Génétique et Evolution des Maladies Infectieuses, UMR IRD-CNRS 2724, Montpellier, France.

Nicolas Ris, INRA, 1382 Route de Biot, 06560 Valbonne, France.

Gabriele Sorci, BioGéoSciences, CNRS UMR 5561, Université de Bourgogne, 6 Boulevard Gabriel 21000 Dijon, France.

Frédéric Thomas, GEMI, 911 Avenue Agropolis, BP 64501, 34394 Montpellier cedex 5, France. IRBV, Université de Montréal, 4101, rue Sherbrooke est Montréal (Québec) Canada H1X 2B2.

Éric Wajnberg, INRA 400 Route des Chappes, 06903 Sophia Antipolis Cedex, France.

Introduction

Frédéric Thomas, Jean-François Guégan, and François Renaud

What are the effects of parasites and pathogens on the ecology and evolution of free-living organisms? This is a huge question! The time when parasites were of interest only to parasitologists is definitively over. The influence of parasites on the evolution of free-living organisms is for some scientists a central research topic and for others an inescapable factor, given that all organisms are affected by parasites. While past studies have mainly focused on the direct effects of pathogens on the fecundity and survival of their hosts, current research shows a surprising number of unexpected consequences for various host traits such as behaviour, morphology, and/or physiology. As is usually the case in ecology, these phenomena can be greatly amplified by cascade effects. Thus, parasites and pathogens can strongly alter trophic chains, demography and the genetic variability of host populations, and competitive interactions between species as well as their invasive potential. All these changes can be favourable or conversely detrimental for biodiversity. The recent acquisition of this knowledge is the result of dynamic research in a continually evolving field. These investigations are being conducted along a continuum from purely fundamental questions to more applied problems, aimed, for instance, at understanding the links between the spread of epidemics and the environmental characteristics of ecosystems. Knowing that most, if not all, ecosystems on our planet are modified by the consequences of human activities, it seems essential to improve our knowledge of the interactions between human activities and parasitism.

This book was initially written for French university students, which is why the contributors are all Francophone scientists. The authors themselves have made a translation to provide a book in English summarizing our current knowledge about the influence of parasites on the ecology and evolution of free-living organisms. We present several topics which stand at the interface between the free-living and parasitic worlds, for instance host population regulation, sexual selection, life-history traits, immuno-ecology, behaviour, hybrid zones, conservation biology, biological control, and the ecology of health. Let us now rapidly summarize the contents of each chapter.

The study of host–parasite relationships is intimately associated with the understanding of host defence mechanisms and the evolution of virulence in pathogens. Study of the immune system has traditionally been approached from the viewpoint of immunologists. Recently a new discipline has emerged—studying the functioning of the immune system in an ecological context, to understand how natural selection has shaped the way organisms invest their energy in immune functions. While immunology mainly focuses on proximate factors, immuno-ecology studies the evolutionary aspects (fitness implications) of energy investment in immune functions. Although it is a recent discipline, immuno-ecology is now central to numerous studies of host–parasite relationships. In the first chapter, Bruno Faivre (lecturer at the University of Dijon), Thierry Boulinier and Gabriele Sorci (researchers at CNRS, Montpellier and Dijon, respectively), and Michel Gauthier-Clerc (researcher at the Station Biologique de la Tour du Valat), provide a global view of the concepts and main findings in immuno-ecology.

The adjustment of life-history traits (for instance growth, age at first reproduction, reproductive investment) is a way for hosts to reduce the impact of parasites on their fitness. This last mechanism is explained in detail by Yannis Michalakis (researcher at CNRS, Montpellier) in Chapter 2. For instance, accelerating development to enable earlier reproduction is a possible mechanism to decrease the detrimental impact of parasites that kill and/or castrate their hosts. These responses can be plastic and concern individuals, but they can also be evolutionary at the population level.

From the time of Charles Darwin, the study of sexual selection processes has been central to evolutionary biology. Since 1982, which saw the publication of the paper by Hamilton and Zuk entitled 'Heritable true fitness and bright birds: a role for parasites?' (*Science* **218**: 384–387), there has been huge scientific enthusiasm for the topic of 'parasitism and sexual selection'. In Chapter 3, Patrice David and Philipp Heeb (researchers at CNRS, Montpellier and Toulouse, respectively), present a review of this work which has generated in only 20 years an impressive number of hypotheses and mathematical models and a vast quantity of empirical data. Their conclusion is clear: parasites have, and through numerous processes, the capacity to alter the ways in which host species choose their sexual partners.

The study of behaviour (both animal and human) is a fascinating topic for many people, both scientists and non-scientists. Parasites can influence many of our behaviours (feeding, sexual, movement, etc.). Frank Cézilly (professor at the University of Bourgogne) and Marie-Jeanne Perrot-Minnot (lecturer at the University of Bourgogne) give us a detailed survey of these phenomena in Chapter 4.

The study of natural hybridization is central to evolutionary biology since it is closely connected to speciation processes. Parasites clearly constitute selective factors that favour isolation of taxa and the potential emergence of new species. Botanists have known for a long time that parasites are important in plant hybrid zones. Studies of animal hybrid zones are fewer, but are rich in information. In Chapter 5, Catherine Moulia (lecturer at the University of Montpellier), and Pierre Joly (professor at the University of Lyon) explore animal

hybrid zones and explain why it would be interesting to carry out more research on them.

The role of pathogens in regulating host populations (Chapter 6) has attracted the attention of a large number of scientists, including both theorists and field ecologists. Serge Morand and Julie Deter (researcher at CNRS, University of Montpellier and post-doctoral fellow at Ifremer, respectively) present this topic from the perspective of both theory and observation. For instance, we see how mathematical models can help to detect particular situations of host regulation by parasites in the field just by examining the distribution of parasite loads in the host population.

Parasites can be used to eliminate species that are dangerous to humans and/or to their activities (biological control). In Chapter 7, Eric Wajnberg and Nicolas Ris (both researchers at INRA) give us an insight into this subject at the interface between fundamental and applied research. We discover in this chapter the wonderful diversity of parasitoid insects, and how this diversity can be used to regulate populations of undesirable species: a nice illustration of how evolutionary biology can be helpful to applied science.

Understanding the complex links between environment and disease is the central topic of the emerging discipline of health ecology. In Chapter 8, Guillaume Constantin de Magny (post-doctoral fellow at the University of Maryland), Jean-François Guégan (researcher at IRD, Montpellier), Patrick Durand, and François Renaud (both researchers at CNRS, Montpellier) introduce this new research field, its challenges, and its implications for human societies. Where do pathogens come from? Why and how do they develop? What are the spatial and temporal dynamic behaviours of diseases? Why are pathogens virulent? How do current environmental changes influence disease evolution? All these questions are topics for which it has become essential to develop research aimed at understanding and predicting the dynamics of pathogens in our changing world (global changes, fragmentation of habitats, intensive agriculture, etc.).

Many recent international meetings have concluded that it is essential to study and to understand the mechanisms that govern the organization and functioning of ecosystems, and hence of biodiversity. Compared with free-living organisms, parasites and pathogens (probably because of their small size) have rarely been considered as real participants in the functioning of ecosystems. How can they be integrated into ecosystem ecology? Are parasites beneficial or detrimental for biodiversity? Should we use parasites in conservation programmes? Camille Lebarbenchon (PhD student), Robert Poulin (professor at the University of Otago, New Zealand), and Frédéric Thomas (researcher at CNRS, Montpellier) show us in Chapter 9 the diversity of mechanisms through which parasites interfere with processes that influence biodiversity.

In addition to providing a wide overview of the ecology and evolution of host–parasite relationships, a second objective of this book was to stimulate readers' curiosity and invite reflection. For this, each chapter provides suggestions for further reading and also questions for discussion. Finally, there is an Appendix that gives a broad explanation of the experimental methods that are mentioned in the chapters (e.g. RT-PCR, ELISA, etc.). We hope that this book will be stimulating, and will encourage students and scientists to pursue research on the ecology and evolution of host–parasite relationships.

Montpellier, 20 March 2008

CHAPTER 1

The evolutionary ecology of the immune response

Gabriele Sorci, Thierry Boulinier, Michel Gauthier-Clerc, and Bruno Faivre

1.1 Introduction

The immune system is without doubt one of the most important physiological functions, crucial for individual survival. Our bodies witness a daily fight between invading micro-organisms and immune defences. Given the importance of the immune system for human and animal health, immunology has a central place within the life sciences, as witnessed by the number of Nobel prizes won by eminent immunologists.

Immunology aims to study the effectors of immunity and to understand the molecular and cellular mechanisms underlying the complex cascades and interactions between effectors. However, even though study of the mechanisms is necessary, understanding the outcome of the interaction between hosts and pathogens cannot be achieved unless we acknowledge the evolutionary forces that have shaped host defences as we see them now. Ecological immunology is a novel discipline born during the early 1990s, with the idea of merging the study of proximate factors (immune effectors) within a larger context of selection pressures acting on such effectors. Ecological immunology therefore aims to understand how **natural selection** has shaped the investment in immune defences. Since the publication of a couple of seminal papers (Folstad and Karter 1992; Sheldon and Verhulst 1996), the number of articles related to ecological immunology has increased monotonically since 2000. One of the most robust and general results of these hundreds of papers is that immune

functioning is at the heart of the ecology of many different species. Because immune defences are costly, investment in immune functioning can only be achieved at the expense of other functions such as somatic growth, reproduction, or the expression of sexually selected signals. Not surprisingly, therefore, investment in immune defences has been considered in the same way as any other life-history trait: its evolution and phenotypic expression follow the rule of maximizing **fitness**.

In spite of the importance of immune defences for individual survival and the strong selection pressures exerted by parasites on host defences, it is common to observe a high degree of diversity both for genes that govern the immune response (i.e. the major histocompatibility complex, MHC) as well as the pattern of allocation to immune defences. In this chapter we will mostly focus on how such variability is generated and maintained. We will then see how the pattern of allocation to immunity may have subtle effects on the evolution of sexual signalling and mate choice. Finally, we will conclude with the suggestion that investment in immune defences can have profound consequences for the **epidemiology** of infectious diseases in the wild.

1.2 Sources of variability

Within host populations, the variability of various traits among individuals is a key issue for the ecology and evolution of host–parasite interactions (Frank 2002). In particular, variability of the immune system can explain part of the variability in **parasite load**s among host individuals, but also part of parasite **virulence**. This will have direct consequences for the dynamics of the interaction by affecting the rates of **morbidity** or mortality within host populations, but also by affecting the evolution of species in interactions, e.g. by affecting the fitness of various parasite strains.

Because of its biomedical implications, an enormous amount of information is available about immunity in humans, domesticated species and animal models (rats, mice, chickens). A large part of this knowledge deals with the fine molecular mechanisms involved in immune responses, but relatively little is known about variability in the immune response, especially in natural populations. Study of the physiological and molecular mechanisms of immunity has nevertheless often relied on information obtained from individuals that expressed some particular variation at the level of the cells, genes, or molecules involved. The fact that there is a strong variability among individuals is thus clear, even if its causes and consequences are rarely fully addressed. It should nevertheless be noted that the variability of the considered traits is then rarely natural, having often been subjected to particular selection regimes in the case of domesticated species (e.g. Parmentier *et al.* 2001), or to particular environmental conditions, like the living conditions of human populations in industrialized societies or the laboratory housing conditions of animal models. After discussing the question of how to quantify among-individual variability in immunity, we will see that the individual genetic, environment, maternal effects, and **genotype**–environment interactions all contribute in a complex way to this variability, and that these effects are at the heart of the questions considered in ecological immunology.

1.2.1 Measuring variability

The variability of immunity can be addressed at different levels due to the complexity of the immune system and the various forms of response to parasites. Its quantification has been debated because a unique measure of individual 'immunocompetence' has often been used in ecological immunology

studies, despite the fact that different components of the immune system can vary independently or can be negatively correlated (Sheldon and Verhulst 1996, Norris and Evans 2000; Schmid-Hempel 2003; Viney *et al.* 2005), and that this variability is due to various factors (see below). The most problematic measures are those that can be affected by the past exposure of individuals to parasites, like the parasite load of individuals or their white blood cell count. If an individual has very few parasites is it because its immune system is performing well or is it because it has had very little exposure to the parasites that are quantified? If an individual has very few circulating white blood cells, is it because it has succeeded in controlling parasitic infections or is it because, to the contrary, its immune system is deficient? The same type of question arises when the concentration of serum **antibodies** is quantified. An approach that alleviates these problems is to experimentally control the exposure of individuals by using an element mimicking exposure to a parasite towards which the immune system will be able to react. Another problem is to figure out what actually constitutes a performing immune system: is it a system that responds strongly following a stimulation, or alternatively is it a system that does not respond to the maximum of its capacity, which can limit the negative effects of too strong a response (Viney *et al.* 2005).

From a practical point of view, experimental protocols of controlled exposure to a known **antigen** are often used in studies assessing the variability in the humoral response (antibody production). The numerous studies conducted with birds have used sheep red blood cells or vaccines (Staszewski and Boulinier 2004). It should be noted that the type of antigen used, and the way it is administrated (e.g. subcutaneously or intramuscularly; with or without an adjuvant) can affect the dynamics of the immune response. To assess non-specific cellular immunity (often referred to as cell-mediated immunity), subcutaneous injection of phytohaemaglutinin (PHA) followed by measurement of the local increase in skin thickness after 24 h has been largely used (Tella *et al.* 2002), although clarification of the specificity of this approach has shown that interpretation of the measure needs to consider that cells other than T-cells are attracted to the injection site (Martin *et al.* 2006). The quantification of the innate humoral response (which is associated with natural antibodies, i.e. those not produced as a response to a specific stimulation) is possible from a sample of plasma (Mauck *et al.* 2005). Approaches enabling the quantification of more refined components of the immune system are rarely used in studies conducted in an ecological context. The numerous studies being undertaken on certain emerging zoonotic diseases, like Lyme disease, West Nile virus or avian influenza, open up the possibility of applying more refined immunological approaches to the study of the ecology of host–parasite interactions in natural populations. In general, cautious interpretation of the biological significance of the different measures of immunocompetence is called for (Schmid-Hempel 2003), so such measures are needed in order to be able to consider the factors affecting this variability, and their consequences.

1.2.2 Genetic variability

What contribution does genetics make to variability in immunity? Relatively few studies have addressed the genetic variability of the immune response in natural populations (Sorci *et al.* 1997a). The bulk of our knowledge thus comes from classical immunological work, but also from selection programmes conducted with domesticated animals (Parmentier *et al.* 2001).

The MHC is identified as a major source of variability in the immune system. The MHC is a complex of genes which codes for a family of cell surface glycoproteins. There are two types of MHC molecules: class I, which are found on virtually all cells with a nucleolus, and class II, which are expressed on **macrophages**, B cells and a few other cells. The MHC is the most variable genetic system in vertebrates. It allows each individual to have a relatively unique response following exposure to an antigen. Indeed, when a macrophage ingests a foreign protein by phagocytosis, the protein is partially degraded into peptides which are then presented at the surface of the macrophage by MHC molecules. If the peptide adjusts well to the MHC molecule, this presentation at the surface of the macrophage can stimulate the response of T cells from the immune

system. Studies have shown that variability in the MHC and other genes known to intervene in the immune response can be associated with resistance against infectious diseases, but knowledge about this topic for natural populations is still developing (e.g. Coltman *et al.* 2001a; Bonneaud *et al.* 2006; Loiseau *et al.* 2008). If some alleles at the MHC are, for instance, positively associated with resistance towards some micro-organisms, other parts of the genome also contribute to resistance against parasites (Frank 2002).

In natural populations, there is little evidence on the **heritability** of resistance to parasites (Sorci *et al.* 1997a), despite its theoretical importance for work on host–parasite co-evolution. In addition to studies conducted with invertebrates (e.g. Grosholtz 1994), a classical study used a design involving cross-fostering of nestlings between barn swallow (*Hirundo rustica*) broods and experimental infestation to estimate the contribution of the father to the correlation between the parasite load of nestlings parasitized by the mite *Ornithonyssus bursa* (Møller 1990a). In another colonial bird (*Rissa tridactyla*), the positive relationship between the number of ticks *Ixodes uriae* found on nestlings of one generation and on nestlings from the next generation (measured on nestlings reared several years after) also suggested a heritable genetic variability in parasite resistance (Boulinier *et al.* 1997). Similar results have also been reported for a system involving endoparasites of mammals (Coltman *et al.* 2001b). From studies that also focused on the heritability of components of the immune system (e.g. Brinkhof *et al.* 1999; Råberg *et al.* 2003; Soler *et al.* 2003b), it is interesting to note that there is still little information on the relationship between heritability in susceptibility to parasites and in components of the immune system (Hõrak *et al.* 2006).

1.2.3 Environmental variability

Environmental conditions can affect the development of the immune system. In addition to quantitative energetic needs associated with the setting up of the immune system, its maintenance and the development of specific responses, qualitative needs in terms of nutrients also make the immune system dependent on the environment in which individuals live (Chandra and Newberne 1977; Gershwin *et al.* 1985; Lochmiller *et al.* 1993). The question of the relative costs of maintaining versus activating the immune system has been debated (Klasing 1998), but it is clear that strong nutritional constraints can negatively affect the capacity of individuals to respond to parasite attacks. Some nutrients, such as carotenoids, can directly affect the immunity of individuals (Chew and Park 2004). Living conditions, and in particular the conditions under which young individuals are reared, can affect different aspects of the immune system, even if not all studies report evidence of that (Råberg *et al.* 2003). In natural populations of the common magpie (*Pica pica*) or the kittiwake it has been shown that the immune system of nestlings coming from replacement broods, and thus laid later in the season when food is supposedly less available, was less developed than that of nestling from first broods (Sorci *et al.* 1997b; Gasparini *et al.* 2006a). Moreover, it is clear that a large contribution to the quality of the rearing environment is due to parental factors (e.g. maternal effects; see below).

Stressful conditions (Apanius 1998) and the production of steroid hormones involved in the stress response and in **sexual selection** processes (see below) are other potentially important environmental factors that can affect the immune system. These factors can vary seasonally and be affected by **trade-off**s (see below and Martin *et al.* 2008). The activity of the immune system will also vary with the age of an individual (Cichon *et al.* 2003).

Finally, the spatial and temporal patterns of exposure to parasites will also directly or indirectly affect the state of the immune system, and will thus contribute to the observed heterogeneity of the immune system among individuals (Frank 2002).

1.2.4 Maternal effects

Among environmental factors, parental effects, i.e. non-genetic effects due to one or both parents of an individual, have the potential to play a particular role in determining variability in immunity among individuals. As we have seen in the previous section, the rearing conditions of offspring can potentially affect the development of their immunity,

notably their level of nutrition and the level of stress to which they are exposed.

A particular transgenerational effect is due to the fact that molecules directly involved in the immune response can be transferred from the mother to the offspring in many species (Gasparini *et al.* 2001; Grindstaff *et al.* 2003). In many species of mammals, a large quantity of **immunoglobulin** G can be absorbed by the ingestion of **colostrum** by the offspring, early after birth. Later, the milk is rich in immunoglobulin A, which is not absorbed but will contribute to the protection of offspring against infectious agents present in the digestive track. In birds, a large quantity of immunoglobulin is transferred to the chick via the egg yolk. These immunoglobulins are degraded relatively fast (within 2 weeks of hatching) but could also help to protect chicks from parasite attacks at a time when their immune system is still not fully functional. This property is used in the poultry industry, where hens can be vaccinated to protect chicks against prevalent viruses. The transfer of antibodies to the egg is a modern alternative technique for antibody production (Schade *et al.* 2005). As the amount of antibody transferred is directly linked to the amount of immunoglobulin circulating in the mother at the time of the transfer, this could be a strong source of variability of immunity among individuals.

In natural host–parasite systems it was shown relatively recently that the quantity of specific maternal antibodies transferred via the egg yolk to the chick in a seabird species was, as expected, a function of the local level of infestation by the ectoparasitic **vector** of the bacterium responsible for the production of those specific antibodies (Gasparini *et al.* 2001, 2002). Work on this topic is developing actively (Boulinier and Staszewski 2008), and it has for instance not yet been demonstrated whether maternal antibodies actually protect offspring in the context of natural host–parasite interactions, as suggested (Heeb *et al.* 1998; Gasparini *et al.* 2001). This point, as well as the possible existence of genetic variability in the ability to maternally transmit antibodies, is the topic of on-going work, with potential insights coming from studies conducted in domesticated animals (Abdel-Moneim and Abdel-Gawad 2006). It should be noted that maternal antibodies have the potential to affect the immune response of juveniles (Gasparini *et al.* 2006b; Grindstaff *et al.* 2006; Staszewski *et al.* 2007), and thus could be at the origin of grand maternal effects, i.e. they could affect the development of offspring from the next generation. The potential importance of the repertoire of antibodies received from the previous generation is illustrated by a hypothesis proposed for the adaptive value of allo-suckling (Roulin and Heeb 1999). Under this hypothesis, offspring of some mammal species could benefit from suckling from females other than their mother because they could benefit from antibodies against a wider spectrum of pathogens. This hypothesis still needs to be explored, but underlines nicely the importance of considering the role of the various factors potentially affecting variability in immunocompetence and their evolutionary and ecological consequences.

1.2.5 Genotype–environment interactions

Interactions among the various factors discussed above can exist and play an important role. In particular it is clear that an individual's history of exposure to parasites, and interactions between the genotype of individuals and the environmental conditions in which they have been living, will be able to affect their immune response. One of the challenges of immuno-ecology will be to determine whether such interactions are important for the ecology of individuals, but also for the ecology and evolution of host–parasite interactions in ecosystems.

1.3 Costs and benefits of the immune response

Identifying the benefits of immune defences is a relatively easy task if we think of the severe consequences, in terms of mortality risk, of both innate and acquired immune deficiencies. In a natural setting it is highly improbable that an animal would survive for long with no immune defences at all. Does this mean that natural selection should favour those individuals with the most effective immune response (or that allocate the most energy to immune defences)? This is unlikely, for at least

two reasons. First, like any other physiological process, the immune response consumes metabolic resources. Maintaining, and in particular activating, the immune response requires energy and, everything else being equal, this energy will no longer be available for other functions such as somatic growth and reproduction (Norris and Evans 2000; Lochmiller and Deerenberg 2000). It is likely, therefore, that an over-investment into the immune function can generate a metabolic cost exceeding the benefits. In such a case, natural selection should set a limit to the optimal allocation of resources to immune defences. Second, an over-investment in immune defences can exacerbate the risk of autoimmune diseases. The immune system has evolved to allow discrimination between self and non-self. This property is, of course, the basis of the function of the immune system: protection from invading pathogens while avoiding attacking the structure (cells, organs) of the host. Now, it is relatively easy to envisage that this ability to discriminate does have a limit: beyond a certain level of immune 'performance', the risk of immunopathology might become unbearable. To conclude, it seems clear that the evolution of immune functioning is driven by the benefits of parasite resistance and the costs generated by a 'hyper' immunity (Viney *et al.* 2005).

1.3.1 Energetic costs and the trade-off between immunity and life-history traits

Any physiological function, including the immune system, requires energy. The metabolic cost of the immune response has been assessed in several vertebrate species (Demas *et al.* 1997; Ots *et al.* 2001; Martin *et al.* 2002; Råberg *et al.* 2002; Ksiazek *et al.* 2003; Eraud *et al.* 2005). Three main experimental approaches have been adopted to evaluate the metabolic cost of the immune response. The approach that has most often been adopted consists of measuring the variation in the basal metabolic rate (the amount of energy consumed by an organism at rest) due to the activation of the immune response (following the injection of an antigen) (Fig. 1.1). Even though some studies have failed to detect any measurable change in basal metabolic rate (Svensson *et al.* 1998), most published studies have reported

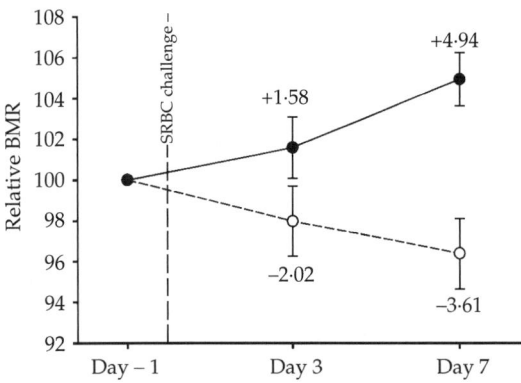

Figure 1.1 Effect of immune activation on the relative basal metabolic rate (BMR) of collared doves. Doves were either injected with a suspension of sheep red blood cells (SRBC) (full dots) or injected with saline as a control (open circles). Relative basal metabolic rate refers to the percentage change relative to day–1 (from Eraud *et al.* 2005).

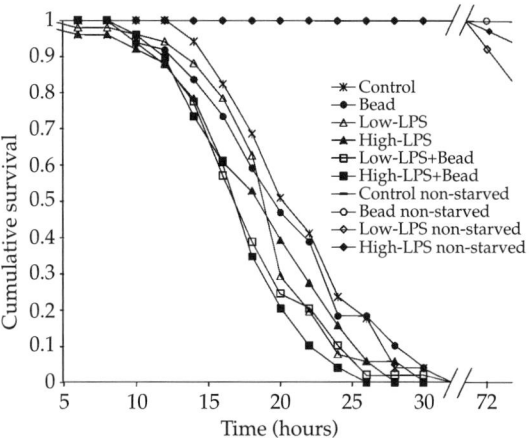

Figure 1.2 Survival of bumblebee workers exposed to different immunological insults (two lipopolysaccharide (LPS) doses and Sephadex beads). Control workers were injected with saline. Bumblebees were subsequently allowed to feed or were starved (from Moret and Schmid-Hempel 2000).

variable degrees of metabolic cost due to immune activation (Martin *et al.* 2002; Eraud *et al.* 2005). Of course, one should look at these results with a degree of caution, since the observed cost must be compared with the total energetic budget to give the real impact of immune activation. Moreover, it is still unclear how an increase of a few per cent in the basal metabolic rate affects Darwinian fitness. That said, we should nevertheless acknowledge that a real infection imposes a tougher challenge than those inflicted by an experimenter.

The second approach takes advantage of artificial selection experiments that have selected lines of mice for high and low antibody production in response to an immune challenge. Ksiazek *et al.* (2003) compared the basal metabolic rate of these selected lines. In agreement with the idea that immune defences are costly, lines selected for high antibody production had a higher basal metabolic rate than lines selected for low antibody production. Finally, a third way to assess the metabolic cost of immunity is to compare the basal metabolic rate of 'nude' mice (lacking **lymphocyte**s) with control lines of mice (Råberg *et al.* 2002). This study provided a surprising result since, contrary to the prediction, nude mice had higher basal metabolic rates than controls. This result might be interpreted as evidence for the absence of a metabolic cost of lymphocyte production; however,

an alternative explanation might be that 'nude' mice compensate for the lack of lymphocytes by over-investing into the innate immune response, thought to be more energetically demanding that the adaptive one (Råberg *et al.* 2002).

The metabolic cost of the immune response has not been measured only in vertebrates. Moret and Schmid-Hempel (2000) performed an elegant experiment in bumblebees (*Bombus terrestris*). In this study, bumblebees were injected with an antigen (lipopolysaccharides) at variable doses, either alone or in combination with Sephadex beads; a group of insects was kept as a control. Half of the bumblebees were then maintained with no food, and the other half with food. Interestingly, bumblebees maintained with food had a similar survival rate whatever the immunological treatment they received (Fig. 1.2). Conversely, when they were deprived of food, bumblebees whose immune system was experimentally activated had significantly lower survival than control individuals (Fig. 1.2). This experiment therefore suggests that the metabolic cost depends not only on the severity of the immune insult but also on the environmental conditions animals face during the challenge.

Studying the evolutionary ecology of the immune response implies that we need to explicitly address the question of how the metabolic cost of the immune defences directly shapes the expression of life-history traits, such as somatic growth and reproduction. Theory about the evolution of life-history traits tells us that these traits have evolved as to maximize Darwinian fitness (Stearns 1992). Life-history traits cannot, however, freely evolve, since several constraints limit their expression. For instance negative genetic correlations, antagonistic **pleiotropic** genes, and competition for limited resources all contribute to constrain the phenotypic expression of life-history traits. Since immune defences are thought to evolve under the same constraints as any other life-history trait, one might predict that an increased investment in immune functioning should be paid for in terms of reduction of investment in other traits. The trade-off between immunity and life-history traits has been explored in several studies in both vertebrates and invertebrates (Sheldon and Verhulst 1996; Norris and Evans 2000; Rolff and Siva-Jothy 2003). The vast majority of published studies indicate that reproductive effort and immune defences are generally negatively correlated (but see, for instance, Williams *et al.* 1999, for a counterexample). The trade-off between reproductive effort and immunity has been investigated by the means of experimental manipulation of both components of the trade-off. Some studies have manipulated the breeding effort by increasing and decreasing the number of offspring parents have to care for (Deerenberg *et al.* 1997) and looked at the consequences of such manipulated effort on immune defences; others have activated the immune response and searched for the impact of such activation on reproductive investment (Råberg *et al.* 2000); finally, a few studies have manipulated both traits simultaneously (breeding effort and immune activation) and assessed the additive and interactive effects on reproductive success (Bonneaud *et al.* 2003). This last study found a particularly interesting result that recalls those previously reported in a completely different biological system by Moret and Schmid-Hempel (2000). Female house sparrows challenged with a lipopolysaccharide injection do pay a cost in terms of reproductive success (number of fledglings produced) but only when the

breeding effort they had to put in was experimentally increased. Like the Moret and Schmid-Hempel (2000) experiment, this study therefore stresses the importance of the environment as a modulator of the trade-off between reproduction and immune functioning.

Breeding success is, of course, not the only life-history trait whose expression is traded against immune functioning. Reproduction is almost universally preceded by a phase of somatic growth which takes more or less time depending on the species. The developmental time usually determines two key life-history traits: age and size at sexual maturity (i.e. the age and size at which individuals reproduce for the first time) (Stearns 1992). Everything else being equal, natural selection should favour individuals that attain such an age at maturity with the shortest delay and largest size. Saying that somatic growth requires metabolic resources is almost trivial, since organisms need to build up a considerable biomass, sometimes within a very short period. In this context, an over-investment in immune defences during the critical developmental phase can be paid for later on in terms of size and age at maturity. The trade-off between growth and immunity has been extensively investigated both in domesticated and wild species. The existence of a negative correlation between immunity and growth is particularly worrisome for animal breeders who aim to select rapidly growing individuals with no (or at least minimal) cost. In wild species, the trade off between growth and immune defences has been explored by the means of original experimental designs. For instance, Soler *et al.* (2003a) have experimentally provided nestlings of the magpie (*Pica pica*) with methionine, an essential amino acid which stimulates the immune response without increasing the total energetic input received by the nestlings. They subsequently measured growth rate and immune protection of methionine-supplemented and control chicks. They found that whereas methionine-supplemented nestlings had a stronger immune response and were better protected against infection with avian malaria, they also paid a cost in terms of reduced growth (Fig. 1.3) (Soler *et al.* 2003a).

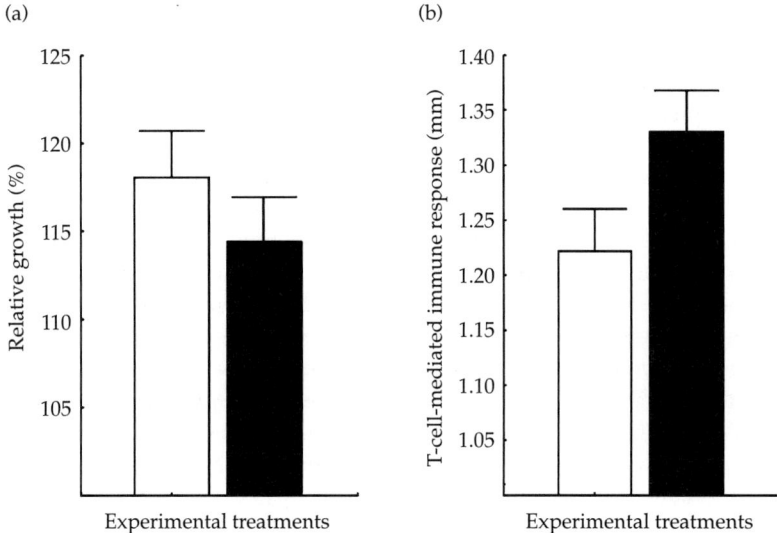

Figure 1.3 Population marginal means of (a) relative growth and (b) the T-cell-mediated immune response for magpie nestlings experimentally fed methionine (black bars) and controls (open bars) (from Soler *et al.* 2003a).

1.3.2 Immunopathology

The immune system is under the control of several regulatory mechanisms that prevent over-expression of the immune response and targeting of the structure of the host. In spite of these regulatory mechanisms, immune disorders can arise at any time during the life of an organism. The selection pressures acting on these regulatory mechanisms (selection to prevent autoimmunity) are supposed to have shaped a number of host traits, including the genetic diversity of immune genes (Nowak *et al.* 1992). As mentioned above, the MHC is one of the most polymorphic gene families of the vertebrate genome. This variability is supposed to allow the presentation of a large spectrum of antigenic peptides to T lymphocytes. There exists, however, a limit to the potential benefits of this diversity, since MHC molecules might erroneously present host peptides and start an autoimmune response. Indeed, some MHC haplotypes are associated in humans with an increased risk of developing auto-immune diseases (McDevitt 1998). Selection operating in the thymus contributes to the elimination of lymphocyte clones that bind to MHC molecules presenting self-derived peptides. In addition, theoretical and empirical work suggests that the constraint

imposed by the mis-recognition of host peptide sets a limit on the optimal MHC diversity (Nowak *et al.* 1992; Wegner *et al.* 2003; Bonneaud *et al.* 2004).

Beyond the role played by MHC genes during the processing of self and non-self peptides, immunopathology can also have a purely environmental source. The activation of the immune response by an invading pathogen implies a series of complex cascading interactions between multiple immune effectors. The intensity (strength and duration) of such cascading events also needs to be regulated to avoid the risk of undesirable side-effects. In many human infections the most severe symptoms are probably not due to the direct effect of the pathogen (pathogen replication within the host) but to the over-expression of the immune response (Graham *et al.* 2005). Fully acknowledging the costs of immunopathology can therefore provide a more reliable picture of the selective forces that shape the evolution of the immune system as well as the strategies adopted by pathogens to optimally exploit host resources.

1.4 Sexual selection

The link between ecological immunology and sexual selection has recently become one of the hottest topics in evolutionary ecology. Yet an association between immunity and the evolution of secondary sexual traits does not seem obvious at a first sight. It was explicitly hypothesized for the first time by Hamilton and Zuk (1982), who proposed that males showing the most developed secondary sexual traits are also those with the best resistance to parasites. The hypothesis states that a host's resistance to parasites is genetically determined, and assumes the following points. First, females choose their mates on the basis of the level of expression of secondary sexual traits. Second, expression of sexual ornaments is affected by parasitic infections. Males which have better resistance to parasites are more ornamented than less resistant males. Finally, females benefit from breeding with more ornamented males, because they ensure an increased fitness for their offspring who inherit resistance genes from their father. The Hamilton and Zuk hypothesis predicts a negative relationship between male ornamentation and parasite burden. Therefore, it clearly gives a central role to immune defences in the sexual selection process.

This theoretical framework, which associates immune defences with secondary sexual traits, has attracted considerable attention, and several studies have investigated the link between these two fitness-related traits. The amount of published work devoted to this topic is impressive and therefore we will focus on a few studies to illustrate the different approaches which have been adopted and the diversity of biological models.

1.4.1 The link between immune defences and sexual ornamentation

A considerable number of observational studies have investigated the relationships between immune functioning and the expression of secondary sexual traits. Some works fully included the evolutionary perspective through comparative analyses, whereas other studies generally focused on one species, assessing the correlations between the expression of secondary sexual traits and various estimates of immune functioning. These estimates of immunity are assumed to mirror the males' resistance to parasites (see Section 1.3.1). For instance, Ryder and Siva-Jothy (2000) reported a positive correlation between the quality of song (which influences female choice) and the number of immune cells in the cricket, *Acheta domestica*. Skarstein and Folstad (1996) found a negative correlation between red coloration and the number of lymphocytes in the Arctic charr (*Salvelinus alpinus*). However, correlative results remain globally contentious. First, confounding factors cannot be excluded from correlative studies. Second, discrepancies have appeared between and within studies. For instance, Gonzalez *et al.* (1999) reported both positive and negative correlations, depending on the season, between bib size and cellular immunity in male house sparrows (*Passer domesticus*), whereas in the blackbird (*Turdus merula*), male bill colour was positively correlated with cellular immune response and negatively with secondary humoral response (Faivre *et al.* 2003a). Similar results were reported for the correlation between comb size and immune responses in male red jungle fowls (*Gallus gallus*) (Zuk and Johnsen 1998). Finally, assays of immune functioning have been criticized on the ground that these measurements do not reliably reflect parasite resistance (see above and Owens and Wilson 1999; Norris and Evans 2000; Adamo 2004). Considering these technical limitations, direct exposure of hosts to parasites may appear to be a more relevant approach. For instance, Lindström and Lundström (2000) showed that male greenfinches (*Carduelis chloris*) with larger yellow patches on their tail feathers produced more antibodies after a controlled infection with the Sindbis virus. However, if this approach seems more realistic, it does not clearly demonstrate that activation of the immune system imposes a cost to the expression of secondary sexual traits (and vice versa), and then does not elucidate the basis of the trade-off between immune defences and sexual signalling.

Recent experiments on the topic have provided less equivocal answers. Briefly, two types of experiments have been conducted independently at the same time: manipulation of the expression of secondary sexual traits and manipulation of immune functioning. Using the zebra finch (*Taeniopygia guttata*), Blount et al. (2003) showed that supplementation with carotenoids enhanced bill redness in males, who became more attractive for females and mounted a higher cell-mediated immune response.

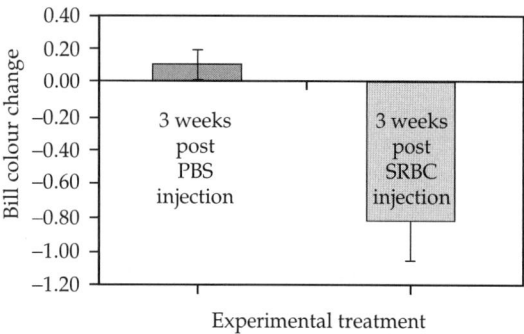

Figure 1.4 The effect of immune activation on the expression of a carotenoid-based secondary sexual trait in male blackbirds. One group of birds received an injection with a suspension of sheep red blood cells (SRBC), whereas a control group was injected with saline (PBS). Bill colour was scored immediately before and 3 weeks post-injection (from Faivre *et al.* 2003b).

Similar results have been reported for the same species by McGraw and Ardia (2003). Activating the immune system and looking at the consequences in terms of expression of a secondary sexual step was the approach followed by Faivre *et al.* (2003b). They showed that activation of the humoral immune response produced a change in bill colour in male blackbirds (immune-activated males having less orange bill than control individuals) (Fig. 1.4). In addition, the amount of colour change was correlated with the intensity of the immune response. Similar results have been reported by Alonso-Alvarez *et al.* (2004) in male zebra finches and by Peters *et al.* (2004) in male mallards (*Anas platyrhynchos*).

1.4.2 The immunocompetence handicap hypothesis

Given that both immune defences and secondary sexual traits share limited resources, it is straightforward to predict that individuals cannot maximize both traits, and that trade-offs should emerge. This leads to the question of the proximal, physiological, mechanisms at the basis of the resource allocation trade-offs. Folstad and Karter (1992) suggested, in their immunocompetence handicap hypothesis, that testosterone might play a key role. The hypothesis is based on the idea that testosterone has a positive effect on males' sexual ornaments, while

it also alters immunity. A high level of circulating testosterone enhances the expression of sexual ornaments in the males of many different species: lizards (Salvador *et al.* 1997), birds (Buchanan *et al.* 2001; Gonzalez *et al.* 2001), mammals (Ditchkoff *et al.* 2001) (see also Chapter 3). However, given that androgens can have an immunosuppressive effect, over-expression of testosterone-based sexual ornaments is paid for in terms of increased susceptibility to parasites. Therefore, only high-quality males can afford a high level of circulating testosterone (which enhances sexual ornaments) and effectively resist parasites and pathogens.

Several observational studies have attempted to provide supports for this hypothesis. In the white-tailed deer (*Odocoikleus virginianus*), antler size was positively correlated with the abundance of helminths (Ditchkoff *et al.* 2001). However, there was no correlation between circulating testosterone and parasite burden. Other indirect evidence was based on the fact that in many species males have higher circulating levels of testosterone, weaker immunity and a higher parasite burden than females (Poulin 1996; Moore and Wilson 2002; Owens 2002). However, several studies have also reported that females may be more parasitized than males (McCurdy *et al.* 1998). Overall, descriptive studies only weakly support the hypothesis suggested by Folstad and Karter (1992) because (1) some correlations remain equivocal and (2) no single study has reported the complete series of correlations needed to provide full support for the hypothesis. In addition, the immunosuppressive effect of testosterone is still a matter of debate (Ros *et al.* 1997; Hasselquist *et al.* 1999).

The best support for the immunocompetence handicap hypothesis probably comes from an artificial selection experiment (Verhulst *et al.* 1999). Chicken lines divergently selected for antibody production showed a correlated response in terms of expression of a secondary sexual trait (comb size) and levels of circulating testosterone. Lines selected for increased antibody production had smaller combs and lower testosterone levels (Verhulst *et al.* 1999). Therefore, these results clearly support the idea that testosterone could mediate the trade-off between immune defences and ornamentation.

1.4.3 Honest signalling

If it is agreed that secondary sexual traits do inform females about the immunological quality of potential partners, how is cheating prevented? A trait is usually considered to be resistant to cheating if its expression and/or maintenance are based on resources which are crucial for immune defences. The point is, then, to identify the nature of such crucial resources. Because immune defences and sexual displays require energy (cf. Section 1.3.1), honesty might be ensured by the limited amount of resources allocated to the two functions (Loyau *et al.* 2005).

Other than energy, two groups of pigments are seen as key resources for both immune defences and secondary sexual traits: melanins and carotenoids.

Melanins give a wide diversity of colours (from jet black to beige) and are crucial for the immunity of invertebrates. Indeed, the encapsulation process induces the deposition of melanin on parasites to neutralize them. Because melanism and immunity share the same physiological pathway, we might expect melanin-based secondary sexual traits to be expressed at the expense of immunity (Rolff and Siva-Jothy 2003). In agreement with this prediction, a few studies have reported correlations between immune defences and melanin-based secondary sexual traits (Siva-Jothy 2000). For instance, Rantala *et al.* (2000) found that in a calopterygid damselfly (*Calopteryx splendens*) males with larger wing spots (a melanin-based signal) had a faster rate of encapsulation of nylon implants.

Carotenoids are ingested with food and are responsible for yellow to red colours in a wide range of animal taxa. Their immuno-enhancing properties (Bendich 1989; Chew and Park 2004), added to their common occurrence as the basis of coloured sexual signals and to their presumably limited availability in the diet (Hill 1999), explain why carotenoid-based ornaments have been extensively studied in recent years (Olson and Owens 1998). Considered to be reliable signals of foraging performance in males (Endler 1980; Kodric-Brown 1985), carotenoid-based sexual traits have more recently been put forward as classical examples of honest advertisements, possibly indicating male quality in terms of immune functioning and antioxidant defences (Lozano 1994, 2001). An extensive literature has been devoted to the topic 'sexual selection–carotenoids–immunity'. For instance, female sticklebacks choose their mates on the basis of the intensity of their red coloration and by this mean, avoided more parasitized males (Milinski and Bakker 1990). Other studies have explored the trade-off between allocation of carotenoids between sexual advertisements and immune functioning (see Section 1.4.1).

Although our understanding of the role played by immune functioning in the process of sexual selection has greatly improved in the last decade, many unanswered questions remain. A deeper knowledge of the fine physiological mechanisms and genetic bases of immunity will probably help us obtain a better picture of the implications of immunity in sexual selection.

1.5 The immune response and the emergence of infectious diseases

The selective pressures exerted by parasites on the immune response influence the composition of the population of potential hosts. In the same way, the process of vaccination modifies the heterogeneity of the host population by creating two subpopulations, vaccinated and unvaccinated hosts. At the same time, the host's immune response exerts selective pressures on the parasites. This selective process usually favours parasite phenotypes that escape host immune response. Therefore the interaction between infectious agents and host immune defences is of prime importance for the establishment and dynamics of epidemics. Grassly *et al.* (2005) compared two infectious diseases and showed that fluctuations in the frequency of these infections corresponded to changes in the immunity of the hosts. Syphilis and gonorrhoea are two sexually transmitted diseases of humans. Gonorrhoea, or gonococcus, is caused by the bacterium *Neisseria gonorrhoeae*. Syphilis is caused by the spirochete bacterium *Treponema pallidum*. The **incidence** of syphilis was at a minimum in the United States in the mid-1950s. Since then, this disease has re-emerged, and undergoes periodic fluctuations. These fluctuations were formerly attributed to social and behavioural changes. The epidemic during the 1970s was explained by the sexual revolution and the liberation of homosexual behaviour, and during the 1980s by poverty, urbanization, and prostitution associated with the use of hard drugs. Syphilis and gonorrhoea are fairly similar in terms of the probability of transmission, the duration of infection, and the probability of case detection, and are treated in fairly similar ways. Yet gonorrhoea did not show these fluctuations in case numbers, even though the social and behavioural changes should have brought them about. The difference is due to the immunity developed by the hosts. When exposed to syphilis, humans acquire partial immunity. When exposed to gonorrhoea, no immunity is developed. The epidemic fluctuations are only possible because the host develops temporary immunity, or dies, removing individuals from the population of potential hosts.

The outbreak of an epidemic depends on the proportions of individuals within the population that are vulnerable, infected, or immune to the parasite. If the infectious period is short in comparison with the duration of immunity, the number of cases will fluctuate according to a periodicity determined by the stock of vulnerable individuals. Additional costs may bring about the transition from one state to the other. Energetic stress may cause a weaker immunity and an increase in the population of vulnerable individuals. Using an experimental approach, the stress of migration was simulated in a wild bird, the redwing (*Turdus iliacus*) (Gylfe *et al.* 2000). These birds may be carriers of the spirochete bacterium *Borrelia burgdorferi sensu lato*, which is transmitted by ticks of the genus *Ixodes*. In humans, this bacterium can cause Lyme disease, which is becoming increasingly common in North America and Europe. The stress due to migration caused a decrease in the birds' defences and an increase in the number of bacteria. Migratory birds may continue to be latent carriers of the pathogen for several months. The infection may be reactivated under the influence of migratory stress. In epidemiological terms, this decline in the birds' defences may result in an increased dispersal of the parasite over long distances and increased transmission during the migratory movements of wild birds.

Many factors may cause a decline in immunity within populations. Economic disparities in human populations result in very variable levels of immune protection and determine the emergence, or otherwise, of diseases. An unstable political and economic environment may result in the collapse of public health systems and consequently in the loss of immunity gained through vaccination. This situation may lead to the re-emergence of diseases. This has been the case with diphtheria in Russia and yellow fever in Africa. Improvements in medical practice have enabled human life expectancy to be extended considerably, particularly for those affected by diseases. The immune system of old or sick people, however, is less capable of fighting against parasites. Treatments for cancer and infection with HIV have also caused an increase in the number of immunosuppressed people. These changes have implications for the epidemiology of some parasites and for the emergence or re-emergence of diseases. The lowering of immune resistance resulting from the spread of HIV has led to the reappearance of a number of parasites and associated diseases, such as tuberculosis. Tuberculosis is caused by the bacterium *Mycobacterium tuberculosis*. More than 95% of cases of tuberculosis occur in poor countries. Owing to inappropriate treatments, tubercular bacilli with multiple resistance to antibiotics have appeared and in some cases are now in the majority (Pablos-Mendez *et al.* 1998). In parallel with this multiple resistance to treatments, frequent co-infection with HIV has resulted in an increase in the number of patients developing a serious, contagious tuberculosis; hence the reappearance of genuine epidemics of this disease (Murray 1998). In the campaign against the pneumococcus in humans, the stimulation of immune defences by vaccination is a reliable and effective method of preventing infection. Vaccination enables the use of direct treatment with antibiotics to be minimized. It therefore enables the frequency of strains of pneumococcus that are resistant to antibiotics to be minimized too. The traditional approach to controlling disease is to eradicate the pathogenic parasite, or at least to minimize its transmission. The elimination of forms that are only weakly pathogenic may, however, result in an increase in the circulation of more virulent forms. In this case, the ecological immunology approach may help us to understand how to improve control strategies by increasing the penalty of virulence for the parasite.

1.6 Conclusion

The aim of ecological immunology is to provide a framework for understanding variability among individuals in immune effectors and consequent parasite resistance. Although ecological immunology is still in its infancy, our knowledge of the selective forces acting on immune defences has substantially progressed in the last decade. Nevertheless, a few key points still need to be addressed. We have stressed above how important it is for evolutionary biologists to be able to measure reliable and informative immune parameters. Studying the evolutionary ecology of the immune response can only be achieved with non-model, free-living species; unfortunately, this also has a number of drawbacks in terms of the immunological tools available for such non-model species. One of the challenges of ecological immunology will be to increase the arsenal of genetic and immunological tools available for use with free-living organisms. This will allow us to specifically address a few points that are still awaiting further study. In our opinion these points are related to the immunogenetics of the immune response (not only the MHC, but also innate immunity genes) as well as better characterization of immunopathology in natural populations. Finally, we believe that much would be gained by explicitly addressing the question of how parasites modify and manipulate host immune responses to their own advantage. How parasites' strategies for avoiding/escaping host immunity affect the evolution of immune defences is still a virtually untouched field.

Important points

• Ecological immunology aims to study the impact of the immune system on the fitness of organisms in their natural context.

• Studying immune defences is obviously a fundamental step for the understanding of host–parasite interactions.

• Moreover, applying evolutionary thinking to the immune system can help us to forecast the potential impact of human activities (vaccination programmes, use of drugs, etc.) on the evolution of pathogens.

Discussion points

• What are the sources of the variability in immune defences among individuals?

• How can we explain the maintenance of a high degree of **polymorphism** in genes involved in the presentation of pathogen peptides to immune cells?

• Can human activity aimed at improving immunity (vaccination programmes) affect the evolution of parasite virulence? And how?

Further reading

• Frank, S.A. (2002). *Immunology and evolution of infectious disease.* Princeton University Press, Princeton, NJ. An excellent overview of the link between the immune system and the evolution of infectious diseases.

• Grassly, N.C., Fraser, C., Garnett, G.P. (2005). Host immunity and synchronized epidemics of syphilis across the United States. *Nature* **433**: 417–421. An example illustrating the role played by the immune system in the epidemiology of an infectious disease.

• Rolff, J., Siva-Jothy, M.T. (2003). Invertebrate ecological immunology. *Science* **301**: 472–475. Ecological immunology applied to invertebrates.

• Viney, M.E., Riley, E., Buchanan, K.L. (2005). Optimal immune responses: immunocompetence revisited. *Trends in Ecology and Evolution* **20**: 665–669. A review of the selection pressures acting on the immune system.

CHAPTER 2

Parasitism and the evolution of life-history traits

Yannis Michalakis

2.1 Introduction

It is a well-known fact that a parasitized host is not just like any other host. Alterations due to parasitism, whether behavioural or physiological, feed back on the host's life-history traits: the traits pertaining to survival and reproduction. By definition, such alterations have a central role in the ecology and evolution of the host, a good enough reason for any research project on host–parasite interactions to include at some stage characterization of the effects of the parasite on its host's life-history traits. Several chapters of this book refer to various aspects of such parasite-induced alterations. We will mention them briefly, and then concentrate on a specific point: the possibility for the host to try to minimize the cost of parasitism by modifying its life-history traits. This possibility to reduce and tolerate the effects of infection has been recognized relatively recently, and is currently a very active field of investigation. Other than a description of the phenomenon, it is important to show that such modifications of life-history traits positively affect host **fitness**. It is also important to explain why such modifications are not observed in the absence of parasites (or in low-**prevalence** populations). Finally it is important to establish whether such modifications of life-history traits have a genetic determinism or must be attributed to **phenotypic plasticity**.

2.2 Parasitism and the reactions of parasitized hosts

As indicated by the term's etymology (from the ancient Greek *parasitos*, a person who eats at

another's table), parasites invite themselves to dinner in or on their host. This 'dinner' has several negative consequences for the host: it may deprive the host of resources it may have benefited from directly (e.g. sugars or lipids), and it may damage host organs (e.g. the liver, or its functional equivalent). Obviously these two possibilities are not mutually exclusive, and result in the modification of host physiology. Such modifications inevitably affect the host's life-history traits: its developmental time, adult body size, fecundity, capacity to survive—all ecologically and evolutionarily fundamental traits. As we will see below, these traits are not independent of each other. In some cases parasites cause the modification of a trait, such as the host's reproductive system, which may result in another modification from which the parasite may benefit directly. It is thus important to ask not only who benefits from such modifications, but also who is responsible for them. Finally, modifications of the host's life-history traits may result from, for example, toxic products of the parasite's metabolism: these are 'inevitable' consequences of parasitism, but the parasite does not incur any direct benefit from them.

2.2.1 Defence reactions against parasitism

Hosts must defend themselves against virulent parasites. Several lines of defence are possible. The first is to avoid encountering the parasite. This behavioural defense is discussed in Chapter 4. However, hosts will not always manage to avoid parasites, and must thus have some ways to defend themselves when such encounters occur.

2.2.1.1 Keep the parasite out of the host

This goal may be achieved either by proposing physical barriers, such as the cuticle of arthropods or the skin of vertebrates, or by eliciting more or less specific recognition mechanisms whose goal is to stop infection before it has even started. We may cite in this category the **gene-for-gene** system governing the specificity of interactions between plants and many of their pathogens; this system lies at the basis of most plant breeding programmes aimed at increasing the resistance of plants to pathogens. In many cases, however, parasites are able to circumvent these defences and penetrate into the host, in which case the host must try to limit the damage the parasite may cause.

2.2.1.2 Impede the parasite's within host development

In this line of defence the host uses a variety of mechanisms in order to limit within-host parasite growth. The immune system of free-living organisms, in a general sense, aims to both stop an infection before it has even started and to reduce as much as possible the within-host parasite growth. The complex machinery constituting the immune system often incurs costs that generate modifications of the host's life-history traits. These issues are further developed in Chapter 1.

2.2.2 Reactions against the parasite's effects: modifications of life-history traits

Another way to reduce the negative effects of parasitism, without directly opposing the parasite's development, is to modify life-history traits. An example of such an adjustment of a life-history trait following infection would be the increase of reproductive effort under some conditions. We may intuitively understand such a modification if we consider that an infected host should invest resources in reproduction given that its future survival is compromised by the infection, or that the costs induced by parasitism will increase disproportionately following, for instance, exponential within-host growth of the parasite population. The hypothesis stating that hosts may respond to parasites through such modifications is relatively recent. It was fist proposed by Minchella (1985) and formalized for the first time by Hochberg and

collaborators (Hochberg *et al.* 1992). This mechanism will be the main subject of this chapter.

We have just briefly mentioned the cascade of defences that hosts may use to counter parasitic attacks: behavioural avoidance, stopping infection through more or less specific physical or physiological barriers, fighting against the within-host development of the parasites, and the adjustment of life-history traits.

These different defence mechanisms are not mutually exclusive, and often, if not always, a given organism may use several or even all of them. We have also seen that life-history traits may be modified as a result of several very different processes: the direct action of parasites which divert host resources for their own development; the pathological consequences of infection; the costs of maintenance and functioning of the immune system. In an adaptive life-history trait modification host fitness is higher following this modification than if the trait had remained unchanged. It is important to be able to identify which process is responsible for any modifications of life-history traits that may be observed following infection in order to be able to interpret them.

In the rest of this chapter we will first briefly discuss the most important life-history traits and their often antagonistic links. We will then provide some examples of adaptive life-history trait modifications and discuss their determinism.

2.3 Life-history traits

Living organisms acquire resources that they find or actively extract from their environment, and then allocate them to different functions: development, survival, and reproduction (Roff 1992; Stearns 1992). The latter function has several components: age of first reproduction and number and quality of the offspring, often reflected in offspring size. (Body size is often a mate choice criterion; the processes we are describing here underlie, at least partially, the mechanisms responsible for the implications of parasitism for sexual selection, described in Chapter 3.) Indeed, body size is often a good indicator of individual quality because it is highly correlated to many other life-history traits, such as juvenile survival or fecundity. Another important point is that most life-history traits change with an individual's age. For example, the reproductive effort early and late in the life of an individual can, and often should, be considered as two distinct functions.

2.3.1 Some key notions

In order to understand the evolution of life-history traits it is often useful to adopt some vocabulary and metaphors from the field of economics. This is because the problem that each organism faces is the allocation of resources to the various functions noted above. **Natural selection** will favour individuals which adopt patterns of resource allocation that maximize fitness. Thus, whenever we mention resource allocation 'strategies' we do not imply that organisms make decisions consciously, but that for different reasons they adopt resource allocation patterns which have been favoured by natural selection.

2.3.2 The problem of resource allocation

The quantity of resource from which each organism may benefit is, in the great majority of cases, limited. This limitation may derive from either an intrinsic limitation of natural resources or the presence of competitors, which may be conspecifics or parasites. Thus, the perfect organism—the 'Darwinian daemon'—with maximal survival and fecundity, reproducing very early in life and living for a long time, cannot exist. Living organisms face **trade-offs** when they allocate resources to different functions: resources allocated to survival will not be available for reproduction. Organisms which develop fast and reach adulthood early will be relatively small adults with relatively low fecundity. Females who have many offspring at their first clutch will have lower chances of surviving to their second reproductive event, and even if they manage to survive will have a smaller second clutch size than other females who will have 'saved' resources during their first reproductive event.

The various life-history traits are antagonistic: everything else being equal their values should be negatively correlated (Fig. 2.1a). It is useful to make two very important qualifications to this last sentence. The first concerns the mechanisms underlying the antagonism between functions, while the second must specify the conditions under which such negative correlations are indeed expected and observed.

2.3.2.1 Mechanisms underlying the antagonism between functions

Trade-offs between functions, or life-history traits, can be of two kinds that we must distinguish,

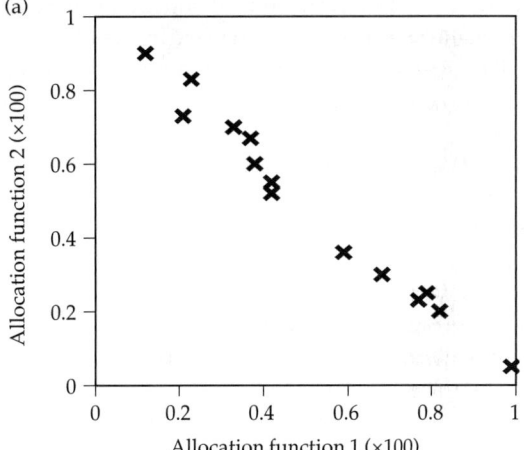

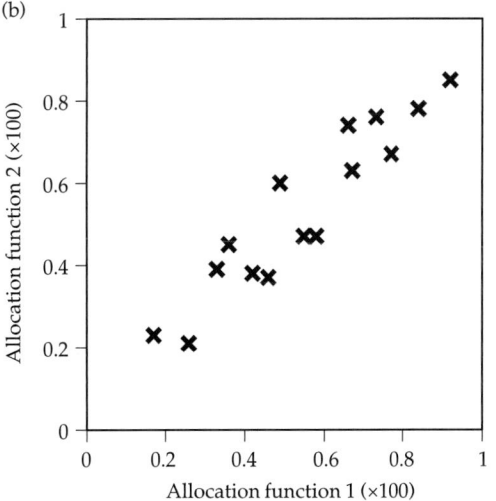

Figure 2.1 Resource allocation to two antagonistic functions. In (a) all individuals dispose of the same total quantity of resources. Antagonism between the two functions is expressed. In (b) individuals acquire different amounts of resources. The antagonism between the two functions is masked by this variability. Resource allocation to the two functions leads to a positive correlation.

because the underlying mechanisms are very different with very different implications. Such trade-offs can be physiological or micro-evolutionary. Micro-evolutionary trade-offs arise whenever selection in increasing the value of a trait causes the decrease in the value of another trait. Micro-evolutionary trade-offs often, but not necessarily, have a physiological basis. For instance, Rice and colleagues (Chippindale *et al.* 2001) showed in the fruit fly, *Drosophila melanogaster*, that genes conferring high fitness in males confer low fitness in females, and vice versa. This trade-off is never expressed at the level of the individual, because each individual is either male or female. It is an intergenerational trade-off which derives in this case from the different roles that individuals of the two genders have in reproduction.

2.3.2.2 *Conditions favouring antagonism between functions*

Physiological trade-offs correspond to cases where two traits are in competition for the same 'materials', for instance amino acids or mineral nutrients, or the same energy source. Such traits are antagonistic 'by construction'. We may note that parasites may confuse the apparent resource allocation pattern of an organism between two traits linked by a physiological trade-off by competing directly with these traits for the same resource.

2.3.2.3 *Cases where the antagonism is not observed*

Trade-offs between life-history traits may not be observed even in cases where they are expected. There are many reasons for this. An anthropomorphic metaphor may help us illustrate one of these reasons. We look at the relationship between money allocated to housing and money allocated to a car by each household. It is obvious that sums allocated to housing will not be available to buy a car, and vice versa. Thus these two budgetary lines are antagonistic because they rely on the same resource. We thus have a good reason to expect a negative correlation between the two. Nevertheless, in general, we do not see a negative correlation but rather the opposite: people who have big houses tend to also have big cars; those who have small houses tend to have small cars. The very large variance in resources available to

be allocated by the different households stops us from observing the expected trade-off: rich people have a lot of money and can buy both a big house and a big car (Fig. 2.1b). It is only among individuals with limited resources that the trade-off could be observed.

The other things that may mask trade-offs are phylogenetic constraints, the fact that some organisms may stock a given resource and thus not be limited by that resource, and **genotype**–environment interactions (Stearns 1992).

Now that we have established how different life-history traits can interact, we can see how hosts may reduce the negative effects of parasites by modifying their resource allocation patterns. The studies that we mention below are examples and not an exhaustive enumeration.

2.4 Adjustment of allocation to reproductive effort: another way to tolerate parasitism

As we have previously seen, parasitism may affect a host's life-history traits, and thus a host's resource allocation patterns, in a variety of ways. We will now concentrate on those modifications that may constitute an adaptive response by the host. Before examining these responses in detail, we must bear in mind that they can be either at the individual level or at the population level. Individual responses are those of the infected individuals themselves. They thus constitute phenotypically plastic reactions of an organism, and are an inducible defence. Such reactions can be revealed by comparison of the traits of infected individuals with those of otherwise equivalent, uninfected individuals. Responses at the population level can in principle be observed even in the absence of parasitism. They are the evolutionary responses of populations that have evolved under parasite exposure. These responses can be revealed by comparing the performances of individuals originating from populations which have evolved under strong parasite pressure, or in the absence of such pressure. An inherent problem with that kind of response lies in the difficulty of establishing causality in the between-population differences. Indeed, individuals originating from different populations may differ in their resource allocation patterns for many reasons, parasitic prevalence being one of the many environmental factors that differ among them. In practice, studies revealing this kind of response are often correlational.

In many cases the cost of parasitism increases as within-host infection progresses. Several mechanisms may be responsible for this: (1) as infection progresses pathological effects, such as the deterioration of host tissues, increase as well; (2) the parasite population within the host often increases over time in a faster than linear manner, often exponentially. This results in an accelerated increase in the costs of parasitism over time, either because of the disproportionately increasing proportion of the resources acquired by the host and diverted by the parasite for its own growth, or because of the increase in deterioration of host tissue resulting directly from the parasite's growth, or both. Whatever the mechanism involved, the important point is that the cost of parasitism increases with time. In such cases, theory of life-history evolution predicts that hosts will be selected to increase allocation towards current reproductive effort at the expense of their survival or future reproduction. It is as if a host which will die soon because of the infection reproduces while it is still able. Theoretical models, applied specifically to host–parasite interactions, confirm this prediction (Hochberg *et al.* 1992).

2.4.1 Mechanisms adjusting allocation of resources to reproductive effort

An individual may allocate more resources to its early reproductive effort in different ways. It may increase the production of offspring, which may be measured either directly, for example by counting the number of eggs laid, or indirectly by analysing an activity related to reproduction, such as the time invested in courtship, to mating itself, or in parental care. An alternative way to accelerate reproduction is to speed up development in order to decrease the age at first reproduction. Such modifications incur energetic costs which may result in either a decrease in future reproduction or longevity, or a decrease in adult body size, itself highly correlated to fecundity. The following sections provide

examples where all the above cited responses to parasitism have been observed.

2.4.1.1 *Increase in fecundity*
Increased allocation to reproductive effort through an increase in fecundity following exposure to parasites was the first observed response to parasitism involving the adjustment of a life-history trait. It was first reported in a paper on snails parasitized by trematodes and was called 'fecundity compensation' (Minchella and Loverde, 1981). These authors showed that the snails *Biomphalaria glabrata* exposed to the trematode *Schistosoma mansoni* right after reaching reproductive maturity respond by producing more eggs immediately after exposure, relative to unexposed snails of a similar age. This increase is accompanied by a cost expressed as a reduced fecundity later in their life. The timing of infection is very important in this system. In general, these parasites castrate their hosts. If the snails are infected before they have reached reproductive maturity, their reproduction is compromised, and it is impossible to observe the response consisting of a shift in the resource allocation pattern (Gérard and Théron 1997). In this latter study the authors observed a similar response in the same experimental system even on individuals which were actually infected, and not only on exposed but uninfected individuals as was the case in the experiment reported by Minchella and Loverde. Thornhill and colleagues (Thornhill *et al.* 1986), in another experiment on the same system, also observed an increase in egg production in both exposed but uninfected and in infected hosts.

Another phenomenon often observed in snail–trematode interactions, as well as in other biological systems, is gigantism (see, e.g., Arnott *et al.*, 2000, for a review on fish–cestode interactions). Indeed, several authors have observed that hosts infected by **castrating parasites** show abnormal increases in body size. We will not discuss gigantism in detail, because its interpretation is still being debated. In particular it is unclear whether the host benefits from this accelerated growth, or whether it is the host, and not the parasite, which controls this growth. Detailed discussion on this subject can be found in Arnott *et al.* (2000), Ballabeni (1995), and Minchella (1985).

Other studies on snail–trematode interactions have yielded analogous results. For instance the freshwater snail *Lymnaea stagnatilis* produces more eggs following infection by the trematode *Trichobilharzia ocellata* than do uninfected individuals (Schallig *et al.* 1991). Between-population variability in fecundity of the snails *Elimia livescens* is negatively correlated to the prevalence of their castrating trematodes (Krist 2001).

Increases in fecundity of infected individuals relative to uninfected individuals have been observed in organisms other than snails. For instance females of the cricket *Acheta domesticus* produce more eggs one day after infection by the pathogenic bacterium *Serratia marcescens* (Adamo 1999).

Other studies have documented increases in processes involved in reproduction, such as courtship or parental care, following infection. The mite *Macrocheles subbadius* parasitizes the fruit fly *Drosophila nigrospiracula* inducing a decrease in survival of infected individuals. Males of this fly increase their courtship behaviour when they are parasitized by these mites: they spend more time courting females (Polak and Starmer 1998). Moreover, they appear to be able to fine-tune their increase in courtship behaviour, since the time spent courting females was shown to increase when the number of parasitizing mites was higher.

Infected hosts can also increase parental care. The great tit *Parus major* may be parasitized by the louse *Ceratophyllus gallinae*. This louse is a nest parasite and feeds on adult or young birds. Male great tits in experimentally infested nests were shown to increase their rate of provisioning relative to males in uninfested nests (Christe *et al.* 1996a). This increase in parental care leads to higher survival of the chicks but incurs a cost to the males: males making a greater investment in nestling provisioning have a higher risk of becoming infected by avian malaria and dying in the following year (Richner *et al.* 1995; Richner 1998). Analogous results were found in the blue tit *Parus caeruleus* (Hurtrez-Boussès *et al.* 1998; Richner and Tripet 1999; Tripet and Richner 1999). Females of the lizard *Lacerta vivipara* react similarly by increasing their investment in their offspring, estimated by total clutch weight as well as the average weight of individual offspring, when they are infected by blood parasites (Sorci *et al.* 1996).

2.4.1.2 Developmental acceleration

As previously mentioned, the second way to increase reproductive effort early in life is to accelerate development in order to decrease the age at maturity. Other snail species respond to infection in this way. For example, the marine snail *Zeacumantus subcarinatus* develops faster when infected by the trematode *Maritrema novazealandensis*. In agreement with this result, its body size, which is in general positively correlated with age at maturity in snails, is negatively correlated to parasitic prevalence in natural populations of these snails: snails develop faster as the risk of infection increases (Fredensborg and Poulin 2006).

A negative relationship between risk of infection, estimated through parasitic prevalence, and age at maturity, estimated through body size at maturity, was found in other snail–trematode interactions as well. For example, in the marine snails *Cerithidea californica* and *Cerithidea mazatlanica* parasitized by several castrating trematodes (Lafferty 1993) or in the interaction between *Potamopyrgus antipodarum* and the trematodes *Microphallus* (Jokela and Lively 1995). A more recent study, however, partially contradicted these findings in this latter system (Krist and Lively 1998). This latter study showed that when snails are infected when they are juveniles they show neither fecundity compensation nor a decrease in the age at maturity. On the contrary, their individual development is slowed down: they become smaller. These contradictory results could be due to a variation of the effects of parasitism depending on the age at which hosts are infected (Krist and Lively 1998). This effect of the age at infection is analogous to that reported earlier for fecundity compensation in the *Biomphalaria–Schistosoma* system (Gérard and Théron 1997).

Developmental acceleration is not specific to snails. Cats respond similarly to infection by the feline immunodeficiency virus (FIV). Orange coat colour is associated in domestic cats with a number of behavioural and life-history traits, and when infected by this virus Orange females [i.e. individuals carrying the Orange allele; in cats there are no orange-coloured females, all orange (or orange and white) individuals are males, because this locus is sex-linked so females carrying the 'Orange' allele bear three colours, orange, dark brown and white—the 'tortoiseshell' coat] reach maturity earlier than non-infected females (Pontier *et al.* 1998). Similarly, females of the amphipod *Corophium volutator* reproduce soon after having been infected by the trematode *Gynaecotyla adunca* and before they have reached the body size which would have maximized their fecundity had they not been infected (McCurdy *et al.* 2001).

The two mechanisms leading to an increase of the investment in early reproduction have been observed in the interaction between the mosquito *Culex pipiens* and the microsporidian *Vavraia culicis*. In a first study on the relative fecundity and survival of infected and non-infected mosquito populations, Reynolds (1970) observed that infected mosquitoes produced more eggs than uninfected ones during the first week and that the pattern was reversed afterwards. Another study, using different populations of both host and parasite, revealed an acceleration of the development of infected hosts. Infected females reached the adult stage faster than uninfected females at the expense of their adult body size, which is highly correlated to fecundity (Agnew *et al.* 1999). This example illustrates the fact that different populations of any given species may respond to the same parasite in different ways. These differences may be due either to differences in the way selection acts on these populations, in the case of this example the experimental protocols, or to the way in which selection has acted on these populations before their reactions were experimentally observed. For example, some populations may show high variability for developmental time while others may show high variability for fecundity. It is also important to bear in mind that the factors maintaining such variation in life-history traits may be independent of parasitism.

2.5 Limits and perspectives

The studies discussed in the previous section require some comments and qualifications. Indeed, parasitic infection may generate a modification of the host's resource allocation pattern towards early reproduction. This change may be incompatible with an increase in parasite fitness. It is nevertheless hard to demonstrate that such modifications are indeed adaptive strategies by the host. To demonstrate this, one would need to use controls corresponding to hosts which do not modify their resource allocation pattern when infected. In practice such controls are not available, such that the adaptive nature of such modifications is inferred rather than demonstrated, alternative interpretations being less plausible.

2.5.1 Variability of life-history trait modifications

We saw that in some cases the infected, or exposed, individuals themselves respond to infection by modifying their resource allocation pattern. In these cases, the reaction is a manifestation of phenotypic plasticity. In other cases, differential resource allocation is observed in the absence of the parasite: this is the case for instance in all studies comparing the age at maturity of individuals originating from populations with different parasite prevalences. In this case, any observed differences must be attributed to factors which are transmissible from one generation to the next. The most obvious candidates are genetic factors, though very often experimental protocols do not allow the exclusion of **parental effects**. Why some species respond by phenotypic plasticity while others show genetically determined responses is a much debated issue which has received little experimental attention. The most common arguments are based on the predictability of the factor causing the response, in our case parasites. Thus, we would expect plastic responses when the presence of parasites is not predictable, while genetically determined responses would be expected in cases where the risk of infection is consistently high over time. This issue deserves further experimental attention. For instance, we could submit populations of a host species responding through phenotypic plasticity to exposure to parasites over several consecutive generations to see how their response evolves. The answer is not trivial. One possibility would be the evolution of a genetically determined response. In other words, we could observe a case of **genetic assimilation**, *sensu* Waddington (Flatt 2005). Another possibility would be that instead of trying to react to parasitism by modifying its pattern of resource allocation to life-history traits, the host could respond by investing more in its immune system. For example, it could respond by developing mechanisms allowing the infection to be stopped before it could build up within the host.

This last possibility illustrates one reason why we may not observe host responses consisting of modification of resource allocation patterns. Other reasons exist as well. Forbes (1993) and Perrin and colleagues (Perrin *et al.* 1996) discussed the effects that different kinds of parasites could have on a host's life-history traits. The key point in this comparison lies in the way parasites affect the relationship between reproductive effort, or the resources allocated by the host to its reproduction, and present reproductive success, that is the benefit the host draws by such an investment. The outcome of these interactions will of course also depend on characteristics of the parasite. Not all trematodes cause the effects we have described:

• The snail *Lymnaea elodes* does not show fecundity compensation when infected by the trematode *Echinostoma revolutum* (Sorensen and Minchella 1998). The authors of this study attributed this to the fact that in this system the infective stage of the parasite is different, and much more energy demanding, than in other systems. The high energy demand of the infectious stage would preclude the possibility of reacting through modifications of life-history traits.
• The trematode *Microphallus papillorobustus* does not cause an acceleration of the reproduction of its amphipod host *Gammarus insensibilis* (Ponton et al. 2005).
• Females of the cricket *Acheta domesticus* react to pathogenic bacterial infection, but do not modify their resource allocation pattern if they are challenged by the **parasitoid** *Ormia ochraea* (Adamo 1999).
• Variation of age at maturity and variation in fecundity of the snail *Helicosoma anceps* are not correlated with trematode prevalence (Krist 2006).

In yet other cases the parasite may take control and manipulate itself its host's life-history traits in order to increase its **transmission**. For example, the fecundity of the beetle *Tenebrio molitor*, an intermediate host of the tapeworm *Hymenolepis*

diminuta, is reduced soon after infection. This reduction in fecundity is accompanied by an increase in longevity. This increase in longevity can be understood as the result of reallocation of host resources from reproduction to longevity. This increase in longevity benefits the parasite directly, however, since it increases the probability of transmission to one of its definitive hosts, the black rat *Rattus rattus* or the brown rat *Rattus norvegicus* (Hurd 2001; Hurd *et al.* 2001).

To summarize, the host reaction consisting of an adaptive modification of the resource allocation pattern may not be observed for a number of reasons: the host may invest in a different kind of response, the population under study may not be variable for the trait under study, or the parasite may control the interaction. It is important to remember that the effects of infection and host responses may vary depending on the age of the host at the time of infection. Finally, all this is relevant only if parasitism is a significant evolutionary or ecological factor in the system under study.

The increase of allocation to early reproduction, at the expense of future reproduction, was proposed as an explanation for the evolution of senescence (Williams 1957). Indeed, pleiotropic genes increasing fitness early in life at the cost of decreases in fitness late in life would be selected for. We may well imagine that parasites acting in such a pleiotropic way would increase the fitness of their hosts, and could thus avoid the costs of an 'arms race' (Michalakis *et al.* 1992). Such parasites have never been described. One reason is that there has not been sufficient effort devoted to finding them. Another reason, the plausibility of which awaits evaluation, is that selection would favour the evolution of vertical transmission of such parasites, assuming this trait is evolutionarily labile, which in turn would generate gender-specific effects (since non-nuclear factors are only maternally transmitted). Thus, the effects of such parasites could evolve very rapidly, the state of a 'pleiotropic parasite' being evolutionarily transitory.

On this occasion we may regret the absence of a recent synthesis of the gender-specific effects of parasites, not only in relation to host response through modification of the resource allocation pattern, but more generally concerning the differential effects of parasites on individuals of the two genders. Indeed, very often the two genders are subject to different selection pressures, not only because of the different role they have in reproduction, but also because they often have different ecological niches. Theses differences result in differences in the relationship between the value of a trait and the fitness it confers to its carrier depending on whether the carrier is female or male. The resulting ontogenetic conflicts may also be expressed through host–parasite interactions. In some of the above mentioned examples host responses were indeed gender specific. Thus, only male great tits increase their provisioning rate, while only female *Culex pipiens* reach the adult stage faster when infected. This area still awaits its synthesis.

The host responses through modification of life-history traits that we have described here are not specific to responses to parasitism. They are compatible with the theory of resource allocation as a function of age-specific risks (Stearns 1992). Similar adjustments have been observed in response to, for example, predation (Crowl and Covich 1990; Reznick *et al.* 1990).

Finally, it is probably useful to specify that this kind of response is not specific to animals; the animal bias of the cited examples only reflects the bias of the author of this chapter. This kind of response is actually common in plants. Gardeners know that the best way to make a plant flower is to stress it.

2.5.2 Role of the evolution of the parasite

This book deliberately adopts a host-centred point of view. Nevertheless, we must not forget that parasites evolve, in most cases faster than their hosts. The patterns we observe in the field or in the laboratory will bear the footprints of such evolution, or perhaps even co-evolution. This aspect has not been addressed by the experimental studies we have discussed so far. Several recently published theoretical studies enable a brief discussion of its potential implications. These recent studies differ from previous theoretical work in two ways: they explicitly take into account the epidemiological dynamics of the interaction, and they explicitly consider parasite evolution (Koella 2000; Koella and Restif 2001; Restif *et al.* 2001; Gandon *et al.*

2002). The first cited study bears on the evolution of host investment in fecundity while the three other studies are concerned with the evolution of age at maturity. They all reach similar conclusions about certain important points. First, in agreement with the earlier models, hosts respond to infection by modifying their resource allocation pattern towards an increase in investment in reproduction. In these models this corresponds to an increase in reproductive effort or a decrease in age at maturity. The prediction of a positive relationship between investment in reproduction and parasitic prevalence is verified (Gandon *et al.* 2002). Second, parasites generally respond to such host shifts by increasing their **virulence**. This may be intuitively understood by considering that increases in a host's reproductive effort, or decreases in its age at maturity, decrease the period over which the parasite can exploit its host. Third, the host's response does not vary monotonically with parasite virulence (Fig. 2.2). As parasite virulence increases, the host's allocation to reproduction increases or age at maturity decreases. After a given level of virulence, however, the value of which depends on the other model parameters, the relationship is inverted: now allocation to reproduction decreases or age at maturity increases. This inflection is due to an epidemiological feedback following an increase in virulence. Indeed, very virulent parasites kill their hosts rapidly, resulting in low prevalence. As a consequence, the selection pressure imposed by the parasites on the host population is decreased, while the effect of the parasites on the infected host individuals is very important.

The other results of these recent studies are model specific. We will only discuss two here, both concerning variation of host life-history traits as a function of parasite virulence. The first bears on the last point we just discussed. Gandon and colleagues (Gandon *et al.* 2002) compared the case where hosts adopt non-conditional responses, i.e. where hosts allocate the same proportion of their resources to reproduction independently of whether they are themselves infected or not, with the case where the host response depends on whether the host individual is infected or not. They showed that infected hosts may follow a different allocation pattern from healthy hosts: allocation

(a)

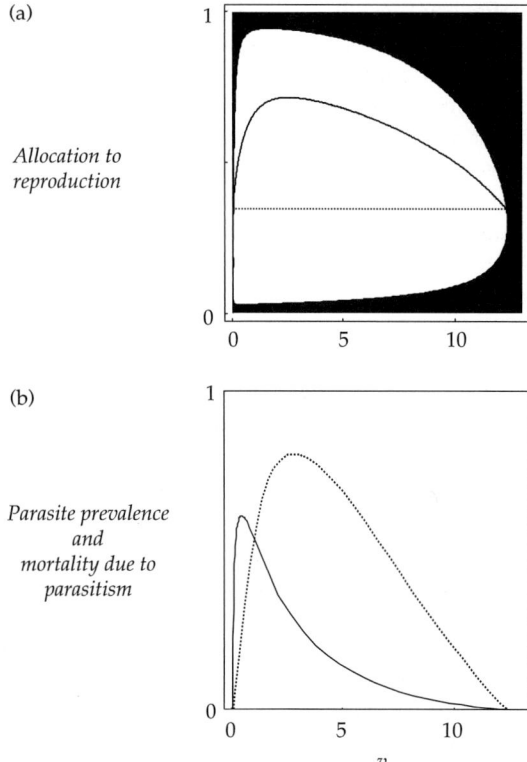

Allocation to reproduction

(b)

Parasite prevalence and mortality due to parasitism

v

Figure 2.2 Effect of parasite virulence (v) on host resource allocation to reproduction (a) and parasite prevalence (b). In (a) the continuous line shows allocation of resources to reproduction. In this case all hosts, infected or not, react similarly. The dotted line shows allocation to reproduction in the absence of parasites. In the black zone the parasite population becomes extinct. In (b) the continuous line represents parasite prevalence while the dotted line represents parasite-induced mortality (prevalence × virulence). Adapted from Gandon *et al.* (2002).

to reproductive effort by healthy hosts varies non-monotonically with parasite virulence, while allocation to reproductive effort by infected hosts always increases with parasite virulence. The mean of the host population is dominated by the allocation of healthy hosts, because as parasite virulence increases prevalence decreases.

The second point concerns the coexistence, under some conditions, and particularly when parasites can be transmitted both horizontally and vertically, of two host strategies (Restif *et al.* 2001). The first corresponds to the 'classical' strategies, these hosts having a lower age at maturity

when the parasite is present. Such hosts respond to infection by decreasing the period over which they are exposed to the parasite, at the expense of their fecundity. The second strategy follows an age at maturity close to the optimal age at maturity in the absence of parasites. Hosts adopting this second strategy accept the risk of being infected with the potential benefit of reaching maturity late, synonymous to high fecundity in this model. The coexistence of the two strategies is observed either beyond a given parasite virulence level or beyond a given host growth rate, which itself affects the evolution of parasite virulence.

The recent models, while confirming the main predictions of the early models, offer new research perspectives. They generate hypotheses on the variation of the host allocation pattern as a function of parasite virulence, of host growth rate, and of the mode and rate of parasite transmission. All these predictions await experimental investigation.

2.5.3 Role of population structure

Most of the studies mentioned in this chapter do not take into account the effects of host and/ or parasite population structure. Predictions for life-history traits were extended in this case by a theoretical study (Kirchner and Roy 1999). These authors showed that in a **metapopulation** selection could favour hosts with a brief lifetime, over the expense of long-lived hosts, because the decrease in host longevity results in a decrease in parasite prevalence. It is important to note that in this model host longevity is not subject to a resource allocation trade-off. Competing hosts differ only in their longevity. In the absence of parasites, long-lived hosts dominate and drive short-lived hosts to extinction. Parasitism displaces this equilibrium only because of its epidemiological consequences. By living longer, long-lived hosts disproportionately suffer from infection. Thus, when the effects of the parasites on host longevity, fecundity, or both traits are severe, the productivity of sites occupied by short-lived hosts is larger than that of sites occupied by long-lived hosts. This in turn allows short-lived hosts to colonize more sites. High extinction rates, at the metapopulation level, generate many empty sites, thus giving short-lived hosts a larger colonization advantage, despite the fact that within each site long-lived hosts are favoured.

The authors of this model noticed that parasites could influence host life-history traits even in the absence of resource allocation trade-offs, when the effects of infection are not shared by hosts which have different life-history trait values. In the model described above, the effects of infection are not shared between hosts with different life-history trait values because of the spatial segregation generated under the conditions of this model. The same authors showed that specificity in host–parasite recognition could generate similar effects (Kirchner and Roy 2001). In this second model, hosts differ with respect to their reproduction rate. Again, this trait is not subject to a resource allocation trade-off. In the absence of parasites the hosts with the lower reproduction rate are selected against and disappear. The same outcome is observed in the absence of any host–parasite specificity, that is whenever parasites are able to infect all hosts. As parasites become more and more specific to a given type of host, however, the dynamics of the system are modified. This time it is possible to maintain a **polymorphism** between hosts having different reproduction rates because of the frequency-dependent selection generated by the specificity of the host–parasite interaction. Hosts with a high reproduction rate increase in frequency within the population, but while doing so they generate conditions favouring the parasites, which infect them preferentially. The negative effects of these specific parasites allow hosts with a low reproduction rate to increase in frequency in their turn, which results in a new change in the composition of the parasite population, and so on until host and parasite populations reach their equilibrium. This process assumes the existence in the host of an association between parasite specificity and life-history trait values. Such an association could be generated through **pleiotropy** or **linkage disequilibrium**. The maintenance of polymorphism is possible only when the effects of parasites are severe and their specificity is high. Importantly, the evolutionary stability of such polymorphisms, that is their capacity to resist invasion by other hosts, has not been evaluated.

The two models by Kirchner and Roy that we have just discussed share some common features. First,

under some conditions they allow polymorphism in host life-history traits to be maintained in the absence of intrinsic resource allocation trade-offs. In other words, any laboratory experiments trying to show that, for example, long-lived hosts reproduce less will fail, simply because such trade-offs are unnecessary under the assumptions of these models. Second, and this is particularly true in the second model, these mechanisms allow the maintenance of polymorphisms that intuitively we would not be inclined to qualify as 'adaptive'. Increased specificity in the host–parasite interaction allows hosts with low reproductive rates to persist in the population. They are not there because they are adapted to an intrinsic environmental condition, other than that they are less infected by the parasites of the 'other' hosts. It is important to keep such points in mind when trying to interpret and understand variation of life-history traits in the field.

Finally, it is important to remember once more that life-history traits are also involved in other types of host defences to parasitism. In particular, the costs of immune responses are often expressed as life-history trait modifications (cf. Chapter 1). Interactions with the other possible reactions to parasitism must be borne in mind during the conception and interpretation of empirical investigations.

Important points

- Hosts may respond to parasitic infections by modifying their resource allocation pattern.
- Such modification consist of increasing reproductive effort or decreasing age at maturity (Minchella 1985; Hochberg *et al.* 1992; Forbes 1993; Perrin *et al.* 1996; Hurd 2001).
- Consideration of epidemiological dynamics and of parasite evolution predict a non-uniform relationship, and in some cases a polymorphism, of these life-history traits as a function of parasite virulence (Koella 2000; Koella and Restif 2001; Restif *et al.* 2001; Gandon *et al.* 2002).

Discussion points

- Why do some hosts not respond to parasitism by modifying their life-history traits?

• How can we demonstrate that such modifications are adaptive for the host?

• How can we experimentally investigate variation of life-history traits as a function or parasite virulence? Or as function of host reproduction rate?

Further reading

• For the evolution of life-history traits see: Roff, D.A. (1992). *The evolution of life histories: theory and analysis.* Chapman and Hall, New York; Stearns, S.C. (1992). *The evolution of life histories.* Oxford University Press, Oxford.

• For the co-evolution of host and parasite life-history traits see: Gandon, S., Agnew, P.A., Michalakis, Y. (2002). Coevolution between parasite virulence and host life-history traits. *The American Naturalist* **160**: 374–388; Koella, J.C. (2000). Coevolution of parasite life cycles and host life-histories. In: *Evolutionary biology of host-parasite relationships. Theory meets reality* (ed. R. Poulin, S. Morand, A. Skorping), pp. 185–200. Elsevier, London; Koella, J.C., Restif, O. (2001). Coevolution of parasite virulence and host life history. *Ecology Letters* **4**: 207–214; Restif, O., Hochberg, M.E., Koella, J.C. (2001). Virulence and age at reproduction: new insights into host-parasite coevolution. *Journal of Evolutionary Biology* **14**: 967–979.

CHAPTER 3

Parasites and sexual selection

Patrice David and Philipp Heeb

3.1 Introduction

In numerous species, such as the peacock and the red deer, males differ morphologically and behaviourally from females. They are often bigger, more brightly coloured, and have more complex songs. Charles Darwin was the first to address the paradox of **sexual dimorphism**: why do males acquire traits that render them more detectable and appear to reduce their survival? These observations seemed to contradict his theory of evolution by **natural selection** (Darwin 1859). To address this contradiction he invented the concept of sexual selection, defined as 'the advantage which certain individuals have over other individuals of the same sex and species, in exclusive relation to reproduction' (Darwin 1871, ch. VIII). Briefly, the extravagant traits of males would increase their reproductive success, thus compensating for their reduced survival. Research in this field has increased considerably in recent decades.

It might seem a strange idea to associate some of the most amazing traits in nature—the song of the nightingale, the feathers of birds of paradise, the antlers of deer—with parasitic worms, haematophagous ticks, and bacterial pathogens. However, since 1980 researchers have uncovered deep relationships between parasites and sexual selection. In this chapter we will discuss these recent discoveries and present their limitations.

Parasites can modify three parameters in their hosts: (1) their ability to pursue/defeat opponents of the same sex (intrasexual selection); (2) their capacity to attract the opposite sex (intersexual selection); (3) their choosiness with respect to sexual partners (Andersson 1994). In Section 3.2 we will discuss the theory of sexual selection. In

Section 3.3 we will describe how parasites, by their direct effect on their hosts, affect male–male competition, mate choice, and reproductive success. In Section 3.4 we will address the most controversial and theoretically challenging issue, namely the relationship between parasites and the genetic benefits (indirect benefits, since a parasite does not modify the genes of its hosts) that females obtain by choosing their mating partners.

3.2 Sexual selection: general theoretical aspects

3.2.1 An issue not only for males

Darwin (1871) discussed the relationship between sexual dimorphism and the reproductive success of males: complex characters, like the antlers of deer or peacocks' tail feathers are a priori costly or represent handicaps for males. They are maintained during evolution because they increase reproductive success, that is the number of matings (and consequently the number of offspring produced) that males can obtain. This increase occurs through two different, and non-exclusive, mechanisms: (1) some characters (like the antlers of deer or their roaring calls) serve as weapons or threats during male–male combat for the control of females; (2) other traits (like a peacock's tail) are aesthetic and serve to attract females.

The evolution of 'fighting weapons' takes place essentially among males and does not raise specific theoretical problems: sexual selection favours an 'arms race' between males that stops when the costs of the characters (the energy lost to produce them or the reduction in mobility associated with them) is greater than the reproductive benefit they provide. Nevertheless, the evolution of ornaments used only to seduce females poses greater theoretical problems. Darwin considered that female preferences pre-existed and caused the evolution of ornaments in males: male peacocks have long blue and green feathers with eyespots because females fancy these ornaments. However, this only displaces the evolutionary problem towards the female side: why should females maintain such preferences? Sometimes males do seem to exploit a pre-existing preference in females that has evolved in another context (e.g. sensory bias; Kirkpatrick and Ryan 1991). But most females are only attracted by the specific ornaments produced by the males of their own species. This specificity suggests that female behaviour and male ornaments have had a lineage-specific co-evolutionary history.

We emphasize that the asymmetry of the system (the female does the choosing, the male is chosen) rests on a classical assumption (which in most, but not all, species, is verified): the energetic investment by the female in each offspring is large in contrast to the male investment which is small. The number of offspring for each female is limited by her available energy, not the number of matings; therefore her best way to increase **fitness** is to maximize the probability of survival of her offspring. The male can potentially sire many offspring, and obtains greater benefits by trying to increase their number. To this end he needs to obtain as many matings as he can. Nevertheless, in cases where the asymmetry of investment is reversed, the logic that usually prevails for the evolution of the secondary sexual characters of males can be applied to females (Amundsen 2000). In the jacana, *Jacana jacana* (a bird living in the swamps of tropical America) females are larger, more aggressive, and more colourful than the males. They fight among themselves to defend territories and monopolize males (Emlen and Wrege 2004). Amundsen and Forsgren (2001) found that in the gobby, *Gobiusculus flavescens*, a small marine fish, males show a preference for brightly coloured females. In both examples, males take care of the young, investing a lot of time and energy. In the following sections we will describe the theory assuming that females choose and males are chosen (the classical case), but it should be kept in mind that the roles of the sexes are sometimes reversed.

3.2.2 The theory of signal evolution

The evolutionary origin of female choice has been the subject of numerous theoretical models. A common characteristic of all these models is to consider male ornaments as a signal that provides information about the males. By using this information females could increase the quality of their offspring. This hypothesis is suggested by the fact that most ornaments (for example the peacock's tail) do not have any direct positive effect on the female or her descendants. The interest of the female in the male ornament arises because it allows her to detect other male characteristics that will lead to changes in the number and/or quality of the descendants produced.

What are these characteristics? 'Direct benefits' models stipulate that male ornaments are an indicator of the quality of the conditions for raising the young: for example if a highly ornamented male provides greater parental care (defence of the young, feeding etc.), or if he is less likely to transmit diseases or parasites to the female or to the offspring. On the other hand, theories based on 'indirect benefits' or 'good genes' state that male ornaments reflect genetic characteristics that can be transmitted to the offspring and increase their survival and/or reproduction. This theory is not incompatible with direct benefits: ornaments can indicate at the same time both the conditions for raising the young and the quality of the father's genes.

In an extreme version (the first to be proposed historically) of the 'indirect benefits' hypothesis it is assumed that the genes that determine the development of the ornament do not have any other effect on the male (besides the cost of the ornament itself). Consider a theoretical population where females preferring males with blue feathers coexist with indifferent females. In this population blue males will sire more chicks than the other males since they have more access to some of the females (the choosy ones). By this simple fact a choosy female is favoured because its offspring will inherit the male ornament. This produces a runaway process because the benefit for the selective females increases with their frequency in the population. This model has been popularized under the name 'Fisher's runaway process' after the first person to have suggested it (Fisher 1958). Later mathematical analyses (for example Kirkpatrick, 1982) showed that this process allowed the evolution of arbitrary female choices and of corresponding male ornaments even if the latter were costly. Unfortunately, these models did not take into account the fact that choice is not free for the females: they dedicate time and energy to choosing a good male, sometimes rejecting some of them and thus delaying their reproduction. In the barn swallow (*Hirundo rustica*) it has been shown that the choice criterion (the length of the tail feathers) led females to prefer less faithful males and males that spend less energy on parental care (Møller 1994), thus constituting a direct cost to the female. When this type of cost, however small, is incorporated in models of Fisher's runaway process, the runaway process stops and the choosy females disappear, out-competed by indifferent females (Pomiankowski *et al.* 1991; Kirkpatrick 1996) since the benefit of choosing is never great enough to compensate for the cost. Currently, scientists agree that, under realistic conditions, the 'Fisherian' benefit, even if it exists, cannot explain by itself the evolution of preference for extreme ornaments. For the theory of 'indirect benefits' to work it is necessary to assume that genes for the ornament have other effects besides the development of the ornament. In technical terms, a **genetic correlation** is required between the ornament and other characters, like, for example, survival or one of its components (resistance to disease, physical strength, etc.). This concept is detailed in Fig. 3.1.

3.2.3 Honest signals and handicaps

As we have seen, the notion of a signal is a central concept in the process of sexual selection. The roaring of deer is used as a warning signal to inform males about the strength of their competitors. Ornaments such as the peacock's tail inform females about the material or genetic benefits that males can transmit to their offspring. A central concept of these hypotheses is the honesty of the

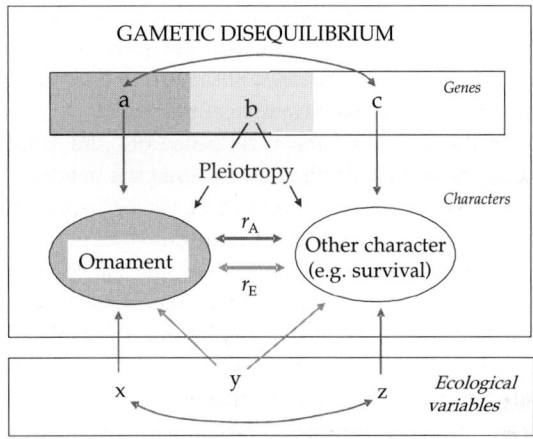

Figure 3.1 What is a genetic correlation? The correlation between two phenotypic traits (here, a sexual ornament and survival) has two components: genetic correlation (r_A) and environmental correlation (r_E). There are two reasons for the genetic correlation: (1) the fact that the same gene (here b) influences both the ornament and survival (pleiotropy); (2) the fact that genes affecting the development of the ornament and genes affecting survival (a and c) co-occur more often than expected by chance within gametes (gametic disequilibrium). Similarly, the environmental correlation can first result from one ecological variable (y) influencing the two characters; alternatively, it can also result from a correlation between two ecological variables (x and z) acting on the two characters.

signal: if male 'cheaters' can display signals of a quality they do not possess, the system cannot evolve. In the case of the deer's roar (or other signals allowing an estimation of the strength of the adversary) real fights are frequent: a weak male disguised as a strong male would pay dearly for his cheating since he will be confronted by opponents stronger than him. In contrast, females cannot directly punish males with deceptive ornaments. What is the process that warrants the honesty of signals for females?

The most influential idea proposed on this question is the 'handicap hypothesis' first proposed by Zahavi (1975). This idea arises from the fact that most signals (sexual ornaments) have a handicapping effect on the individual bearing them: they require energy to be produced or to be maintained, they render the individual more detectable by predators, and/or complicate their movements. Theory assumes that the costs of the signals guarantee their honesty: a mediocre male could produce and carry a cheap signal but would not have the resources to produce costly ornaments. This idea has gained a large currency in the scientific community, but as we will see it is not as simple as it seems.

In the simplest version of the model, the 'pure epistatic handicap', it is assumed that the handicap plays the simple role of a selective sieve (Maynard-Smith 1985). It is assumed that 'good genes' exist (those favouring survival) as well as 'bad genes'. Only males with the 'good genes' can survive with the handicap imposed by the ornament: by choosing a male with an ornament the female ensures that he is carrying good genes. A preferential association or **gametic disequilibrium** appears between the genes for handicap and the genes for survival. This association is due to a particular type of selection where the cost of the handicap is reduced when the individual is also carrying good genes for survival (e.g. **epistasy**). Theoretical models have shown that this mechanism does not work; in the long run handicapped males, with or without the 'good genes', do not compensate for the survival costs associated with the handicap by the number of partners obtained.

Today we tend to prefer more complex forms of the handicap concept that theoretically allow both the handicapping trait and female preference to be maintained in evolutionary time (Maynard-Smith 1985; Iwasa *et al*. 1991). The central idea is as follows: the same genes determine at the same time the ornament–handicap and fitness components (survival or fecundity). The genetic correlation between the handicap and these components is based on this double influence or **pleiotropy**. This correlation is often expressed by supposing the existence of a synthetic variable: the physical condition of the individual. This covers a number of variables positively correlated among themselves (health state, metabolism, muscular mass, hormonal state, nutritional status, etc.), assumed to have a positive effect on survival and fecundity, and thus on individual fitness. If the ornamental character of males is condition dependent, so that it can only develop properly in individuals in good physical condition, it will honestly reflect their genetic qualities.

It has to be noted that in the hypothesis of condition-dependence male ornamentation reflects both the genetic and the environmental variations of their condition. For example, a male who developed in a food-rich environment and protected from parasites would have a better condition and develop more beautiful ornamentation: the correlation between ornament and condition is at the same time genetic and environmental. Only the genetic part is important for promoting the evolution of female preference in the context of indirect benefits (e.g. 'good genes'). In the context of direct benefits, both components, genetic and environmental, are important since it is the phenotype (and not the genotype) of the male that determines its ability to improve the conditions under which the young are raised.

3.2.4 Which genes should one choose to make fitter offspring?

One particular hypothesis implicitly underlies the 'good genes' theories described up to now: the males' fitness genes must have an **additive effect**. This means that if these genes improve the survival of the male that carries them they will have the same effect on the offspring independently of the genes received from the mother. Recently, particular attention has been directed to the fact that some fitness components are not transmitted in an additive fashion. More precisely, it is often a specific complementarity between the maternal and the paternal genes that seems important. The simplest example is provided by **inbreeding depression**: crossing two similar genomes (from related individuals like a brother and sister) often leads to the production of descendants with low fitness that carry defects and genetic diseases. In these conditions, the female does not have any benefit in choosing exclusively males in good physical condition. Each female should benefit by choosing males as different from her as possible so as to protect her offspring from the negative consequences of inbreeding (the 'inbreeding avoidance' hypothesis; Pusey and Wolf 1996). This type of choice has been observed in a number of species (for example Palmer and Edmands 2000; Stow and Sunnucks 2004; Wheelwright *et al*. 2006; cf. Section 3.5). Signals that allow a female to recognize these males are not the same as condition-dependant ornaments: the information they carry has to be relative to the 'choosing' female, not absolute as in the 'good genes' hypothesis. The two signalling systems can coexist in the same species, but the interaction between the two remains poorly studied.

3.3 Parasites and male–male competition

In comparison with the rituals by which one sex attracts the other, the competition between individuals of the same sex shows us the 'dark side' of sexual selection. Competition among males for access to females, or territories allowing them to attract females, can involve violent, sometimes fatal, fights. Thus, many secondary sexual characteristics have certainly evolved as structures used in male–male competition (Howard and Minchella 1990). The violence of intraspecific competition can even extend to juveniles. In lions, a male that takes over a group of females kills the cubs fathered by other males, thus removing the oestrus inhibition associated with lactation. Conversely, in jacanas, females seek to destroy the clutches of other females so as to obtain a greater share of male incubating time for their own eggs (Emlen and Wrege 1989).

Given that parasites can affect the health and physical strength of their hosts, parasite loads are expected to have an immediate effect on the outcome of competition between individuals of the same sex for access to mates of the opposite sex. For example, in the lizard *Scelopus occidentalis* infested by *Plasmodium mexicanum*, parasitized males are competitively inferior to healthy males and achieve lower reproductive success (Schall and Dearing 1987). This reduction in fitness of infected individuals when facing healthy ones has also been observed in insects such as the damselfly *Enallagma ebrium* infested by the ectoparasite *Arrenurus* spp. (Forbes 1991).

Parasites thus have a direct influence on the outcome of male–male fights. They can also have an effect through their action on the development of certain organs or specific behaviours. In many species, males exchange coded signals (for example sounds like the roaring of red deer or the songs of birds) allowing them to assess the strength of their congeners. These signalling systems, if they are honest, are beneficial to all males: the coding allows two males of unequal strength to avoid fighting (the subordinate giving away the resource to the dominant). Indeed, knowing that the outcome is clear from the start, a fight would constitute a useless and risky waste of energy for both males. Real fights will take place only when the information obtained through reciprocal evaluations does not allow the individuals to clearly predict the outcome (Maynard-Smith 1982). In this context, parasites can alter the competition between sexes when they act on the expression of characters used in reciprocal evaluation.

In the North American wild turkey, *Meleagris gallopavo*, male–male competition is intense. The outcome of fights establishes a dominance hierarchy on which the access to females depends. In a series of experimental tests Buchholz (1997) showed that male dominance was positively correlated with the size of the carbuncle (a fleshy appendage that hangs from the beak), and that males used it to assess their opponents. The size of the carbuncle was negatively correlated with coccidia (unicellular parasites of the gut) loads and positively correlated with male condition. In this example, parasites affect both the capacity to win fights and the signal used in evaluation. Furthermore, in a preliminary study, Buchholz (1995) had shown that females based their mate choice on the same signal. These multiple roles of signals are frequent (e.g. see the example of bird songs below).

Parasites can also stress their hosts, affecting the development of organs used by males as weapons in their fights. Markusson and Folstad (1997) showed that the asymmetry of reindeer antlers was positively correlated with the **prevalence** of parasites.

Nevertheless, other physiological mechanisms (for example hormones) are known to affect the frequency and outcome of fights between males. Male aggressivity in many vertebrates depends on a hormonal cocktail, of which testosterone is a major component (Mougeot *et al.* 2005). Folstad and Karter (1992) suggested that the mobilization of the immune system during an infection would require a lowering of testosterone titres since this hormone has an immunosuppressive action. This immunosuppression was demonstrated in the Scottish grouse, *Lagopus lagopus scoticus* (Seivwright *et al.* 2005): an artificial increase in testosterone tends to increase the rate of parasitism by the nematode *Trichostrongylus tenuis*. In natural conditions, an infection would lead to a decrease in the general state of health and energetic resources of the males, but would also lead to males redistributing their energy expenditure in favour of the immune system and to a reduction in agonistic behaviours via a diminution of testosterone.

The same mechanism could influence the production of signals involved in reciprocal male evaluation such as feather colour, or bird songs, also influenced by testosterone. Bird songs are among the most studied natural signals (Catchpole and Slater 1995). These complex signals have multiple functions (Garamszegi 2005). In the context of sexual selection they play a role in both male–male interactions (territory and mate defence) and mate choice. For this reason, the influence of parasites on bird song affects both frameworks of sexual selection. We discuss this aspect here in the context of male–male competition, but it has to be kept in mind that these results can also be applied in the context of direct (Section 3.4) or indirect (Section 3.5) benefits obtained through mate choice. A recent example was provided by the study of Spencer and collaborators (Spencer *et al.* 2005). In this study they demonstrated that young canaries infected by the blood parasite *Plasmodium relictum* later develop adult songs that are less complex than those of uninfected canaries. Infected individuals also showed weaker development of the areas of the brain implicated in song production. Experimental approaches allow a better understanding of the links between parasites and song properties in males. Thus, Garamszegi *et al.* (2004) have examined the existence of a **'trade-off'** between the immune response and song characteristics in the collared flycatcher, *Ficedula albicollis*. To demonstrate this trade-off they simulated an infection by injecting a neutral **antigen** (sheep red blood cells) in a number of males, whilst a second control group was injected with a sterile solution. Garamszegi *et al.* (2004) showed that the immune response to sheep red blood cells was costly and led to a decrease of body mass, testosterone levels, and song frequency. Nevertheless, the influence of parasites on song, in the context of sexual selection, remains poorly understood.

3.4 Parasites and mate choice: costs and direct benefits

During reproduction, individuals have to come into physical contact thus increasing the risk of transmission of parasites and pathogens (Kulkarni and Heeb 2007). Because by definition parasites have a negative effect on their hosts, sexual selection can favour individuals that avoid mating with infected partners (Andersson 1994; Kavaliers *et al.* 2005): This can occur for three reasons: (1) the individual avoids contamination by the parasites of the partner (if the parasite can be directly transmitted); (2) the individual also avoids the transmission of parasites to its offspring (if the partner takes part in raising them or is likely to contaminate the nest, for example with ectoparasites like fleas and ticks); and (3) if the partner takes part in raising the young, a partner with fewer parasites will be in better health and will provide better parental care. This preferential choice supposes that the parasite burden of a host can be directly assessed (through visual, tactile, or olfactory detection of the parasites) or indirectly through a signal by a sexually selected trait sensitive to infections.

Avoidance behaviours constitute the first line of defence for potential hosts (Hart 1990). As we will see in Section 3.5, this avoidance can also lead to indirect benefits associated with genes for resistance to parasites carried by the partner (the 'Hamilton and Zuk hypothesis'; Hamilton and Zuk 1982). Able (1996) pointed out that a large number of experimental studies showing the existence of preferences for uninfested partners directly concern transmissible parasites. In this case, the direct effects on the female and the offspring, in comparison with indirect effects of the parasite, could constitute important selection pressures for the maintenance of female preferences. Able (1996) proposed three hypotheses, all characterized by a greater reproductive success of unparasitized males (Fig. 3.2): (1) the hypothesis of parasite avoidance; (2) the hypothesis of a contagion indicator; and (3) the Hamilton and Zuk hypothesis.

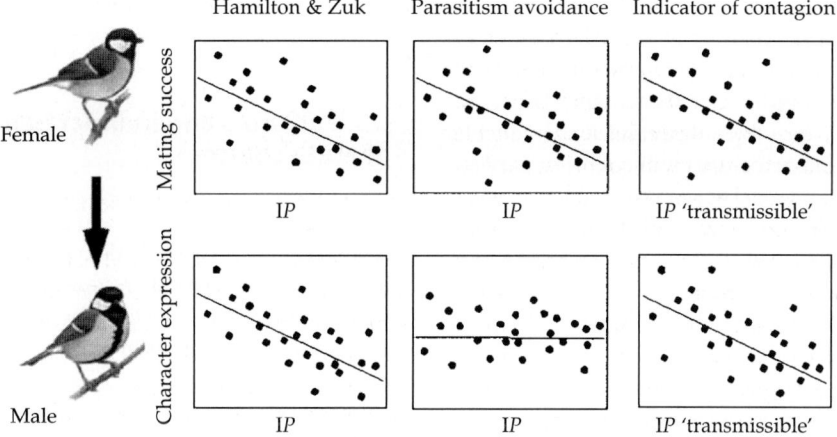

Figure 3.2 Why are non-parasitized males better? There are three hypotheses (IP is the intensity of parasitism). In the three hypotheses, male mating success decreases with parasitism. In the parasitism avoidance hypothesis, females directly detect parasites and avoid parasitized males. In the other two hypotheses, females focus on sexual traits (e.g. colours) that are influenced by the intensity of infection. According to the Hamilton and Zuk hypothesis, choosy females obtain genetic benefits (males that are genetically resistant transmit resistance genes to their offspring). In the indicator of contagion hypothesis, the benefit is direct: females avoid contact with parasitized males thereby protecting themselves from infection. This scenario is limited to situations involving parasites with direct transmission (IP 'transmissible').

The parasite avoidance hypothesis suggested by Borgia (1986) is the simplest. The female directly detects parasites on males and avoids them. In this scenario the development of secondary sexual signals does not need to be sensitive to parasitism. In this way, in the satin bowerbird (*Ptilonorhynchus violaceus*, a passerine living in Australia), females can detect ectoparasites present on the dark plumage of males; however, the intensity of male coloration does not vary with the intensity of parasite infestation (Borgia 1986; Borgia and Collis 1989). In a similar way, in the domestic mouse the female is able to detect the odours associated with parasites or with the metabolic changes they cause in the host. She detects the presence of intestinal parasites via the smell of male urine and then shows a strong avoidance reaction (Ehman and Scott 2001, 2002; Kavaliers *et al.* 2005).

In the contagion indicator hypothesis, secondary sexual signals help females detect the presence in males of directly transmissible parasites and thus avoid contagion. A correlation between sexual signals and all parasites is not necessarily expected. Parasites with indirect transmission (for example those with complex life cycles, like the parasitic trematodes where various phases of development

occur in different host species) or those parasites that need a transmission **vector**, like malaria, will not necessarily act on the development of sexual signals since the female cannot be directly infected by a male with such parasites. Finally, the Hamilton and Zuk hypothesis assumes that the signals indicate to the female which males she has to choose to transmit resistance genes to her descendants (see below for details): This hypothesis concerns all types of parasite (Hamilton and Zuk 1982).

In many animal species, sexual partners stay together to raise their young. Sexual selection does not necessarily stop at fertilization of the egg(s) and can continue until the offspring become independent (Trivers 1972). Individuals thus have an interest in choosing a partner that will not infect their young in the nest. In species with male parental care females should maximize their fitness by choosing to reproduce with healthy males (Milinski and Bakker 1990). Such a choice can be particularly important for hosts that have developed antiparasitic strategies that imply an increase in parental investment in response to the presence of parasites (Perrin *et al.* 1996). For example, the blue tit, *Cyanistes caeruleus*, responds to infestations of the nest by the hen flea, *Ceratophyllus gallinae*, by increasing the rate of feeding of the young. This increase in parental effort will partly compensate for the deleterious effect of the ectoparasites on the survival and condition of the young (Tripet and Richner 1997). It is thus important for female blue tits to choose a partner that will not bring parasites into the nest and/or that will be ready to increase his parental investment in response to the presence of parasites. This last study suggests that parasites not only play an important role in mate choice but also in parental investment strategies.

Up to now we have discussed how parasitism modifies the attractiveness of a potential partner to a choosing individual. However, there are cases in which it is the infected individual making the choice. It has been shown in certain cases that a parasite infestation modifies the decision-making processes involved in mate choice (Milinski 1990). In the broad-nosed pipefish, *Syngnathus typhle*, the male takes care of the eggs and the young. Given this high level of male parental investment, males are expected to exercise selection for a female partner, thus inverting the asymmetry typical of most other species (see also the examples of the jacana and gobby described above). Pipefish males that are not infected by trematodes of the genus *Cryptocotyle* sp. show a preference for females that are not parasitized; however, this preference is not shown by infected males (Mazzi 2004). Females, whether they are infected or not, are not choosy and accept any male. This study suggests that the parasite only has a direct effect on the choice process in the sex which has more to lose in terms of investment by making a bad choice. (In this case the 'good genes' theory may apply, because fish are infected by free-swimming parasite larvae that are not directly transmitted by contact between two host individuals.) The meaning of a decrease in the selectivity exercised by parasitized individuals is difficult to assess. Is it because a sick or weak individual has little to lose by being in contact with diseased congeners? Is it a simple secondary effect of parasitism in which a diseased individual has reduced sensitivity or loses behavioural flexibility? Could it be an adaptation that limits the damage in parasitized males if the choice is costlier for parasitized males than for non-parasitized males? We could also consider the possibility that

the modification of mate choice by the parasite is a manipulation of the host by the parasite.

3.5 Parasites and indirect benefits of mate choice

3.5.1 Choosing parasite-resistant males

Resistance to parasites is defined as the ability of a host to avoid parasitic infections, or to limit parasite loads. Many studies have shown that a large proportion of the variation in longevity and fecundity in host species is related to parasites. Hence, differences in resistance among individuals may be a major determinant of variation in fitness. Many examples illustrate the existence of genetic variation in resistance to parasites within host species. Some of these concern specific resistance to a single parasite; For example, humans who are heterozygous for the genotypes A/S at the haemoglobin gene are more resistant to *Plasmodium falciparum*, the agent of malaria (Allison 1954). Resistance may also be more general; for example the encapsulation reaction, which allows insect larvae to neutralize **parasitoid** wasps, can be triggered by many different kinds of objects injected into the **haemolymph**. In *Drosophila*, artificial selection experiments have documented a large **genetic variance** for the encapsulation reaction, involving both a general resistance component and components specific to particular parasitoid species (Fellowes *et al.* 1999).

By detecting and avoiding parasitized males, females can increase the probability that their offspring will be genetically resistant to parasites (an indirect benefit). This ties in with the direct benefit of avoiding contamination of themselves and their young (in the case of directly transmitted parasites; see above).

3.5.1.1 The Hamilton–Zuk hypothesis and the lek paradox

A major difficulty is that parasites themselves are often difficult to detect within hosts. In 1982, Hamilton and Zuk, in what became a very influential paper, proposed that sexual ornaments acted as signals of resistance to parasites. According to this hypothesis, heavily infected males (being in bad physical condition) were incapable of producing such elaborate ornaments as healthy males. The underlying model was the condition-dependent handicap model (cf. Section 3.2.3), in which it was assumed that condition was primarily dependent on the infection status of hosts.

One of the main strengths of this idea is to provide a solution to a fundamental problem of good genes theories, the so-called 'lek paradox' (Kirkpatrick and Ryan 1991). This paradox was named for the lekking behaviour (males grouping together in an arena and displaying to attract females) observed in several bird species. The central tenet of the lek paradox is that the driving force for good genes sexual selection is the heritable fitness variation within a population. By definition, genes that affect fitness are under strong selection and should either rapidly invade populations (if they are advantageous) or be eliminated (if deleterious). Therefore it would seem hard to maintain a large additive genetic variance for fitness. Parasites offer a solution to this problem. Unlike other components of the environment, parasites co-evolve with their hosts, and continuously tend to bypass every resistance mechanism invented by their hosts. Through this so-called '**Red Queen**' process hosts are under fluctuating selection pressures. Alleles that at one time provide resistance to parasites may later become obsolete. Such a cyclical system could maintain, at least for some evolutionary time, enough genetic variance in fitness to feed a good genes sexual selection process. We should mention, however, that cyclical co-evolution will only apply to specific resistance to one kind of parasite. More general forms of resistance, although they may be reflected by sexual signals as well, pose the same problem as parasite-independent components of fitness, as they are under consistently directional, not periodically changing, selection.

3.5.1.2 Testing the Hamilton and Zuk hypothesis

Hamilton and Zuk (1982) made two predictions. The first was at the intraspecific level and stated that species that had been more exposed to parasites during their evolutionary history should have developed male sexual ornaments or signals more often, and thus show a stronger sexual dimorphism, than less exposed species. The second was at the intraspecific level: in species with ornamented males, ornament size (or beauty) should be correlated with parasite load, i.e. parasites should inhibit the development of ornaments (Fig. 3.2).

Among North American passerine birds, species with high prevalences of blood parasites had on average more brightly coloured feathers and more complex songs, which seemed to confirm the first prediction. Some, but not all, later studies found similar patterns in other groups (Møller 1990b). Importantly, studies that controlled for **phylogenetic** relationships (and hence, lack of statistical independence) among species did not confirm the interspecific positive correlation between parasitic prevalence and intensity or complexity of sexual signals (Ward 1988, 1989; Read and Harvey 1989). Some studies even found a negative correlation (Lefcort and Blaustein 1991). However, the main problem does not reside in the data but in the test itself. The Hamilton–Zuk hypothesis predicts that two processes, contact with parasites and heritable variation in resistance, encourage the evolution of sexual signal in hosts. But is the current prevalence of parasites a good indicator of these two processes? The cyclical dynamics of host–parasite co-evolution, which is the central tenet of the Hamilton–Zuk hypothesis, certainly results in a large temporal variation in prevalence or parasitic loads, but not necessarily in a global increase of these quantities (Clayton et al. 1992). Highly parasitized species might be those that never managed to evolve specific resistance, and are thus poor candidates for the evolution of sexual signals of resistance. Thus, interspecific comparisons, although they were Hamilton and Zuk's first argument, do not provide strong evidence for the Hamilton–Zuk hypothesis after all.

On the other hand, the intraspecific prediction can be directly tested. An abundant literature has flourished on this theme, especially in birds, and has established several important points: (1) parasites negatively affect host fecundity and survival; (2) parasitized males have dull colours and small ornaments in comparison to healthy ones; and (3) females are particularly keen on ornaments that are very sensitive indicators of parasite load (examples, among others, Milinski and Bakker 1990; Pruett-Jones et al. 1990). Fewer studies go as far as showing that parasite loads are transmitted from father to offspring, which denotes heritable genetic variation in resistance within host populations (e.g. Hillgarth 1990; Zuk *et al.* 1990; Box 3.1). Although negative results also exist (and have probably been less broadcast than positive ones), the accumulation of all these data in the space of a few years suggests that the intraspecific predictions of Hamilton and Zuk (1982) are often met. A well-known caveat is that very few studies have simultaneously tested all the points mentioned above. However, such studies do exist and are quite convincing (Box 3.1). In summary, it seems probable that one of the benefits that females gain by choosing ornamented males is a genetic resistance to parasites.

Can we now consider that the Hamilton–Zuk theory has been definitely validated? Certainly not! The hypothesis states that resistance to parasites is the major, or only, reason why sexual ornaments evolve. It is one thing to show that such resistance does play a role, but quite another to show that it is the only, or most important, factor. As mentioned above the Hamilton–Zuk hypothesis belongs to the category of condition-dependent handicap models. Although parasite load is an important component of body condition in animals, it is certainly not the only one; for example nutritional status is very important. To demonstrate the validity of Hamilton and Zuk's argument, one has to show that male ornaments reflect genetic resistance to parasites much more than they do any other form of heritable variation affecting the offspring's survival or fecundity. To our knowledge, this has never been done. On the contrary, in many studies that verified one or several of the cited predictions, brightly coloured males differ from dull males in many other condition variables apart from parasite load: they may differ in body size, metabolic reserves, or physical abilities. Of course one can always argue that these differences are consequences of differences in parasite loads; however, when two variables are correlated, nothing allows us a priori to interpret one as the cause or consequence of the other. Both may be consequences of a common underlying variable; for example male age may affect both infection status and the state of the secondary sexual ornaments. In this case, a correlation between the expression of sexual signals and parasite load could be due to the effects of age

rather than to a genetic variation in resistance to parasites (Thomas *et al.* 1995).

Controlled experiments unfortunately have not solved this problem. These kinds of studies have been successfully used to show that many physiological variables change following infections (as we all know by experience), and that good body condition increases the general level of resistance of an organism to infections. However, we still do not know whether genes other than specific resistance genes explain the variation in body condition and sexual ornaments in nature. This leads us to a second problem: the specificity of resistance genes. Any gene that ameliorates body condition can be considered as a general resistance gene. However, the important genes in the Hamilton–Zuk scenario are specific resistance genes that cyclically co-evolve with their particular parasite (cf. Section 3.5.1.1). Unfortunately, no study has ever proved that genes involved in resistance-mediated indirect benefits are specific.

Finally, Poulin and Vickery (1993) suggested several important limits to the correlation between resistance to parasites and the development of sexual ornamentation: (1) spatial aggregation (hosts are not uniformly exposed to parasites—few hosts commonly bear very high parasitic loads); (2) too strong a pathogenic effect of the parasite; or (3) too low an abundance of parasites. Poulin and Vickery (1993) concluded that the necessary conditions for the evolution of ornaments as signals of host resistance to parasites are quite restrictive and probably not often met together in natural systems.

3.5.1.3 Trade-offs
Because of the large impact of the Hamilton–Zuk hypothesis, researchers became interested in the complex links between parasite loads, sexual ornamentation, and body condition, and tried to document the physiological mechanisms underlying these links. Many authors assumed that sexual ornamentation and the ability to fight parasites were linked by physiological trade-offs. Such models assume that the immune function and the development of sexual ornamentation are in competition for some underlying physiological variable or resource. An individual that allocates a lot of energy to sexual ornamentation may pay a price in the form of depressed immunity. Conversely, an

Box 3.1 Female barn swallows prefer males having genes that make their offspring resistant to ectoparasites

In the barn swallow (*Hirundo rustica*), males differ from females by having a red throat and a deeply forked tail. During the 1990s, studies performed by Møller (see Møller, 1991 for review) showed that tail length was a key determinant of reproductive success in males and of their attractiveness to females. In this monogamous bird species, males with long tails are less likely to remain single than their short-tailed congeners, they also obtain females in better condition, have more extra-pair copulations with other females, and have more faithful females themselves.

Nests as well as adults are parasitized by haematophagous ticks (*Ornithonyssus bursa*), with sometimes several hundreds or thousands of parasites per bird. To test the prediction of Hamilton and Zuk, Møller combined two experimental approaches: the manipulation of infection levels (adding ticks or killing them with insecticides) and 'cross-fostering'. This last technique consists in exchanging several juveniles, just after eclosion, between two nests in order to separate genetic (family) and environmental (nest quality, parental care) effects. Results revealed that:

- Ticks have a negative effect on the body condition of chicks: juveniles from nests in which 50 ticks were added had a higher parasitic load, a smaller size, and lower weight at 15 days compared with chicks from control nests.
- Ticks have negative effects on the development of secondary sexual traits in males. Tail size in males changes every year (in the moulting season). In males from nests treated with insecticides, tail growth was higher than in untreated nests. Tail growth was reduced when additional ticks were added to the nest.
- Parasite loads are genetically heritable. Chicks having the same parents display parasitic loads that are similar whether or not they stay in the same nest, while chicks with different parents, even when raised in the same nest, do not necessarily have similar parasitic loads. Similarly, the parasitic load of a chick is positively correlated with that of its biological father, not the social father.
- Finally, secondary sexual traits in males are a good predictor of their genetic resistance to parasites: juveniles from males having well-developed secondary traits (long tails) have fewer parasites, whatever nest they were raised in. However, unrelated chicks adopted by long-tailed males enjoy no particular advantage.

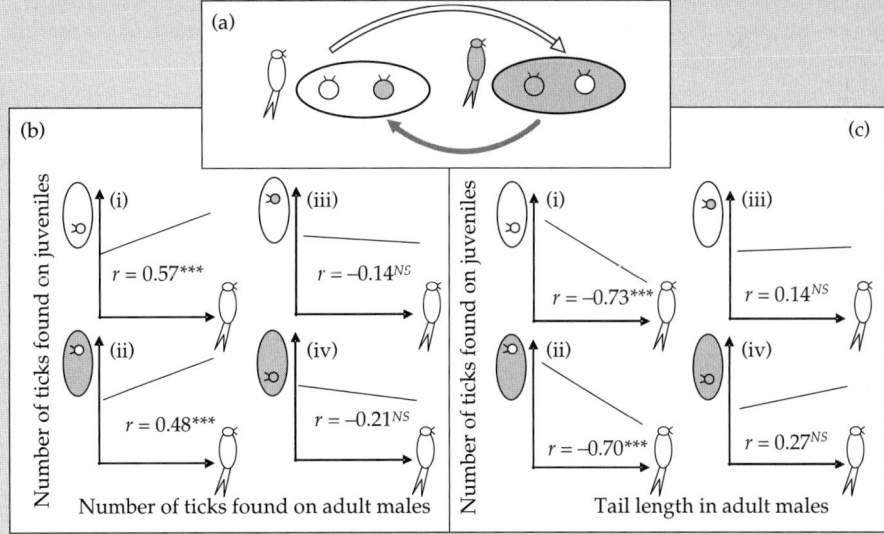

'Cross-fostering' experiments in barn swallows. (a) Design of the experiment. (b) Relationship between the level of parasitism of males with that of: (i) their own chick in their own nest; (ii) their own chicks in other nests; (iii) unrelated chicks raised in their own nest; (iv) brothers of unrelated chicks left in their nests of origin. All the correlations (measured on 24 individuals) are given with a *P* value (***$P < 0.001$; $^{NS}P > 0.05$). (c) The same figures showing the relationship between tail size of males and the parasitic load of chicks. Modified from Møller (1991).

individual that mobilizes its resources to eliminate a parasite would lose its attractiveness (Faivre *et al.* 2003b). This constraint would work as an internal handicap associated with ornamentation: a weak male could not afford a bright ornament because this would pave the way for deadly infections. Two types of physiological variables have been intensely studied: carotenoids and hormones.

Carotenoids

Carotenoid pigments are often involved in brightly coloured male ornaments (feathers or crests) (Lozano 1994). The orange colour of the beak in male blackbirds (*Turdus merula*), the red markings of male guppies (*Poecilia reticulata*), the red crests of cocks (*Gallus gallus*), or red feathers of the passerine bird *Carpodacus mexicanus*, all of which are attractive to females, can be modified by manipulating the dietary intake of carotenoids (Rotschild 1973; Kodric-Brown 1989; Hill *et al.* 2002; Faivre *et al.* 2003b). However, carotenoids also seem essential for the correct functioning of cells in the immune system (Bendich 1989). In many species [cocks, great tits (*Parus major*), guppies, or sticklebacks (*Gasterosteus aculeatus*—another fish with red markings)], males with high parasite loads have smaller, or less intense, red or yellow patches (Zuk *et al.* 1990; Houde and Torio 1992; Folstad *et al.* 1994; Horak *et al.* 2001). This suggests (but does not demonstrate) that the use of carotenoids by the immune system makes them less available for ornamental coloration. However, these results remain ambiguous because a correlation does not mean a direct physiological causality. For example, Saino *et al.* (1999) observed a correlation between carotenoid titres in the blood and the intensity of the red patch in the throat of male barn swallows. Although this apparently supports the idea that carotenoids are a limiting resource for ornament development, a similar correlation was observed between carotenoids and male tail length. Males' tails (an important sexual signal in this species) do not display any red or yellow coloration, and therefore cannot be directly dependent on carotenoid-based pigments.

Hormones

Some hormones, especially androgens like testosterone, may be able at the same time to stimulate

the development of sexual ornamentation and have an immunosuppressive action (Folstad and Karter 1992). In this case, the two functions are not in competition for hormones (because they do not actually consume hormone molecules), but increasing one function implies a decrease in the other. This hypothesis gained credence when Verhulst *et al.* (1999) published the results of an artificial selection experiment in chickens. In chicken strains selected for increased immunocompetence, testosterone levels and the size of male ornaments (the red crest) had decreased, demonstrating a negative genetic correlation between these quantities and the immune response. Following this study, many studies on birds, reptiles, and mammals, have tried to elucidate the link between testosterone and ornamentation on the one hand and between testosterone and immunity on the other. Roberts *et al.* (2004) have meta-analysed 36 studies that experimentally manipulated testosterone levels; their conclusion is that the existence and generality of these links are far from being proven. Positive results concern only a few species, and only some aspects of the immune response, which are not consistent across studies. In addition, the positive effect of testosterone on male aggression during fights is much more firmly established than its effect on sexual ornamentation (Mougeot *et al.* 2005).

Not only do the two hypotheses (carotenoids and testosterone) have ambiguous empirical support, they also share an intrinsic flaw, because they implicitly transfer the problem onto a hidden variable, which makes them very difficult to test (Fig. 3.3). In both hypotheses, the physiological constraint is assumed to impose a negative relationship between ornaments and immunity. If males differed among themselves only in the proportion of resources allocated to each function, females should rather choose unadorned males because they would be healthier. Thus, male must differ with respect to another variable, such as the total amount of resources they can use (e.g. the total amount of carotenoids). As shown in Fig. 3.3, the genetic correlation between ornaments and resistance to parasites can be either positive or negative depending on which character is more variable: the allocation strategy or the total resource uptake

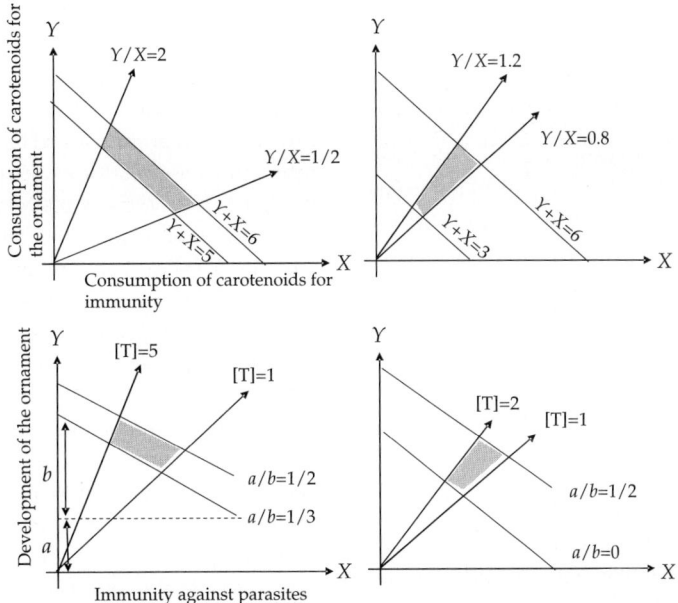

Figure 3.3 How trade-offs between functions may generate positive or negative correlations. Top: competition between two functions (immunity and ornamentation) for ingested carotenoids. The ratio Y/X represents the strategic choices made by an individual (i.e. how carotenoids are invested into each function), while $X + Y$ represent the total amount of carotenoids obtained from food. If individuals are very different in their allocation strategy but not in their global ability to obtain carotenoids (left panel), a negative correlation will arise between the two functions (grey area): individuals with developed ornamentation will tend to have a less efficient immune response. If individuals vary a lot in their ability to obtain carotenoids but are similar in the way they allocate them to the two functions (right panel), the correlation will be positive.

Bottom: Two functions inversely influenced by a given hormone (testosterone). a and b respectively represent the testosterone-independent and the testosterone-dependent components of the immune system activity. High quantities of testosterone [T] increase the development of the ornament and decrease the testosterone-dependent component of immunity. If individuals mainly differ for [T] (left panel), immunity defences and ornament development are negatively correlated. If individuals mainly differ for a/b (right panel), the correlation is positive. Arbitrary units.

(Shykoff and Widmer 1996). This is actually a simple application to sexual selection of a classical problem in models of joint evolution of several traits linked by trade-offs (Van Noordwijk and De Jong 1986).

A similar problem arises with the testosterone hypothesis (Fig. 3.3). It is assumed that males vary in their ability to withstand immunodepression, and this ability is what females 'want' to transmit to their offspring. This ability can be of two kinds: either males possess testosterone-independent mechanisms that allow them to limit parasite loads even when testosterone-dependent mechanisms are impaired; or they are able to survive more easily with high parasite loads (they are parasite-tolerant). The first process (testosterone-independent immunity) is a hidden variable that has never been properly quantified. The second (**tolerance**) has a strange property—it predicts that females should seek heavily parasitized males because they are probably more tolerant; this is counter-intuitive and is contrary to most of the observed data. Whatever the data may be, the theory always seems to have an explanation: parasitized males show that they are parasite-tolerant, and healthy males that they are resistant. We may conclude that our inability to validate or invalidate the theories does not lie with the data, but with the imprecise formulation of the theories themselves.

3.5.2 The major histocompatibility complex, parasites, and mate choice

The major histocompatibility complex (MHC) is a group of glycoproteins which allow self-recognition during the vertebrate immune response; this recognition allows cells to cooperate within an organism so as to destroy alien, infectious cells within the body. For example, proteins of MHC class II allow **macrophages** to present antigens, exposing a fragment of alien protein attached to the MHC II protein, on the cell membrane. Using receptors that specifically bind the MHC II–antigen complex, **lymphocyte** cells can interact with macrophages and start to divide and become active. They are then able to attack any cell that presents the antigen without the MHC II, and is therefore identified as an alien cell (Roitt *et al.* 1998). The polymorphism of MHC II genes is extremely high: two individuals taken at random practically never have the same genotype; this is why they identify each other's cells as alien and reject allo-grafts, in the absence of an immunosuppressive treatment.

In the early 1980s, female mice were found to be capable of comparing their own MHC with that of other individuals, by smelling the urine of other mice. Females were more likely to mate with males with MHC very different from their own (Potts et al. 1991). The same type of process has also been suggested to occur in fish (Landry *et al.* 2001) and humans (Wedekind *et al.* 1995). This choice behaviour is not an absolute preference (as in the case of coloured ornaments) but a preference relative to the choosing female (cf. Section 2.4). The consequence of this choice is to increase offspring heterozygosity. If heterozygosity at MHC genes increases survival, or if it is correlated with other advantageous characteristics, this mating strategy, focused on partners as different as possible from oneself, will be beneficial. Thus, many researchers have tried to find out whether MHC heterozygosity is truly advantageous. This hypothesis had particular appeal because **heterozygote** advantage (or **overdominance**) is a very powerful way of maintaining polymorphism, and could therefore explain the high polymorphism of MHC genes. Alternatively, the MHC could play the simple role of an indicator of kinship, allowing a female to

avoid breeding with her relatives, thus avoiding inbreeding depression in her offspring. The diversity of MHC genes is such that in nature two individuals who share MHC genes are almost certainly kin. These hypotheses (MHC overdominance and inbreeding depression) are not exclusive: it may be that MHC heterozygosity makes an individual more resistant to parasites and the relative disadvantage of being homozygous for MHC combines with other, more general, components of inbreeding depression due to recessive deleterious alleles elsewhere in the genome.

Does MHC heterozygosity by itself provide protection against parasites? The results are not clearcut. In some cases, a particular MHC allele can increase resistance to a particular parasite, but there has been no direct proof that heterozygosity *per se* is favoured (Apanius *et al.* 1997; Wedekind *et al.* 2005). In the long run, selection for particular alleles (not heterozygosity) could also maintain the MHC polymorphism if new or rare alleles, to which parasite populations have not had time to adapt, tend to make individuals more resistant to parasites. In some species females seem to prefer males that carry some particular MHC alleles rather than avoiding alleles similar to their own (Ekblom *et al.* 2004). Studies on sticklebacks by Reusch *et al.* (2001) suggest that females choose their mates on the basis of their 'internal diversity', i.e. the number of different alleles that they carry in a set of several MHC II genes. The female choice strategy is not yet clear: it could result in maximizing internal MHC diversity in males (Reusch *et al.* 2001) or in favouring intermediate internal diversities (Milinski 2003; Wegner *et al.* 2003). However, these data suggest that MHC could sometimes be the target of absolute choices, not necessarily choices that are relative to the genotype of the choosing female. The MHC would in such cases enter the framework of good-genes theory, together with other genes with additive effects, such as those of sexual colour ornaments.

3.5.3 Mate choice and resistance to parasites: a true solution to a false problem

Figure 3.4 summarizes all the mechanisms described in this chapter. The role of parasites in the evolution of mate choice has been a very active

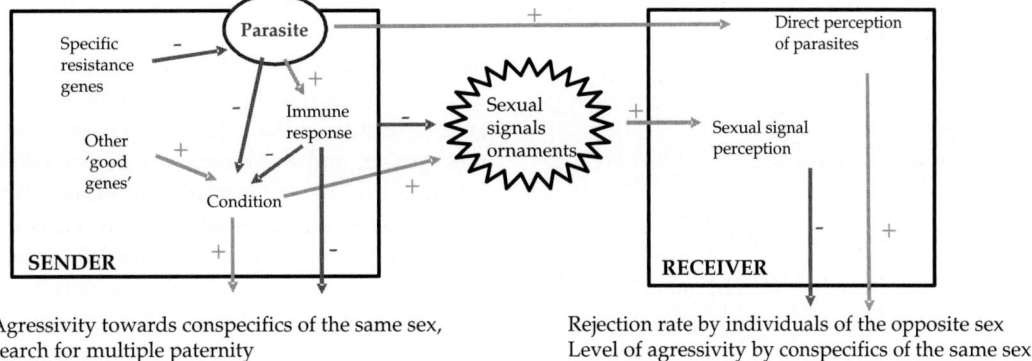

Agressivity towards conspecifics of the same sex, search for multiple paternity

Rejection rate by individuals of the opposite sex
Level of agressivity by conspecifics of the same sex

Figure 3.4 How parasites can alter reproductive success. Parasites may have direct effects on the reproductive success of their host (sender) or an indirect effect through the behaviours of other individuals when they detect the presence of parasites (receivers). Parasites consume their host's resources and elicit an immune response. Through these two effects they decrease the host's body condition. The reduction in body condition, as well as the activation of the immune system, can lead to reduced levels of aggressivity towards individuals of the same sex, lower parental care, and/or reduced development of the secondary sexual traits. The direct perception of the presence of parasites, or the perception of their effects on secondary sexual traits, inform receivers that the sender is parasitized. Receivers of the opposite sex will then be less likely to copulate and/or those of the same sex will be more aggressive. Figure modified from Garamszegi *et al.* (2004).

area of research, and the corresponding literature since 1980 is abundant compared with that for other branches of evolutionary ecology. This is because parasites seem to provide a solution to several problems: they provide a reason to maximize offspring survival by choosing mates with 'good genes' that make them resistant, as well as a mechanism (the Red Queen hypothesis) to indefinitely maintain substantial variation in these genes and, in general, in fitness.

Researchers have mostly focused on the following question: will the offspring of choosy females be genetically more resistant to parasites than those of females who randomly choose their mates? Thanks to a considerable research effort, especially in vertebrates, we now know that the answer is yes, they often will. However, we are still far from having completed the research programme started by Hamilton and Zuk (1982).

First, is the Red Queen hypothesis necessary to maintain genetic variation for fitness? This question is not quite settled. A short time after Hamilton and Zuk's paper was published, it became clear that genetic variance for fitness traits is not usually low, even in artificial populations relatively protected from parasites. Actually, the **heritability** of fitness traits is low because survival and fecundity are very sensitive to environmental

variation. However, their genetic variance, relative to the mean, is very substantial, probably because these characters are 'synthetic traits' that integrate many physiological variables and can therefore be affected by many different **mutations**—most of them deleterious (Houle 1992). Hence, the lek paradox is no longer a problem: if sexual ornaments are condition dependent, their genetic variance reflects variance in fitness, which is large and continuously re-created by mutation (Rowe and Houle 1996; Kotiaho *et al.* 2001). Although it is clear that parasites deeply affect the survival and fecundity of hosts in nature, the proportion of genetic variance for fitness due to specific resistance genes remains essentially unknown.

In addition, demonstrating that resistance to parasites is correlated with male ornamentation is not enough. In the framework of the condition-dependent handicap theory, we need to know, among all the condition variables that affect the development of ornamentation, what proportion of the variance can be attributed to specific resistance genes.

As we have seen, a similar problem arises with the MHC hypothesis. It is quite clear that parasites have a role in maintaining the genetic diversity of the MHC. However, it is not at all clear that the main motivation behind MHC-based female choice has something to do with the immune function of the MHC. It is difficult to evaluate, within the total fitness benefit females obtain through their choice, what proportion directly relies on MHC and what proportion is related to inbreeding depression on other genes.

In conclusion, the emphasis since 1980 on host–parasite relationships has had a deep impact on our understanding of how and why complex mate choice behaviours arise during evolution. On the whole, this area has offered many convincing examples of indirect benefits, and often spectacular validations of the condition-dependent handicap theory. However, we need not be overly enthusiastic: there is actually no definite reason to believe that host–parasite cycles are necessary for the evolution of cock crests, peacock feathers, red markings in guppies, or the extravagant displays of birds of paradise, from either the empirical or the theoretical point of view. Parasites very probably have a role in sexual selection, but they are not necessarily its main or unique cause. The actual problem is to evaluate the extent to which they contribute.

Important points

• Parasites decrease the success of their hosts in competition with their same-sex congeners for access to mates. They also decrease the attractiveness of their host towards individuals of the opposite sex.

• These two effects can be direct, because parasites weaken their hosts, or they can be indirect. In the latter case the parasite can modify a host's sexual signal, such as a patch of colour or a morphological ornament, which allows evaluation of the host's bodily condition, especially in relation to parasite load, by conspecifics.

• Such signals may serve either as elements of dissuasion (in competition among individuals of the same sex) or as attractive elements (for the opposite sex).

• The hypothesis of Hamilton and Zuk (1982) is a major landmark in the study of the role of parasites in sexual selection. It states that sexual signals allow females to recognize and choose males that are genetically resistant to parasites, and will later transmit this resistance to their offspring.

• The existence of a positive genetic correlation between resistance to parasites and sexual signals, one of the main predictions of Hamilton and Zuk's hypothesis, has been documented in several studies. However, the degree of generality of this mechanism, and its importance as the major or unique cause for the evolution of sexual signals, is far from being demonstrated. In many cases, females, by avoiding parasitized males, may obtain more immediate benefits than transmitting resistance genes to their offspring. For example, they may simply avoid contagion. Moreover, sexual signals can act as indicators of many more kinds of genetic variation in fitness than just resistance to parasites.

Questions for discussion

• When we observe in a species that females avoid mating with parasitized males, how can we know empirically if they directly detect the parasites or if they base their choice on a coded signal?

• In the same situation, how can we know if the benefits they obtain are direct or indirect?

• What protocols would allow us to detect a **phenotypic correlation** between a sexual signal and parasite resistance? How can we show that there is a genetic correlation between these two variables?

• How can we define a sexual signal?

• What roles are played by physiological trade-offs in sexual selection ? How do they interact with parasites?

• How can we know whether resistance to a parasite is specific or not?

Further reading

• Andersson, M. (1994). *Sexual selection*, Princeton University Press, Princeton, NJ. A classic, that reviews both the theory and the natural history of sexual selection, including the role of parasites.

• Darwin, C. (1871). *The descent of man, and selection in relation to sex*. John Murray, London. The historical root of all research on sexual selection (with somewhat boring descriptions of natural history!). The sexual selection theory is concisely stated in Chapter VIII, while the initial (I–VII) and final (XIX–XXI) chapters present an audacious application of this theory to the origin and diversity of human phenotypes.

• Judson, O. (2002). *Dr Tatiana's sex advice to all creation.* Metropolitan Books, New York. A best-seller with a very entertaining and hilarious account of sexual selection theory, but which includes very good scientific references and incisive critical appraisal of the current hypotheses. A unique work in the popular scientific literature.

• Møller, A.P. (1994). *Sexual selection and the barn swallow.* Oxford University Press, Oxford. A very complete case study of the barn swallow, illustrating almost every aspect of sexual selection theory, using exceptionally rich field data.

CHAPTER 4

Parasites and behaviour

Marie-Jeanne Perrot-Minnot and Frank Cézilly

4.1 Introduction

The occurrence of parasitic species in the environment is a major evolutionary constraint for a whole range of living organisms. From this viewpoint any form of behaviour may potentially contribute to reducing or increasing the risk and/or consequences of infection. Each individual is regularly exposed to parasites during its life cycle, be it when feeding, breeding, interacting with its fellows, or even right from the time it develops as an embryo. Likewise, the fate of parasites, through their survival, development, or dispersion in the environment is more often than not tied to the behaviour of their hosts.

This chapter analyses the complex linkages between host behaviour and parasitism. We start by looking at the behavioural strategies hosts put in place to reduce the risk of infection, which are known as **prophylactic** strategies. We then review how, once infected, hosts can alter their behaviour further to counter infection or reduce its pathogenic effects by what are known as **therapeutic** strategies. To that end we will propose a simple classification of host behaviour based upon the timing of the host-defence strategy (before the parasite becomes established, depending on the host's ability to perceive the risk of infection, or afterwards depending on the perception of infection itself), on the way the parasite is passed on, and on whether it lives in or on the host. Finally we examine the phenomenon of 'parasite manipulation', whereby a host's behaviour is modified to the advantage of the parasite alone.

4.2 Why study (and how to study) the behaviour of parasitized hosts?

4.2.1 Why?

Until recently the words 'parasites' and 'behaviour' did not seem to belong together and were rarely found together in books on parasitology or ethology. 'Parasites' invariably evoked fixed life forms, endowed with a rudimentary nervous system and poorly developed sensory capabilities. 'Behaviour', by contrast, was restricted to the set of motor actions of an individual (cf. Campan and Scapini 2002). However, things began to change when parasitologists and ethologists started to place their research within an explicitly evolutionary framework. The way we understand behaviour has been progressively refined to cover 'the collection of decisive processes by which individuals adjust their state and situation according to variations in their environment, both abiotic and biotic' (Danchin *et al.* 2008). In this definition 'decision-making' does not necessarily imply the involvement of elaborate cognitive processes. At most the emphasis here is on the fact that organisms are regularly faced with several alternatives, the consequences of which are not equivalent for their survival and reproduction.

From this perspective, the effect of parasites on their hosts' behaviour becomes a subject of study in its own right. For if parasites, by definition, increase their phenotypic **fitness** at the expense of that of their hosts, it is to be expected that their hosts should consequently have developed defence mechanisms over the course of evolution. Alongside immune system defences, host behaviour is a second means of preventing parasite infestation or of countering its effects. However, within the co-evolutionary process linking parasites to their hosts, it also happens that parasites have gained a firm hold, particularly through their effects on the behaviour of their hosts.

And so the study of host behaviour within evolutionary parasitology provides insight into the diversity of antiparasite defence strategies. Vice versa, allowance for the influence of parasites in behavioural ecology may suggest adaptive interpretations for certain altered or specifically produced behaviours in response to infection.

4.2.2 How?

For behaviour to be considered as having an adaptive function of controlling parasitic infection through avoidance, reduction, or elimination, it must fulfil two criteria. The first is that it must respond to a stimulus indicating a risk of infection or signalling actual infection: the behaviour must be specifically produced or modified depending on the risk of infection or the parasitic infection status. The second criterion is that this behavioural response to the perception of risk of parasitic infection or to actual infection must increase the individual's fitness. The same reasoning may be applied to the parasite: the change in the host's behaviour after infection must meet a number of criteria for it to be considered adaptive for the parasite (Poulin 1995; see Section 4.4.2). Accordingly our analysis of the various effects of parasitism on host behaviour follows a **hypothetico-deductive approach** (Cézilly *et al*. 2008). Our basic hypothesis is that the behaviour of infected hosts has been shaped over the course of evolution by the process of **natural selection**. This process has either selected behaviour conferring on the host effective protection against parasitism, or has favoured the ability of parasites to alter their hosts' behaviour to their own advantage. However, beyond this simplistic dichotomy, the reader should keep in mind that, in many cases, the observed behaviour will constitute some sort of compromise between opposite selective forces, and therefore will not correspond to an optimal value for either the host or the parasite. In addition, since our interest is primarily

in behaviour shaped by natural selection acting on host–parasite systems, the behavioural signs of infection akin to a straightforward pathogenic effect with no adaptive consequences for host or parasite are not covered.

Adjustment of a host's behaviour depending on the risk of infection may differ depending on whether the parasite lives in or on the host. In the case of **ectoparasites** that search the environment for hosts, natural selection will favour active avoidance behaviour by hosts (trying to escape from parasites or to repel them) or parasite elimination (e.g. removing them mechanically or chemically). Different strategies will be deployed against **endoparasites** that are usually passed on to the host through routine behaviour such as feeding or interacting with conspecifics. Avoidance of infection, or the limitation of its effects, will then depend on the ability of the hosts to modify their behaviour depending on the risk of infection, or to engage in specific defensive behaviours.

4.3 Pre-infection behaviour: parasite avoidance strategies

In many host species parasites are picked up in the normal course of their activities, such as during feeding. In most cases, the conduct involving the risk of parasitic infection is essential to the host's survival and reproduction so that the host cannot completely forego the behaviour despite the risk of infection involved. Under these circumstances, natural selection is supposed to favour individuals whose behaviour ensures the best **trade-off** between satisfying a need and minimizing the risk of infection. Alternatively, or concomitantly, organisms may develop preventive behaviour allowing reduced exposure to parasites.

4.3.1 Avoidance through modulation of behaviour involved in parasite transmission

While ingesting parasite-contaminated foodstuffs is a major source of infection, other forms of behaviour may be considered 'risky'. Sexual activity, social interactions, or choice of habitat are all forms of behaviour that may have non-negligible consequences in terms of infection. Several studies have shown that organisms have the ability to adjust the level of expression of such behaviour depending on the risk of infection they perceive in the environment.

4.3.1.1 Foraging and avoidance of infection

Foraging behaviour generally obeys the principle of maximizing the quantity of energy (or nutrients) taken in over a given time (Giraldeau 2008). However, some animals are able to modulate their feeding behaviour depending on the quality of the food available in their environment. For example, the influence of the risk of parasitism on the exploitation of food resources has been particularly well documented in domesticated herbivorous ungulates. The parasites involved are usually worms, whose eggs are released into the environment via the faeces of infected individuals, thus fouling the pasture. The contamination of grassland by sheep droppings increases the risk of ingestion of eggs of the intestinal **nematode** *Ostertagia circumcincta*, the agent causing ovine gastroenteritis. The presence of droppings in some areas of the pasture induces selective grazing behaviour: sheep graze almost exclusively in uncontaminated areas (Fig. 4.1). There is a resulting trade-off between feeding on the less-grazed land (land with denser and better-quality forage) and exposure to ingestion of nematode eggs (Hutchings *et al.* 2001). However, the benefit of avoidance differs depending on whether the animal is healthy, parasitized, or immunized by previous infection. It is expected that this compromise

is resolved differently depending on the parasitic status of the sheep. This is exactly what is observed in the field (Hutchings *et al.* 2001): when uncontaminated spots begin to be overgrazed, immunized sheep are the first to exploit the contaminated patches, where grass is more abundant (Fig. 4.1). Such behaviour, of avoiding infection by selective grazing outside defecation zones, is not peculiar to domesticated species. It has been observed in wild ungulates too. In the dik-dik, *Madoqua kirkii*, individuals graze significantly more in areas of their territory away from mounds of dung where densities of strongilid nematode larvae (intestinal parasites) are highest (Ezenwa 2004a).

4.3.1.2 Influence of parasitism on sexual behaviour

The resurgence of sexually transmitted diseases provides a dramatic illustration of the importance of sexual activity in the infection dynamics of many parasites. During (more or less protracted) interaction between mates various parasites may be passed from one to the other either by contact or by transfer of seminal fluid. An essential behavioural component in avoidance is the choice of an uninfected mate (see Chapter 3).

For external parasites, the parasitic status of a potential mate may be evaluated directly through perceptible signs of infection. For example in the sage grouse, *Centrocercus urophasianus* (order Galliformes), females tend to avoid any louse-infested male suitors. Research on the Wyoming Plains shows that females directly assess the

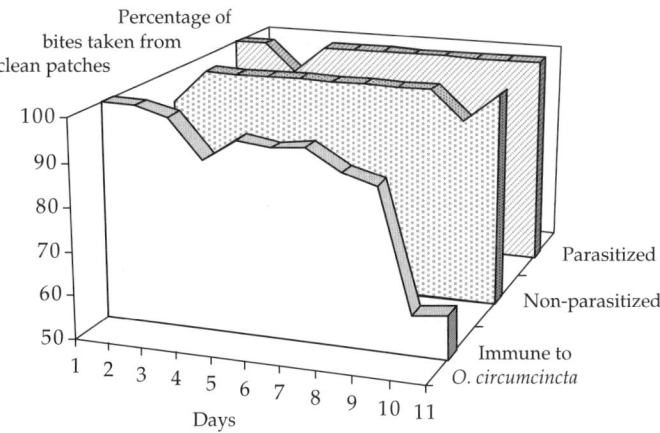

Figure 4.1 Effect of parasitic status on grazing selectivity in sheep. Selectivity is measured by the proportion of bites from clean patches (swards not contaminated by faeces) to the total number of bites taken from the experimental plot. After Hutchings *et al.* (2001).

parasitic status of males during their display, which involves males swelling the yellowish air sacs on either side of their neck: infested suitors thus reveal the haemorrhagic spots on their inflated air sacs caused by lice. The mere presence of fictive spots on a male's air sacs is enough to induce rejection behaviour in the female (Spurrier *et al.* 1991). In the case of endoparasitic infection, indicators of parasitic status are necessarily indirect. Behavioural ecology studies have shown that in many instances the degree of development of various extravagant secondary sexual characteristics (such as the tail of the peacock, *Pavo cristatus*) provides reliable information about the health of individuals and that females of various species are able to rely on such information when choosing mates (Danchin and Cézilly 2008; see also Chapter 3).

The possibility of recognizing the parasitic status of a potential mate is not necessarily given to all host species, and also varies with the type of parasite. Many infections remain without any perceptible consequences for a long time, and the occurrence of extravagant secondary sexual characteristics is far from universal. And yet the benefits in terms of survival of avoiding infection by certain sexually transmitted parasites, such as immunodeficiency viruses, are obviously still high. Until now the existence of behavioural defences against sexually transmitted diseases (STDs) has been referred to mainly in primates and rats (Nunn 2003) and remains difficult to test experimentally. An indirect approach is to resort to a mathematical analysis known as the 'comparative method' (Harvey and Pagel 1991; Cézilly *et al.* 2008). This type of analysis makes it possible to judge whether, at the interspecific level, there is a positive association between behavioural avoidance of a parasitized mate and the risk of being infected by a sexually transmitted parasite independently of phylogenetic kinship relations binding different species. It can be used to establish whether over the course of evolution the influence of the risk of contagion on the appearance of preventive behaviour has arisen independently in different species. Such a study has been made from data available for non-human primate species (Nunn 2003). In primates, behaviour designed to prevent contagion by STDs includes pre-copulatory genital inspection, post-copulatory genital grooming, and

post-copulatory urination. In addition, the risk of infection is assumed to increase with the degree of sexual **promiscuity**: as most Old World primate species and populations are infected by STDs, particularly viruses, the more promiscuous they are, the higher the risk of **transmission** (Nunn 2003). However, a comparative analysis by Nunn (2003) showed no significant association between any of the above-mentioned behaviours and the degree of sexual promiscuity. None of these behaviours seem, then, to have evolved in response to the risk of infection by a sexually transmitted parasite. Moreover, it has also been suggested that in theory monogamy should be favoured when the risk of contagion by STDs increases. However, there is no empirical evidence to support this prediction (Nunn 2003; Cézilly 2006). One would expect, however, that the evolution of pre-copulatory behavioural defences to STDs is limited by their negative impact upon reproductive success, particularly though missed mating opportunities arising for high selectivity towards mates (Nunn 2003). The evolutionary influence of STDs on the sexual behaviour of species therefore remains an open question.

4.3.1.3 *Group life, territoriality, and site fidelity*

Group life
Sociality is generally associated with the permanent or periodical gathering of individuals into groups of variable size and composition depending on species. The increased density, proximity of hosts, and diversity of behaviour associated with social interactions are all factors favouring the transmission of certain parasites. Social species are therefore exposed to more intense selection favouring the placing of behavioural or immune-response barriers to parasite transmission (Altizer *et al.* 2003). The selection pressures exerted by parasites are stronger when the size or density of these groups is greater and/or promiscuity within the group is high. Selection pressures also depend on the type of parasite, the way it is dispersed and transmitted, and its distribution in the environment and in (or on) its hosts. The benefit of parasitizing social species in terms of transmission is greater for parasites with simple cycles than for parasites with complex cycles (Combes 1991). The links

between group size and infection also depend on the mode of parasite transmission (Coté and Poulin 1995; Nunn and Altizer 2006). The **prevalence** and **intensity** of 'contagious' (i.e. transmitted by contact with an infected individual or its faeces) parasites are indeed positively correlated with host group size in social species as taxonomically diverse as greenflies and monkeys (Coté and Poulin 1995). Parasites, or their **vectors**, that localize their hosts via stimuli whose intensity increases with the number of hosts in proximity, should also be favoured by group life. In fact, the expected positive correlation between parasite prevalence and group size is confirmed in many host–parasite pairings, particularly among primates in relation to malaria (Nunn and Heymann 2005). On the basis of this same relationship, one can predict an increase in wealth of parasite species in host species evolving towards sociality. Thus, by using the comparative method, Tella (2002) has shown that the transition from a solitary lifestyle to coloniality in birds has led to increased species richness and prevalence in parasites transmitted by vector organisms.

In territorial species, those forming permanent or ephemeral groups, it is expected that behaviour leading to the avoidance of contaminated habitats or exclusion of infected individuals will have evolved. For example, the simple attitude of avoiding contact with parasitized conspecifics may allow a healthy individual to reduce the risk of contamination provided it can evaluate the parasitic status of its fellows. In bullfrog tadpoles, *Rana catesbeiana*, the proximity of an individual infected by an intestinal parasite, the yeast *Candida humicola*, increases the risk of infection through ingestion of contaminated water or faeces. Healthy individuals are able to detect and avoid infected conspecifics (Fig. 4.2) particularly through the perception of chemical signals (Kiesecker *et al*. 1999).

In rodents, healthy individuals use body odours to differentiate infected individuals from uninfected individuals (familiar or otherwise). This recognition entails avoidance and the expression of aversive behaviour towards infected individuals, be they competitors or potential mates (Kavaliers *et al*. 2005). The cognitive processes relating these social chemical signals and the recognition of parasitic status derive in rodents from partly known hormonal and neuronal mechanisms (Kavaliers *et al*. 2005). Lastly the aggression of resident individuals towards immigrant individuals seeking to join the group, as has been observed in some primate species, has also been interpreted as a strategy to limit possible risk of contagion (Freeland 1976; Loehle 1995).

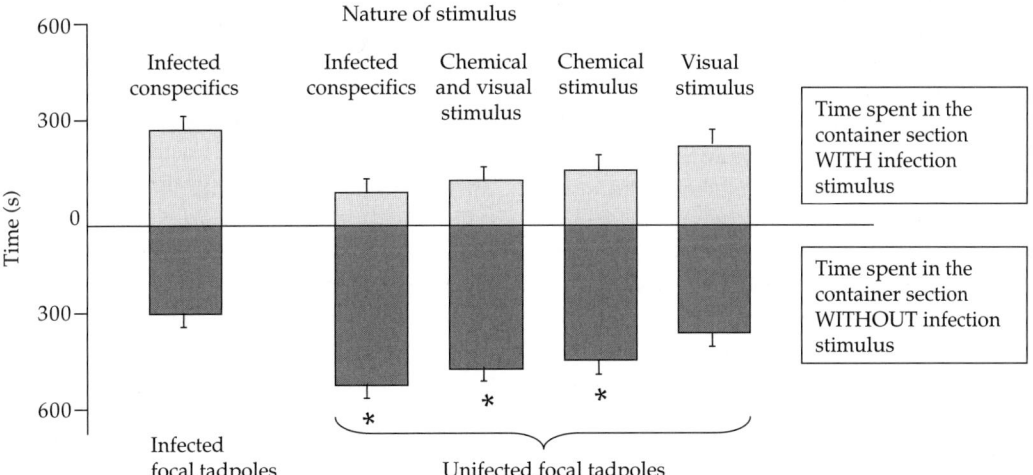

Figure 4.2 Mean time (seconds) spent by *Rana catesbeiana* tadpoles in each of two sections of a container during a test period of 10 min: (1) the container section with an uninfected conspecific or with no stimulus (lower dark part) and (2) the container section with an infected conspecific or with visual and/or olfactory stimuli indicating the presence of an infected conspecific (light upper part). After Kiesecker *et al*. (1999).

The negative effect of group life on the risk of parasite infection that different parasite species or vectors represent may, however, be attenuated by collective strategies aimed at reducing the attractiveness of hosts. The choice of closed sleeping sites (hollow trees or thick tangles of epiphytes or creepers) by some neotropical monkeys has been interpreted as behaviour selected in response to the risk of transmission of the malaria agent, *Plasmodium brasilianum*. On this hypothesis, the use of shelters when resting limits the diffusion of odours and so the risk of being detected by the *Anopheles* sp. mosquito, a nocturnally active vector of *P. brasilianum* (Nunn and Heymann 2005). A comparative analysis of 16 primate genera with varied behavioural strategies has shown a connection between the choice of sheltered or open resting sites and the prevalence of malaria: sleeping in a closed microhabitat reduces the risk of malaria infection, although further studies are needed to increase the power of the tests (Nunn and Heymann 2005). In addition, sleeping behaviour is associated with group size, which remains the host behavioural trait that most influences the prevalence of malaria, with a positive correlation between the number of individuals forming the group (from 3.1 to 37.5 on average depending on genus) and the prevalence of *Plasmodium* (from 0 to 33.3%) (Nunn and Heymann 2005). This finding confirms the decisive role of group size in risk of infection.

Territoriality and site fidelity

Several host species defend a territory with relatively stable boundaries all year round. Others return each year to the same breeding site, either an isolated site or a traditional colony site. In both cases, permanent or regular occupation of the same site facilitates the maintenance and growth of parasite populations associated with the host species. This is particularly true of parasites whose mobility is limited, such as ectoparasites (mites, lice, and fleas, or monogeneans) or parasites transmitted by contamination of resources (food, water) such as endoparasites released in their hosts' droppings.

Various studies have shown that an individual that uses a habitat (e.g. a nest or burrow) permanently or periodically can include the healthiness of the site as a criterion in its habitat-selection process and so reduce the risk of infection for itself and its offspring. For example, the great tit, *Parus major*, prefers an empty nest site not contaminated by the blood-sucking flea, *Ceratophyllus gallinae*, to a contaminated nest site, even if there is a ready-made nest there (Oppliger *et al.* 1994).

Keeping the habitat and its resources healthy involves selective defecation behaviour, allowing occupants to avoid infection through ingestion of parasites released by faeces (be they bacteria, protozoans such as coccids, or helminths). The evolution of selective defecation behaviour is particularly expected in species that build nests or burrows as well as in territorial species. The available empirical data confirm the existence of such a phenomenon. Burrows dug by several small mammal species contain special chambers serving as **latrines**, separated from the space where food is stored. This is the case notably in the mole-rat, *Spalax ehrenbergi*, which systematically reserves a defecation station in its tunnel network (Zuri *et al.* 1997). Similar selective defecation behaviour is observed in several ruminant species. It consists in concentrating faeces in a particular part of the habitat, often on the boundaries of the territory (Ezenwa 2004a). The use of latrines, in the form of faecal mounds deposited at various points in the territory, has also been reported in two species of lemur in Madagascar (Irwin *et al.* 2004).

However, even if such behaviour might have evolved in response to the risk of infection, constraints inherent to the social behaviour of species should not be overlooked. For instance the increase of parasite transmission associated with either group living or territoriality can probably be only partially compensated by 'hygienic' behaviours. Two studies corroborate this. First an interspecies analysis comparing rates of infection rates by strongilid nematodes in territorial bovids with selective defecation behaviour (faecal mounds) and non-territorial bovids shows that the former had significantly higher infection rates than the latter (Ezenwa 2004b). Secondly, a study of Grant's gazelle, *Gazella granti*, shows that the infection rate of territorial males was higher than that of non-territorial males and females because of greater exposure to faecal mounds deposited on the territorial boundaries (Ezenwa 2004b). The use of latrines by other mammals must be interpreted with care. When it detects a risk of intrusion on its territory from another male, the solitary mole-rat, *S. ehrenbergi*, moves its latrines nearer to the unwanted neighbour (Zuri *et al.* 1997). In social species, such as prosimian primates, for which olfactory communication is important in sending signals, the primary function of latrines may be to defend group resources (food, resting site) either by deterring intruders or by leading them to avoid conflict upon encounter with a local group member (Irwin *et al.* 2004). Thus **hygiene behaviour** such as selective defecation may have evolved under constraints other than parasite risk, particularly in direct relation to the marking of territory.

Lastly, when initial choice or maintenance are no longer sufficient to contain the risk of infection, an alternative may be to abandon the territory, nest, or burrow. The same study of the great tit's choice of nest site depending on contamination by fleas also showed that, when tits have no choice and all nest sites are contaminated, 60% of them desert the ectoparasite-infested nest even during breeding (during incubation or after hatching). By comparison only 4.5% of tits that brood or raise their offspring in healthy nests desert the nests (Oppliger *et al.* 1994). By the same token, migratory behaviour may also be interpreted as seasonal or occasional avoidance of an environment where the risk of parasite infection is too high (Combes 2001).

Human history is punctuated by great epidemics (plague, influenza, etc.) and by permanent risks of disease linked to contamination of habitat (especially water) and the presence of vectors. Sedentarization and more recently the intensification of worldwide trade, together with certain heavy infrastructure, have probably contributed to an accentuation of health risks. In Africa the building of a dam resulted in increased prevalence of parasites whose vectors, at one stage of their life cycle, are restricted to the aquatic environment (Zakhary 1997; cf. Chapter 8). Vice versa, emigration of human populations may have been a behavioural response to contamination of the habitat. In the West African savannahs some villages were deserted due to the presence of great numbers of simulium flies. These blood-sucking flies, whose larvae develop in water, are the vectors of the

thread worm, *Onchocerca volvulus*, which causes river blindness or onchocerciasis. In the 1970s this disease could affect more than 50% of men older than 35 years (Mouchet 1984). However, other selection pressures, such as predation or degradation of an overexploited habitat by growing populations, may be alternative or concomitant explanations for such behaviour, making it difficult to interpret such behaviour through a direct connection with parasitism.

Grouping as a means of defence against parasites

Despite the risks of contagion through contact, living in groups may be an effective way of reducing the risk of parasite infection. A group is able to provide the same defence function against parasites as against predators (Combes 1991; Mooring and Hart 1992).

Individually, the benefit of living in a group compared with a solitary existence may result from a reduction in the probability of parasite 'attack'. This 'dilution effect' is effective only if the probability of detection and so of encountering the parasite does not increase with the size of the group ('encounter-dilution effect'; Mooring and Hart 1992). Another behavioural defence allowed by grouping is for an individual to secure protection from parasites or mobile vectors by taking up a position in the centre of the group or close to other animals (Mooring and Hart 1992). This defence is known as the 'selfish herd effect' (Hamilton 1971). There is empirical evidence that these defences afforded by group life—which are familiar antipredator strategies—work as protection against parasites, especially in ungulates (Mooring and Hart 1992). They might also explain shoaling behaviour in certain fish, such as schooling in the stickleback, *Gasterosteus* sp. This behaviour seems to be correlated with the risk of infestation by an ectoparasitic crustacean, the isopod *Argulus canadensis*. As the number of attacks by the parasite is not correlated with the size of the school of fish, the average number of attacks per individual falls as school size increases (Poulin and Fitzgerald 1989). In some mammals the tendency of hosts to form larger groups is a response to the increased presence of ectoparasites (flies, ticks, etc.) probably so as to reduce the likelihood of infestation or lower the parasite burden by dilution (see several references in Altizer *et al.* 2003).

Group living may also allow the emergence of collective-protection behaviour such as clustering that reduces the body surface directly exposed to parasites. Herds of cattle change shape with the intensity of attacks by *Hybomitra* flies that are potential parasite vectors (Ralley *et al.* 1993). Individuals line up (with the dominant individual best protected in the centre) and continue grazing in response to fly attacks. If the level of attacks increases, the herd shape changes to afford more protection, but as a downside grazing is interrupted (Ralley *et al.* 1993).

All the studies of the connections between the evolution of parasite transmission strategies and the evolution of sociality (group size, promiscuity, dispersion, **territoriality**, etc.) suggest that parasitism may have shaped sociality. However, no study to date has shown that group size has diminished in response to parasite risk (Altizer *et al.* 2003).

4.3.2 Prophylactic behaviour specifically related to the risk of infection

Host behaviour specifically involved in avoiding infection by reducing the probability of encountering parasites, or of them becoming established, is known as prophylactic behaviour. It consists in avoiding contamination of the habitat by parasites, lowering a host's attractiveness for parasites or vectors that actively localize their hosts using (often olfactory) stimuli, or preventing parasites from becoming established.

The specific prophylactic behaviour most commonly observed consists in using plants with repellent or toxic effects, mechanically or chemically preventing a parasite from becoming established (Hart 2005). This practice wards off ectoparasites and their vectors or maintains hygiene in a nest or burrow by preventing parasites from accumulating. When ingested, some plant products might protect individuals against worms or microbes becoming established.

Three criteria must be met for the incorporation of such plants into an individual's immediate environment or into its diet to be considered as

prophylactic behaviour (Hart 2005). First, the repellent or toxic (antihelminthic, antibiotic, insecticide, or acaricide) properties of the plant used must be experimentally demonstrable. Thus several medicinal plants, a specific part of which (root, stem, leaf) is consumed by animals, have antimicrobial properties, a rough surface, or a bitter or astringent taste (Hart 2005). Analysis of medicinal plants used in herbariums in the Middle Ages has demonstrated a range of anti-inflammatory, antimicrobial, immunomodulatory, or analgesic effects (Hart 2005).

The second criterion is that there is a correlation between the frequency of the prophylactic behaviour in the population under study (frequency of performance or proportion of a particular plant in the diet) and the risk of infection (measured by the prevalence in the same population at the same time) (Hart 2005). The most telling example is the ingestion of unchewed leaves by chimpanzees coinciding with the rainy season when the risk of infection by strongilid nematodes is highest. Increased gut motility caused by the passage of whole leaves of plants such as madder, *Rubia* sp., is thought to prevent the establishment of these intestinal parasites (see several references in Hart, 2005). Although a therapeutic function may also be invoked (based on the observation that individuals showing signs of infection, such as diarrhoea, display this whole-leaf swallowing behaviour), the increased frequency of this behaviour during seasons where there is a risk suggests a prophylactic function (Hart 2005).

Lastly, to satisfy the final criterion, the ectoparasite or endoparasite burden in individuals having consumed such plants must be lower than that in individuals without prophylaxis. This last criterion is especially difficult to check in the wild, without manipulating the environment or the diet of a control group of individuals. This is why a correlational approach is generally followed, comparing several populations with different risks of infection or a single population exposed to parasites on a seasonal basis and against which the prophylactic behaviour is directed.

Some bird and rodent species have been observed to supply and replenish supplies in the habitat of plants (branches, leaves, or berries) with repellent or toxic effects on ectoparasites. Such defence

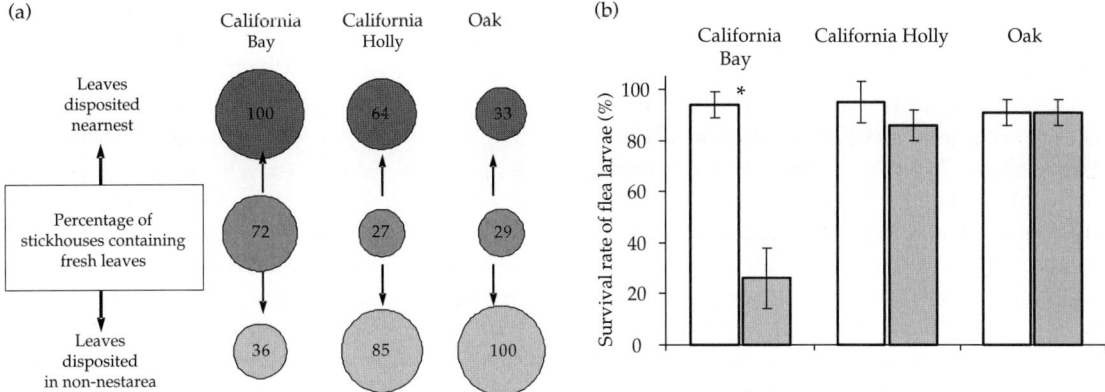

Figure 4.3 Nest fumigation in the dusky-footed wood rat, *Neotoma fuscipes*. (a) Percentages of stickhouses containing fresh laurel leaves (*Umbellularia californica*, Lauraceae), California Holly (also known as Toyon; *Heteromeles arbutifolia*, Rosacae), and/or oak (*Quercus* sp.) (figures in central circles), and, for those containing each plant, the percentage of stickhouses in which leaves were found near the nest (top circles), and/or at a distance from the nest (bottom circles). (b) Survival rate of flea larvae after a 3-day incubation with California bay leaves, Toyon, or oak, depending on whether the leaves are whole (blank bars) or torn (stippled bars) (*$P < 0.0001$). After Hemmes *et al.* (2002).

behaviour is called 'fumigation' of the nest or burrow, although the active plant matter is not released by combustion but by simple volatilization. This hygiene behaviour maintains hygiene in the nest or burrow and offspring-rearing area. The supply of certain plants with bactericidal, acaricidal, or insecticidal properties has been observed in many bird species, from finches to falcons, especially in species reusing nests from one breeding season to the next (Lozano 1998). The blue tit, *Cyanistes caeruleus*, brings from one to five aromatic plants to the nest from the time of hatching and for the next 13 days. These concoctions protect the brood from mosquitoes (and probably from other biting insects) through their repellent effect but possibly also by 'masking' the odour of the chicks making it more difficult for vector insects to detect them by smell (Lafuma *et al.* 2001). Rodents too fumigate their burrows. The dusky-footed wood rat, *Neotoma fuscipes*, builds its nest on the ground or in trees in Californian forests where oaks predominate. These constructions made with plant debris at the forks of trunks and branches act as resting sites, breeding sites (nests), and larders. The rats place fresh leaves in or close to the nest and renew them regularly (Hemmes *et al.* 2002). Nearly three out of four of the constructions contain California bay laurel, *Umbellularia californica*, leaves and inside the

constructions the leaves are invariably found close to the nest and much less often outside (Fig. 4.3a). This distribution contrasts with that of other plants that are used rather as food sources, such as oak leaves and California holly (Toyon) leaves (Fig. 4.3a). The rats nibble the laurel leaves they arrange in their nests, a behaviour that probably favours the diffusion of certain compounds like monoterpene volatile oils (Hemmes *et al.* 2002). The ripped leaves release compounds that are toxic to flea larvae, while the same intact leaves, or oak and holly leaves, have no effect on the same parasites (Fig. 4.3b). This prophylactic behaviour contributes to the avoidance of an accumulation of ectoparasites in these constructions that are reused for several generations (Hemmes *et al.* 2002).

Plants may also be used to protect plumage or fur from attacks by ectoparasites, vector insects, or bacteria. Many bird species and several mammal species (e.g. monkeys, coatis) rub their bodies against vegetation or use leaves, stems, or fruits to this effect. Some birds and mammals practise 'anting', rubbing their bodies with ants, millipedes, or other crushed invertebrates, all having antiparasitic properties (Lozano 1998; Nunn and Heymann 2005). However, multiple functions have been attributed to this behaviour, such as relief of irritated skin or upkeep of feathers and fur

(Lunt *et al.* 2004) so that limiting these adaptive functions to parasite defence is an oversimplification or even mere speculation.

The ingestion of plant products of prophylactic value is akin to self-medication behaviour. This has been suspected in several species of non-human vertebrates for more than 30 years (Lozano 1998). From rhinoceroses to primates, via elephants, the ingestion of stems, leaves, and roots with stimulating, antihelminthic, laxative, or antibiotic properties or simply a high tannin content, seems anecdotal. But the repeated evolution of such behaviour must raise questions about the selective forces at play and the potential to respond to them either through the evolution of innate dietary behaviours including the ingestion of bitter, astringent, or tannin-rich plants, or by individual learning or cultural transmission (Lozano 1998; Hart 2005; Nunn and Altizer 2006). The absorption of small quantities of bitter substances is observed in several mammal species, including primates and humans. Quinine, used as an antimalarial treatment, is extracted from *Cinchona* bark. It was imported into Europe from Peru by the Jesuits in 1627; Peruvian Indians used the plant against malaria (Lederberg 2000). But absorbing bitter substances with prophylactic values is not the preserve of humans. When mice have a choice of drinking straight water or a chloroquine solution, the latter accounts for 10–40% of the total fluid intake in 24 hours. Mice infected by *Plasmodium berghei* (the specific agent of murine malaria) develop less intense parasitaemia and have a higher survival rate so long as they have access to chloroquine (Fig. 4.4; Vitazkova *et al.* 2001). Despite this advantage, infected mice show no initial preference for the chloroquine solution, nor any conditioning or reinforcement in the course of experiment compared with healthy mice. Whatever their parasite status they display significant aversion to the chloroquine solution (Vitazkova *et al.* 2001). It seems, then, that in mice as in other mammals an innate 'repeat sampling' behaviour of unpleasant plant substances has evolved in connection with their prophylactic properties relative to protozoan (plasmodium, amoebae, etc.) and other parasites (Vitazkova *et al.* 2001; Hart 2005). The risks of poisoning (some of these plants contain toxins) are probably limited by the low doses consumed (Vitazkova *et al.* 2001) but

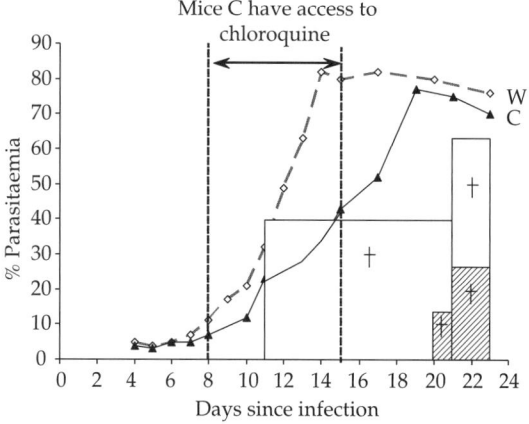

Figure 4.4 Changes in the percentage of parasitaemia over time and in the mortality rate of mice infected by *Plasmodium berghei* depending on whether they had access (C) or not (W) to a 1 mM chloroquine solution (choice between chloroquine solution and water) between days 8 and 15 post-infection. The curves represent the parasitaemia rate and the blocks aggregate mortality (blank, control mice W; shaded, experimental mice C). Initial numbers were eight mice per group. After Vitazkova *et al.* (2001).

may also be reduced by another equally surprising aspect of dietary behaviour: geophagy.

Another behaviour akin to self-medication is the ingestion of earth (geophagy) or of 'stimulating' substances (Lozano 1998). In four species of macaques, gorillas, and chimpanzees practising geophagy, it has been shown that the earth eaten contained three similar types of clay, one of which enters into the composition of a product sold to stop diarrhoea. Many functions, with no direct connection to parasite risk, have been attributed to such behaviour: the ingestion of minerals or the maintenance of intestinal pH values (Lozano 1998; Aufreiter *et al.* 2001), the ingestion of intestinal symbionts (Aufreiter *et al.* 2001), or the neutralization of toxins from foodstuffs, such as alkaloids (Aufreiter *et al.* 2001; Vitazkova *et al.* 2001). Clay-rich soils, nonetheless, have adsorption properties that can provide protection against bacterial toxins, and help form a mucoprotective barrier against intestinal pathogens or inhibit the development of amoebae through competition with ferrous ions (Aufreiter *et al.* 2001). These functions could explain selective consumption by chimpanzees in Tanzania of the earth making up termite mounds (Aufreiter *et al.* 2001).

4.3.3 Limits to effective avoidance

Several types of behaviour that are routine in an individual's life but risky in terms of infection are theoretical targets for natural selection. However, there remain few proven cases, compared with the ubiquity of parasites, where such behavioural adjustments are effectively involved in antiparasite defences. This is probably because of the costs associated with the modulation of behaviour: being choosy about food, mates, or breeding sites entails a cost that is higher when competition is stronger. This cost cannot be offset by the benefit of avoiding infection unless the perception of risk of infection is reliable, and the probability of infection and the **virulence** of the parasite are high. Alternatively, or concomitantly, behaviour may have evolved specifically in the host species in reaction to the risk of infection to avoid encounters with parasites or prevent them from becoming established. Host behaviour to avoid parasites is less frequent than parasite behaviour to reach their hosts (not dealt with here). This can be explained in evolutionary terms in two ways: (1) In the 'arms race' between host and parasite to meet or avoid one another, the result is generally more vital for the parasite than for the host (Combes 2001). Indeed, selection pressure on the parasite to find and become established on (or in) a host is greater than that on the host to avoid the parasite or defend itself against it. (2) Avoidance is difficult because the systems for recognizing parasites by hosts are limited by the very nature of parasites life cycles and strategies of transmission (contrary to the perception of predators), and because the costs are potentially high. This last point leads us to consider the evolution of post-infection behavioural defences as perhaps more probable, the perception of infection by a parasitized host being possible from the very reactions triggered in response to infection.

4.4 Post-infection behaviour: eliminating or limiting pathogenic effects

Once a parasite is established on or inside its host, infection signals could activate specific behavioural defences. Whereas behaviours preventing infection are usually part of an animal's behavioural repertoire associated with feeding, resting, interacting with conspecifics, or reproducing, most behavioural strategies for parasite removal have specifically evolved in response to parasitism. Selective constraints on such behaviours differ between ectoparasites and endoparasites. In both cases, three types of behaviours can be identified based on their consequences: (1) removing the parasite; (2) clearing the infection by killing the parasite; or (3) limiting its proliferation or its development to a non-virulent state. Removal mainly concerns ectoparasites, although some endoparasites such as intestinal helminths can also be removed. The behavioural defences against endoparasites mainly consist in modifying internal physiological conditions to limit the parasite's survival, development, or proliferation.

4.4.1 Behaviours to limit or clear infection

Elimination behaviour is generally 'mechanical' and may be likened to cleaning behaviour. The simplest behaviour consists in driving off or removing ectoparasites. Leg, tail, and muzzle swipes and so on in response to attacks by flies and other biting insects are directly correlated with the intensity of attacks. For example, cattle sprayed with insecticide (cypermethrine) produce significantly less defence behaviour than untreated cattle (Ralley *et al.* 1993).

The most widespread cleaning behaviour is toileting or grooming. We speak of autogrooming or allogrooming depending on whether the infected individual grooms itself or seeks the cooperation of its conspecifics to groom it, particularly for parts of the body that are difficult to reach. Grooming behaviour may be 'scheduled', that is carried out regularly independently of ectoparasite infestation, or induced by skin irritation or bites by ectoparasites (Mooring *et al.* 2004). This mechanical removal is very widespread behaviour in vertebrates and is incontestably effective, especially against ectoparasites such as ticks that are not very mobile. In ungulates, the evolution of complex grooming behaviour and of allogrooming is associated with an increased risk of infection: a comparative analysis of 60 ungulate species showed that this grooming behaviour is, at the interspecies level, correlated with life in forest habitats and with host body size, two decisive parameters for the risk of tick infestation (Mooring *et al.* 2004). The hygienic benefit of grooming has been evidenced in primates as well, although a social function might also explain the evolution of allogrooming (Nunn and Altizer 2006). The costs and benefits of such behaviour must be weighed up (Nunn and Altizer 2006). Cost may be evaluated in time spent on cleaning, reciprocity of the service rendered, or in energy expended. In the mouse-eared bat, *Myotis myotis*, an experimental infestation with the blood-sucking mite *Spinturnix myoti* increased the time spent grooming—behaviour that is induced by infestation stimuli such as cutaneous irritation—by two or three times, depending on the parasite burden. Consequently, time allocated by this bat to resting is halved and oxygen consumption increased by 11–21.3% depending on the mite burden (Giorgi *et al.* 2001). However, despite grooming, the **parasite load** was not reduced while the experiment lasted (Giorgi *et al.* 2001). The actual effectiveness needs to be evaluated too, given that some ectoparasites such as lice may detect grooming by the associated movements and so move away (Clayton 1991).

Other toileting behaviour to remove ectoparasites consists in coating plumage or fur with crushed leaves or invertebrates (see Section 4.3.2) or having mud or dust baths. One means of partial 'de-parasitizing' is to count on a dilution effect by rubbing shoulders with uninfested individuals to which ectoparasites might emigrate. It seems that the adoption behaviour of nest changing in alpine swift, *Apus melba*, chicks has an adaptive function of reducing the chick's parasite load. The chicks that move most are those with the highest burden of blood-sucking flies, *Crataerina melbae*, a burden that falls on average by 42% if they are transferred to a de-parasitized nest (Bize *et al.* 2003).

Cleaning of a habitat that is used permanently or periodically, when contaminated, is also part of an antiparasite defence. For example, nest-cleaning behaviour is known in some birds, including the great tit, *Parus major*. Nest-cleaning time is taken from night-time resting (which falls from 73.5% for females occupying an uninfested nest to 48.1% for females in flea-infected nests) (Christe *et al.* 1996b). The daily foraging effort for the clutch by females in parasitized nests is thus maintained at a comparable level to that of healthy females (Christe

et al. 1996b). The use of fumigation of nests or burrows by some birds or rodents, presented above as a prophylactic strategy (Section 4.3.2) may alternatively or concomitantly be a curative behaviour if the secondary compounds introduced in response to the presence of parasites are toxic for them. The distinction is not always easy to draw in the wild.

Such hygiene behaviour has been known for more than 40 years in social insects and in particular in bees. Bees have developed a whole repertory of behaviour to eliminate or limit propagation of parasites such as bacteria, viruses, or ectoparasitic mites of the genus *Varroa*. For example, *Apis dorsata* and *Apis mellifera* engage in hygiene behaviour consisting in prospecting and cleaning outside cells, or removing a dead pupa from a chamber whose cap is damaged in response to haemolymph discharge (Woyke *et al.* 2004). This hygiene behaviour limits the propagation of varroa, bacteria, or viruses to the rest of the brood. In bees, other antiparasite defence behaviour differs depending on whether the species builds nests in the open air and emigrates regularly or is more sedentary and builds nests in cavities (Woyke *et al.* 2004). In the first case, illustrated by *A. dorsata*, behavioural defences combine hygiene behaviour and migration behaviour to found a new healthy colony (mites cannot survive for more than a few days on an adult bee). In response to the death of a pupa in a cell whose cap is intact, the worker leaves the comb sealed, so preventing parasites from spreading to neighbouring combs. However, the domestic bee *A. mellifera* displays the opposite hygiene behaviour: unlike *A. dorsata* it opens the combs whose caps are not damaged but in which the pupa is dead. As *A. mellifera* is more sedentary, the cost of this parasite-eradication behaviour by removing the pupa (at the risk of reducing productivity of the colony and of spreading the contamination) is probably offset by the benefits of increased hygiene (Woyke *et al.* 2004). In the case of social insects, this hygiene behaviour can control infection of the colony by ectoparasites and endoparasites. In a way, these eusocial species achieve at the colony scale what humans do when they amputate a gangrenous limb to prevent the proliferation of bacteria associated with necrosis.

4.4.2 Behaviour to kill the parasite or slow its proliferation and development

An endoparasite can be killed, its development slowed, or its proliferation contained either by modifying the environment exploited by the parasite (e.g. body temperature, pH, nutritional resources) or by bringing it into contact with toxic substances (e.g. antihelminths, antibiotics).

In ectotherms, one response to infection is to choose habitats, an orientation to sunlight, or a daytime activity period that allow an increase in body temperature. Such a behavioural response is adaptive if the rise in temperature creates harmful or even lethal conditions for the parasite. This behavioural thermoregulation has been reported in species belonging to taxonomic groups as different as insects, crustaceans, scorpions, annelids, amphibians, fish, and reptiles (Ouedraogo et al. 2004). For example the migratory locust, *Locusta migratoria migratoroides*, responds to infection by the entomopathogenic fungus *Metarhizium anisopliae* by moving along a thermal gradient. By raising their body temperature above 40°C by repeated 'sunbathing' for a total of at least 4 hours

per day, the infected locusts considerably increase their survival rate compared with infected locusts that do not practise thermoregulation (Fig. 4.5). However, as soon as infected locusts are deprived of their heat source, fungal proliferation takes over and host mortality increases as a result (Fig. 4.5). Behavioural thermoregulation is therefore an effective defence against proliferation of the entomopathogenic fungus (Ouedraogo *et al.* 2004).

The development and proliferation of parasites that transit through or become established in the gut, **haemolymph**, or blood may be limited by the ingestion of certain plants containing toxic substances that modify the habitat of the parasites (e.g. pH, bio-availability of nutritive elements) or that stimulate immune defences (Lozano 1998; Hart 2005). Such self-medication for therapeutic purposes is suspected in several mammal species. In primates various forms of self-medication behaviour, such as geophagy (Aufreiter *et al.* 2001) and the ingestion of bitter foodstuffs, or of rough whole leaves with a mechanical effect to clear the gut (Lozano 1998; Hart 2005), have been observed in response to symptoms of illness (for a review

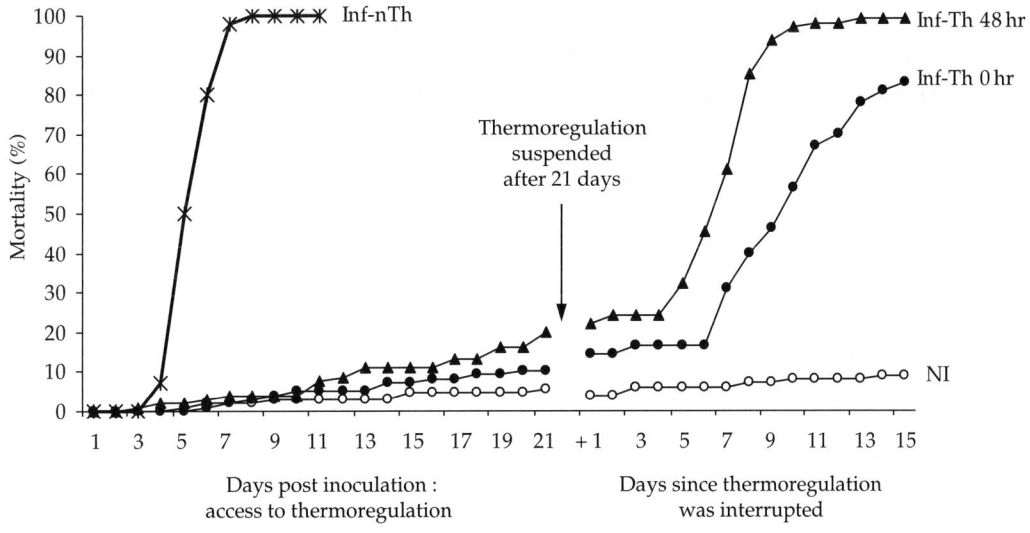

Figure 4.5 Consequences of infection by the entomopathogenic fungus *Metarhizium anisopliae* var. *acridum* (inoculated at 2.103 blastospores per adult male) on the mortality of *Locusta migratoria migratoroides*. Mortality is calculated over the 21 days post-inoculation with (Th) or without (nTh) thermoregulation, and during the days following the interruption of behavioural thermoregulation (no heat source available). Two thermoregulation regimes were applied: access to heat for 0 hours (Inf-Th 0 hr) or 48 hours (Inf-Th 48 hr) after inoculation. NI, non-inoculated control. After Ouedraogo *et al.* (2004).

see Nunn and Altizer, 2006). However, most such behaviour seems to have a prophylactic (preventive and unconnected with parasite status) rather than curative basis, although both functions may sometimes be invoked jointly (Hart 2005; see Section 4.3.2).

The evolution of diet in the genus *Homo*, marked especially by a growing consumption of meat since the early Pleistocene, was probably accompanied by an increased risk of infection. It is likely that this evolution was accompanied from a very early date by self-medication behaviour based on plants with anti-inflammatory, analgesic, immunomodulatory, or antimicrobial properties (Hart 2005). Whether in humans or in other mammals where self-medication is suspected, analysis of the role of learning (by imitation or cultural transmission) or of natural selection in the evolution of self-medication lies outside the scope of this chapter, and may be found in a recent summary of the topic (Hart 2005; see also Nunn and Altizer 2006).

4.5 Manipulation by parasites

The influence of parasites on their hosts' behaviour is not limited to attempts their hosts make to avoid infection or to counter its effects. Actually, the ability, shared by several parasite groups, to modify their hosts' behaviour to their own advantage has been one of the most spectacular discoveries of evolutionary parasitology in the last 30 years. This phenomenon is known as 'parasite manipulation' (see also Chapter 9).

The phenomenon of parasite manipulation is regularly cited to illustrate the concept of the 'extended phenotype' developed by the British evolutionist Richard Dawkins (1982). Dawkins argued that an organism's phenotype is not limited to its bodily envelope. So, for him, beaver dams would be part of the extended phenotype of beavers. One can discuss the point of drawing a distinction between the result of the animal's action on the physical environment and of dam-building behaviour that might be readily connected with the animal's phenotype in the narrow sense. However, the relevance of the concept of the extended phenotype can be easily understood when the influence of an organism's genes is no longer manifest through its

effects on the abiotic environment but through the phenotype of another life form. In such a case, the concept of the extended phenotype draws attention to the fact that phenotypic manifestations in an individual of one species may derive from the action of natural selection on the genome of another species. Thus the adaptive logic of alteration of the phenotypes of infected hosts is often to be sought in the advantage it bestows on their parasites.

4.5.1 Examples of manipulation of host behaviour by parasites

Infected hosts often display what is judged to be aberrant behaviour, whether it is atypical of the ordinary behaviour of the species or whether it contributes to reducing the fitness of the infected individuals, by reducing their reproductive success or lowering their chances of survival.

The phenomenon is particularly obvious in the hosts of various **parasitoids**. One of the most spectacular cases is observed in the spider *Plesiometa argyra* that lives in palm plantations of Costa Rica. This orb-weaving spider normally weaves a circular web, the primary function of which is to capture prey, at least so long as its abdomen has not served as an oviposition site for a wasp of the genus *Hymenopimecis*. As in many hymenopteran parasitoids, the larva of this species develops in its host by sucking the haemolymph. In the first days after infection, the spider behaves normally and continues to weave its usual web. But one night, its weaving behaviour suddenly changes (Eberhard 2000). It then sets about undoing its web to make a new, radically different one. Under the effect of a substance diffused by the parasitoid larva in its abdomen, the spider begins weaving a tougher, rudimentary web, connected to vegetation by a few threads. Once the new weave is completed, the larva mutates and kills and devours the spider before becoming a pupa in its own cocoon, now suspended on a web ready to withstand tropical rains. In a similar register, the behaviour of the buff-tailed bumble bee, *Bombus terrestris*, is largely disrupted by infection by a dipteran larva of the *Conopidae* family. Infected workers bury themselves in the ground just before dying, something healthy workers never do (Müller 1994). In both cases, the modified behaviour has no adaptive consequence for the host. It benefits the parasitoid alone. The cocoon suspended from its threads affords excellent resistance against bad weather and provides the parasitoid hymenopteran with shelter against tropical rains that are particularly deadly for insects (Eberhard 2000). The burial of the bumble bee ensures the dipteran pupa a particularly suitable wintering site. Fatter and heavier adult flies emerge from pupae buried in the ground, with fewer wing malformations than flies from pupae that have hibernated on the ground surface (Müller 1994).

The ability to modify the behaviour of host species is not reserved to parasitoids alone. It is found especially in various helminths with complex life cycles and particularly in the thorny-headed worms (Acanthocephala). These are heteroxene cycle parasites, i.e. they involve at least two hosts. They are known for inducing spectacular changes in behaviour and more generally in the entire phenotype of the various arthropod species that serve as their **intermediate hosts**. These changes can be arranged into two main categories depending on their consequences for the hosts' behaviour. A first possible consequence of infection by an acanthocephalan parasite is a marked reduction in host fecundity that may even extend to complete castration. In female hosts, total or partial castration is a physiological phenomenon involving degeneration of eggs (Bollache *et al.* 2002). The males, however, are not made sterile by infection. It is simply that their behaviour is modified so that they are less able to copulate with healthy individuals: either their competitive capacity is reduced or they are less inclined to react to the presence of a sexually receptive female (Zohar and Holmes 1998; Bollache *et al.* 2001). In nature this phenomenon leads to a lower frequency of pairing among infected hosts than healthy individuals (Fig. 4.6). Such a phenomenon, known as 'behavioural castration', is observed in many host–parasite associations. It is generally considered that it helps to increase the parasite's phenotypic fitness. Like any living organism, the host has only limited energy resources that it can attribute either to growth and maintenance of its physical integrity, or to reproduction. Any reallocation of the host's resources towards the parasite

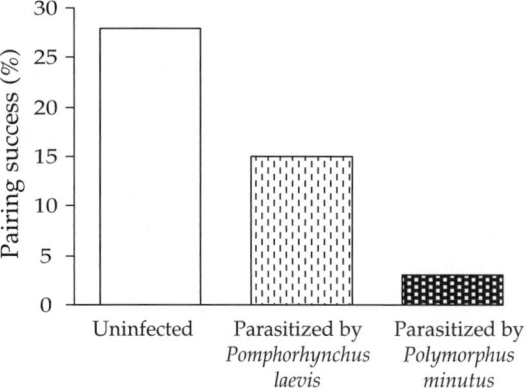

Figure 4.6 Pairing success of *Gammarus pulex* males with their parasite status: healthy, parasitized by acanthocephalans *Pomphorhynchus laevis* or *Polymorphus minutus*. Pairing success is represented by the proportion of males in pre-copula (pre-copulatory grip) in the natural population among all males sampled. After Bollache *et al.* (2001).

would ensure optimal developmental conditions for the parasite within the host.

The second possible consequence of infection by an acanthocephalan parasite is even more drastic: it implies the death of the host as a result of increased vulnerability to predators. In this instance, it is not any predator but usually a species that is a final host for the parasite and in which it can complete its developmental cycle and reproduce. Typically, the behavioural alteration induced by the parasite in its host results in an increased likelihood of an encounter with the final host, leading to increased trophic transmission of the parasite to the final host. Hechtel *et al.* (1993) first showed that the acanthocephalan *Acanthocephalus dirus* reverses the escape reaction of the aquatic isopod *Caecidotea intermedius* when it meets a potential predator such as the creek chub, *Semotilus atromaculatus*, which is also a final host of the parasite. Whereas uninfected isopods move away whenever they perceive the predator, parasitized isopods go in the direction of the predator. Recently, Perrot-Minnot *et al.* (2007) went further by showing that (1) the acanthocephalan fish parasite *Pomphorhynchus tereticollis* reverses the antipredator behaviour of infected *Gammarus pulex* in response to chemical cues from the bullhead, an appropriate final host, and (2) that infected prey are more exposed to predation by bullheads in the wild.

There is nothing anecdotal about the phenomena described above. The ability to alter the breeding behaviour of species acting as intermediate hosts or to modify their antipredator behaviour extends very widely beyond the acanthocephalans. Many examples are found in various parasite species with complex cycles, whether trematodes, nematodes, cestodes, or protozoans (cf. Moore, 2002, for an overview of the phenomenon of toxoplasmosis). A growing number of studies also suggest or demonstrate that parasitic manipulation extends to vector-borne parasites. Indeed, a number of blood-sucking insects (tsetse flies, mosquitoes, sand flies) do manipulate the phenotype of their vector (biting rate) or of their host (alterations of odour profiles making the infected host more attractive to vectors) in ways that could increase transmission (Lefèvre *et al.* 2006).

4.5.2 Manipulation: a reaction for the host's benefit or a simple by-product of infection?

It is tempting to label as 'manipulation' any alteration of an intermediate host's phenotype by its parasite whenever it leads to an increased chance of trophic transmission to a final host. However, it is best to remain prudent and not to jump to conclusions about the adaptive value of the effect of a parasite on its host's behaviour. In effect, a priori, the modification of the host's behavioural phenotype could equally well be an adaptive strategy put in place by the host to counter the effects of the parasite. This argument does seem debatable when the result of the change of the infected host's behaviour is to hasten its capture by a predator and to lead ultimately to a drastic reduction in the host's fitness. Smith-Trail (1980) nonetheless proposed that the intensified predation on infected hosts might be the consequence of a 'suicidal' strategy of infected hosts, serving to limit the spread of the parasite in the population. Other than that it appeals in part to the logic of group selection (an incorrect way of understanding how the natural selection process operates, cf. Danchin *et al.* 2008) this reasoning becomes quite irrelevant when considering trophically transmitted parasites with a complex cycle, in which the intermediate host's would-be 'suicide' ensures the parasite can reach its final host and reproduce.

Beyond any adaptive significance from the standpoint of host or parasite, the parasite's effect on the host's behaviour might just be an inevitable consequence of physiological disturbances due to infection (cf. Edelaar *et al.*, 2003, for a good illustration of this phenomenon). Indirect action of the parasite on its host's behaviour is particularly likely in parasites with simple cycles (Kavaliers and Colwell 1995). For example, an experimental study has recently shown that a parasitic protozoan, *Crithidia bombi*, that lives in the gut of the bumble bee, *Bombus impatiens*, causes severe disruption to its host's cognitive abilities (Gegear *et al.* 2006). In particular, the ability to associate a flower's colour with its sucrose content is largely reduced in infected bumble bees. This marked effect of the parasite that contributes to a reduction in its host's ability to survive does not seem, however, to have any significant effect on its own fitness.

The foregoing example shows that parasites may cause a change in their hosts' behaviour without deriving any particular advantage. It has therefore been proposed to subordinate the use of the concept of 'parasite manipulation' to certain criteria to distinguish what may be considered a truly adaptive character from a simple by-product of infection (Poulin 1995). There are four criteria. The first emphasizes the 'complex' character of the host's changed behaviour. By this principle, the change in the behaviour of the intermediate host must not be a trivial phenomenon. For clearly a simple weakening of the host as a consequence of the pathogenic effect of the parasite may be enough to make it more vulnerable to various types of predators, some of which may be exploited by the parasite to complete its cycle. Yet this first criterion is not easy to apply because of a lack of clear knowledge about the mechanisms behind the manipulation (see below) and because of the recurrent difficulty in defining complexity in biology (Bonner 1988).

A second criterion stipulates that the effect of the parasite must have some purpose, that is, the change in behaviour should, from a functional viewpoint, lead to facilitating the ingestion of the infected intermediate host by the final host. This apparent 'purposefulness' is nothing other than the result of natural selection which, over the course of evolutionary time, has selected among various effects of a parasite species on its host those that best contribute to the completion of the parasite's life cycle. For example, the acanthocephalan *Polymorphus minutus* changes the geotactic behaviour of its host, the amphipod crustacean *G. pulex* (Cézilly *et al.* 2000). A healthy *G. pulex* generally displays positive geotaxis, staying at the bottom of the water column. Conversely, a *G. pulex* parasitized by *P. minutus* regularly ventures towards the water surface. It then becomes easy prey for many water birds that act as final hosts for the parasite. By contrast, the other acanthocephalan parasites, *Pomphorhynchus laevis* and *Pomphorhynchus tereticollis*, the completion of whose cycle depends on their intermediate host being ingested by a fish, do not change the geotaxis of *G. pulex* in any way. However, they reverse its reaction to light: whereas healthy gammarids display clearly negative phototaxis, infected gammarids are attracted by light (Cézilly *et al.* 2000; Tain *et al.* 2006). Both of them also reverse the odour-evoked escape behaviour of *G. pulex* in response to fish cues (Perrot-Minnot *et al.* 2007; Kaldonski *et al.* 2007). Such alterations in behaviour result in a reduction in the use of refuges by parasitized gammarids, increasing their exposure to predation by fish (Perrot-Minnot *et al.* 2007; Kaldonski *et al.* 2007). This is a remarkable example of behavioural adjustment by the effects of three parasites, for which reaching their final host through the same intermediate host involves following radically different pathways.

The third argument for the adaptive character of manipulation rests on the idea of evolutionary convergence, that is, on the fact that the same solution has been found for the same problem several times and independently in the course of evolution. Proving this ideally involves studying the effect of the same parasite on different host species belonging to the same clade so as to judge whether the parasite's ability to manipulate its hosts is ancestral or arose independently several times in the course of evolution. It has thus been shown that the manipulation of the behaviour of different species of cockroaches by the acanthocephalan parasite *Moniliformis moniliformis* is independent of phylogenetic relations among the various host species (Moore and Gotelli 1996). There are still a few such comparative studies that suppose that

one can test the effect of the parasite on the behaviour of a large range of hosts. Nevertheless, the argument of evolutionary convergence may also be invoked when two parasite species that are very remote phylogenetically but depend on the same host species to complete their cycle induce the same behavioural change in their respective intermediate hosts. Let us return to the example of the change in geotaxis induced by the acanthocephalan *P. minutus*. The same manipulative strategy is observed in the trematode *Microphallus papillorobustus* (see Chapter 9) which, like *P. minutus*, exploits an amphipod crustacean as its intermediate host and a water bird as its final host (Helluy 1983, 1984). Because divergence between the evolutionary lineage leading to trematodes and that leading to acanthocephalans is particularly old and most certainly precedes the evolution of complex cycles in both groups, it is reasonable to think that the similarity between the effects of both types of parasite on their intermediate hosts is an example of evolutionary convergence and therefore that the effects observed are indeed adaptive.

The final criterion to be observed before concluding that a manipulation is adaptive is to check whether the manipulation translates effectively and directly into increased fitness of the parasite, that is, increased chances of surviving and reproducing. In the case of parasites with complex cycles, many laboratory studies and some in natural conditions have sought to quantify the effect of infection on trophic transmission to the final host (Lafferty and Morris 1996; Bakker *et al.* 1997; Voříšek *et al.* 1998; Seppälä *et al.* 2004; Lagrue *et al.* 2007). The results are in most cases quite demonstrative. As a general rule, predators consume a proportion of infected prey that is significantly higher than the prevalence of infection (Table 4.1). However, some studies show that behavioural changes accompanying infection do not necessarily translate into an increased risk of predation of the intermediate host by the final host (Webster *et al.* 2000; Seppälä *et al.* 2006).

4.5.3 Parasite manipulation mechanisms

Although the ecological and evolutionary aspects of parasite manipulation are reasonably well understood (Moore 2002; Thomas *et al.* 2005), exploration

Table 4.1 Studies demonstrating overconsumption of intermediate hosts infected by different predator species constituting an appropriate final host

Parasite species	Intermediate host species (prey)	Final host species (predator)	Type of experiment	Prevalence (in intermediate host)	Proportion of infected prey consumed	Reference
Acanthocephalans						
Polymorphus paradoxus	Gammarus lacustris	Anas platyrhynchos (mallard)	Predation experiment (lab.)	50%	78%	Bethel and Holmes (1977)
Pomphorhynchus laevis	Gammarus pulex	Gasterosteus aculeatus	Predation experiment (lab.)	50%	61.50%	Bakker et al. (1997)
Pomphorhynchus laevis	Gammarus pulex	Cottus gobio	Prevalence measurements (wild)	0.38%	17%	Lagrue et al. (2007)
Pomphorhynchus terreticollis	Gammarus pulex	Cottus gobio	Prevalence measurements (wild)	1%	10.2%	Perrot-Minnot et al. (2007)
Acanthocephalus dirus	Asellus aquaticus	Semotilus atromaculatus	Predation experiment (lab.)	50%	63% (light substrate)–88% (dark substrate)	Camp and Huizinga (1979)
Plagiorhynchus cylindraceus	Armadillidium vulgare	Sturnus vulgaris	Predation experiment (wild)	50%	59%	Moore (1983)
Cestodes						
Cyathocephalus truncatus	Gammarus lacustris	Salvelinus alpinus	Prevalence measurements (wild)	0.49%	3.72%	Knudsen et al. (2001)
Trematodes						
Euhaplorchis californinesis	Fundulus parvipinnis, killifish (2nd intermediate host)	Various species of egret, heron	Predation experiment (wild)	64%	98%	Lafferty and Morris (1996)

of the underlying physiological mechanisms is only in its early stages. However, a better understanding of this aspect of the problem is a major step forward in understanding the phenomenon as a whole.

The phenomenon of parasite manipulation regularly expresses itself as the change in a **taxis**, that is, a movement or an orientation reaction of the host in relation to various stimuli, such as gravity in the case of geotaxis, light in the case of phototaxis, or chemical signals like those emitted by a predator in the case of chemotaxis (Heymer 1977). Such behaviour is usually modulated by different neuropeptides (Harris-Warrick and Marder 1991).

One of them, serotonin (5-HT) was implicated in the phenomenon of manipulation in different host–parasite associations (Terenina et al. 1997; Overli et al. 2001; Poulin et al. 2003) in particular in connection with the reversal of phototaxis in amphipod crustaceans parasitized by trematodes (Helluy and Thomas 2003) or acanthocephalans (Helluy and Holmes 1990; Maynard et al. 1996; Tain et al. 2006). The pioneering work of Helluy and Holmes (1990) first showed that the injection of 5-HT in amphipods of the species *Gammarus lacustris* mimicked the inversion of phototaxis caused in this species by the acanthocephalan parasite *Polymorphus paradoxus*. Helluy and Thomas (2003) then showed that in the brain of *Gammarus insensibilis* infected by the trematode parasite *M. papillorobustus*, some areas displayed markedly higher serotonergic activity than in healthy individuals. More recently, by combining behavioural observations with pharmacological and immunocytochemical approaches, Tain and his co-workers (Tain *et al.* 2006) have been able to establish, from a comparative study of the effects of three acanthocephalan parasites on their common intermediate host, the amphipod crustacean *G. pulex*, that the reversal of phototaxis is functionally related to increased serotonergic activity in the brain of infected individuals.

While that work has shown the physiological pathway involved in the phenomenon of phototactic manipulation, how the manipulating parasites operate remains a mystery. It seems unlikely that the parasites can synthesize and release into their hosts quantities of 5-HT similar to those injected into healthy gammarids to reverse phototaxis. It is more likely that the parasites excrete substances than can interact with the host's nervous system. It has recently been suggested (Adamo 2002) that the manipulation could have evolved from the ability of parasites to overcome the host's immune defences by virtue of a connection between the nervous system and the immune system. Thus the active defences developed by parasites to counter attacks of the host's immune system are thought to have side-effects on the nervous system and the host's behaviour, and these effects could in some cases be beneficial to the parasite. This scenario has not yet been confirmed by empirical data but is currently one of the most promising lines of research for understanding the evolution of parasites' ability to manipulate their hosts' behaviour. Alternatively, studies using differential screening of either the proteomes or the transcriptomes of uninfected, infected unmanipulated, and infected manipulated prey or vectors, have recently been highlighted as a way to reveal the molecular pathways involved in parasitic manipulation (Biron *et al.* 2005; Ponton *et al.* 2006; Lefèvre *et al.* 2008; see also the appendix on Methods). The proteomic analysis of excretory/secretory products from infective macroparasites might also contribute to unravelling the mechanisms underlying parasitic manipulation.

4.5.4 Unanswered questions

Despite the diversity of host–parasite associations that have been studied, the phenomenon of manipulation remains imperfectly understood. Several areas of doubt persist, providing new directions for future research.

First, most studies of manipulation have been made on infected hosts taken from the wild so that it is always possible that the change in behaviour observed is not the consequence but rather the cause of the infection and therefore is not necessarily adaptive. However, a simple argument goes against this idea: the change in behaviour generally arises only when the parasite has reached a sufficient degree of development in its intermediate host to be able to infect the final host (Bethel and Holmes 1974; McCarthy *et al.* 2000). In any event, resort to experimental infestation (see Franceschi *et al.* 2008) is a promising avenue for the future, especially for understanding how genetic and environmental factors interact to determine the ability of parasites to manipulate their hosts' behaviour.

Secondly, the effectiveness of the change in a host's behaviour in the case of trophic transmission to a subsequent host deserves further attention. The very great majority of experiments conducted have confirmed that a predator, acting as the suitable final host for the parasite, consumed more parasitized prey than healthy prey when given a choice. Now, the disrupted behaviour of the host induced by infection may equally well increase the host's vulnerability to other types of predator that are not suitable final hosts for the parasite, but rather a dead end (Mouritsen and Poulin 2003). Future work should therefore attempt to show that manipulation of the host's behaviour leads to an increased likelihood of the intermediate host being consumed by a suitable final host rather than by any other type of predator.

Thirdly, experiments conducted to show the greater vulnerability of infected hosts to predation need to be improved upon. In most cases, parasitized hosts are present in equal numbers with healthy hosts, which amounts to mimicking a prevalence of 50%. But in nature the prevalence of most manipulating parasites seldom exceeds

5–10%. The influence of parasite prevalence on effective manipulation by parasites is still to be confirmed, then. In the same vein, it would be necessary to pair more often the quantification of predation on infected hosts under experimental conditions with a prevalence measurement of parasitized hosts in the diet of final hosts compared with the prevalence of infection in the environment (cf. Moore 1983; Lagrue *et al.* 2007; Perrot-Minnot *et al.* 2007).

Important points

• Parasites exert considerable influence over the behaviour of their hosts.
• In the course of evolution, hosts have put in place behaviours with prophylactic effects to prevent infection or therapeutic effects to counter infection. However, it is not always easy to distinguish between the two types of effect in practice.
• Some parasites, especially those with complex cycles, change the behaviour of their hosts. These changes often increase the parasite's phenotypic fitness at the expense of that of its host(s). Such manipulation of the host is essential in the trophic transmission of some parasites from an intermediate host to a final host.

• It is often difficult to discern which behavioural changes are adaptive and which are merely side-effects of infection, and, if adaptive, who benefits. The outlined criteria can be used to experimentally establish the adaptive significance of altered host phenotypes.

Questions for discussion

• Make forecasts about the possible ways for parasites to resist their hosts' defence behaviour (pre- or post-infection), depending on the mode of transmission and the parasite life cycle. Discuss why such resistance might not arise.
• What can we gain from knowing the physiological mechanisms of manipulation from the perspective of understanding how the phenomenon arose and has been maintained over the course of evolution?

Further reading

• Jog, M. and Watve, M. (2005). Role of parasites and commensals in shaping host behaviour. *Current Sciences* **89**(7): 1184–1191.
• Sorci, G. and Cézilly, F. (2008). Inter-specific parasitism and mutualism. In: *Behavioural Ecology* (ed. E. Danchin, L.-A. Giraldeau and F. Cézilly), pp. 615–643. Oxford University Press, Oxford.

CHAPTER 5

Parasitism and hybrid zones

Catherine Moulia and Pierre Joly

5.1 Introduction

Hybridization between animal **taxa** was underestimated for a long time and its contribution to the dynamics and evolution of biodiversity has been understated. The first studies and analyses explained hybrid zones as transient phenomena and barriers preventing genetic exchanges between two taxa. Then, further studies describing long-lasting hybridizations with substantial genetic exchanges revealed that, depending on the case, the **fitness** of the hybrids could be inferior, equal, or superior to the fitness of the parents. The issue became how to determine the selective factors favouring or disfavouring hybrids, and among them the internal (genetic) or external (environmental) factors. Of the biotic factors, parasitism represents a major evolutionary constraint which may influence hybridization favourably or unfavourably.

In this chapter we analyse the role of parasitism in the evolution of host taxa by means of two examples:

- The increased susceptibility to pinworms of complex hybrids of mice, *Mus musculus*, observed though a comprehensive study, from field observations to experimental analysis, of the immunological and evolutionary aspects of the phenomenon.
- The increased resistance of hybrid water frogs to lung helminth parasites in a specific case of hybridization: **hybridogenesis**.

Those two examples highlight the importance of parasitism in evolution. The consideration of parasitism at the level of the host taxa as well as at the level

of the interactions between taxa, in such a frequent and important phenomenon as hybridization, is essential for predicting the evolution of biodiversity.

5.2 Hybridization and the fitness of hybrids

Defining hybridization is as problematic as defining **species**, as one is so intimately related to the other. In theory, hybrid zones are either the result of a divergence in **sympatry**, or the result of the meeting of two taxa, each having diverged but not yet having reached the level of a species as strictly defined (hybridization implies that the individuals have at least partial interfecundity). Natural hybridization is not an anecdotal biological phenomenon as cases are numerous and concern all the main groups of living creatures. In 1985, a team had already recorded over 150 hybrid zones which had become the subject of more or less detailed studies (Barton and Hewitt 1985).

Hybrid zones may correspond to different types of situation. In some cases, only first-generation hybrids (F_1, F_2, or backcrossing) are present: there is little crossing between the two genomes because meiotic recombination happened only for the first or second generations and the hybridization is mostly limited in numbers and area of distribution. Conversely, some hybrid zones are made of 'complex' hybrids, resulting from multiple and repetitive recombinations. For each zone an **introgression index** is defined, evaluating the crossing level of the parental genomes in each hybrid category.

---

* kindly translated by Dr Nathalie Le Brun

5.2.1 Hybridogenesis: a stable or transitory phenomenon?

Hybrid zones are generally understood from one of two opposing points of view:

• Starting from the assumption that the observed divergences accumulated by the taxa are all causes of incompatibility, analysis of hybrid zones must identify the mechanisms which limit the **gene flow** between two taxa, and thus lead to speciation with complete isolation of the genome. Therefore, hybridization is part of the speciation process, in particular with the selection of features favouring the definitive separation of the taxa (like the choice of sexual partner) (Howard 1993; Burke and Arnold 2001). Moreover, among certain species complexes, hybridogenesis—or **hemiclonal** reproduction—is a mechanism which maintains the integrity of the parental genome and limits genetic exchanges despite the existence of crossings. In effect, hybridogenesis corresponds to hybridization without genetic recombination, with the elimination of one of the germ lines before meiosis. Consequently, the hybrids always remain **heterozygous** for all loci. Up to a certain point, we can consider them as permanent F_1 hybrids.
• Conversely, the other approach considers that the mix of two genetic pools constitutes a source of new variation (DeMarais *et al.* 1992; Grant and Grant 1996). Analysis of hybrid zones reveals how new adaptations or new species may appear (Rieseberg 1997; Rieseberg *et al.* 1999; Arnold 2004).

Whichever point of view is adopted, hybridization is generally considered to be a transient phenomenon, bound to disappear either because of isolation of the taxa or the emergence of a new species (Burke and Arnold 2001). Meanwhile, we observe that most hybrid zones appear to be stable, at least on the scale of our perception of time (Moore 1977; Barton and Hewitt 1985; Arnold 1997). For the majority of cases of hybridization this stability seems to result from the incidence of multiple factors having opposing effects on genetic exchange between taxa. Therefore, the two positions described earlier, where the factors acting on the hybridization are clearly favouring or disfavouring gene flow, represent the two

extreme cases among the many possible situations. Between those two extremes most of the hybrid zones represent a relatively durable barrier acting as a more or less efficient filter of gene exchange between the two parental taxa (F. Bonhomme, personal communication).

5.2.2 The fitness of hybrids

The future of the hybridization will depend on the fitness of the hybrids, which depends on **exogenous** (extrinsic or external to the hybrids) and/or **endogenous** (intrinsic) factors. In short, we have two important theories that are in total opposition: the theory of ecotones giving the major role to exogenous selection (Moore 1977) and the theory of tension zones where selection is primarily endogenous depending mainly on genetic processes (Barton and Hewitt 1985; Hewitt 1988).

5.2.2.1 The theory of ecotones
This theory indicates that within hybrid zones the hybrids possess a higher fitness than the parental populations. Those hybrid zones are localized in areas of transition between ecologically contrasting regions (Anderson 1948) and, according to this theory, the hybrids are better adapted than the parental taxa to ecological conditions which are intermediate between the optimum conditions for those taxa. Also, the hybrids seems to be better adapted to new ecological conditions resulting from natural or anthropogenic disturbances, which are adverse for the parental taxa (Moore 1977).

5.2.2.2 The theory of tension zones
This theory relies on intrinsic **counter-selection** factors. The hybrids possess a lower fitness than the parental populations, and the relative stability of the zone results from the equilibrium between counter-selection of hybrids (which would lead to their extinction) and parental migration (which leads to new hybridization events) (Barton and Hewitt 1985). In this case, the decrease in hybrid fitness could be due to incompatibilities between alleles either within a same gene (locus) (Dobzhansky 1936; Muller 1942) or within multigenic systems. Indeed, in parental populations which are (or were) relatively isolated, mutations would have caused

a divergence between the two sympatric populations. New mutations which then accumulate in a population are exclusively selected in the genetic background of this population. The further apart the populations have diverged (and therefore the greater number of new alleles that have appeared), the less likely it is that the new alleles, which have established interactions with the genome, are compatible with the alleles of another genome (Fig. 5.1). In hybrids, the mix of two genomes would create allelic associations with little functionality. The term used is 'disruption of co-adaptation' and the resulting **disgenesis** would lead to decreased viability or fertility of the hybrids.

Therefore, determining the fitness of the hybrids in a hybrid zone is one of the key elements for understanding how the zone functions. But an actual comparison in natural conditions between hybrids and parents, and between hybrids themselves, is often a complex and difficult process.

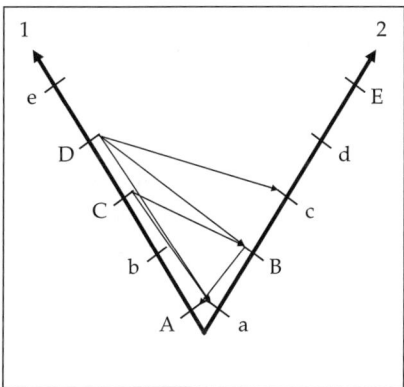

Figure 5.1 The emergence of genetic incompatibilities by allopatric divergence. The two bold arrows represent two allopatric populations (1 and 2) coming from an ancestral population whose alleles, represented by lower-case letters, are fixed. Time flows in the direction of the arrows. Incompatibilities between the new alleles and the native ones occur at each new allelic substitution in each population. The first substitution happens in population 1: the new allele A is fully compatible with the ancestral background. The second substitution of b for B in the population 2 causes incompatibilities with allele A. Then the substitution of c for C happens in population 1: the new allele C is incompatible with either a or B; and so on; D is incompatible with a, B, and c. This diagram shows that the later a substitution occurs, the higher its incompatibility. Therefore, the incompatibilities between alleles seem to increase with time in a nonlinear fashion. Modified from Orr (1995).

Nevertheless, many markers can be used to estimate the fitness of hybrids. The variables to focus on are the viability of the hybrids, their growth rate, their developmental process, and their fertility and/or fecundity (e.g. Bert *et al.* 1993; Nürnberger *et al.* 1995). In general, the performance of hybrids is inferior to that of the parents (Arnold and Hodges 1995; Jiggins and Mallet 2000). But a small group may have a higher fitness, at least in some environments (Barton 2001).

The current ideas about hybrid fitness and the future of a hybridization may be summarized as follows: not all hybrids are superior or inferior to their parents but they can be assigned to categories of different fitness. The future of the hybridization will depend not only on the proportions of those categories and on the habitats and niches helping (or not) the establishment of the hybrid, but also on the genetic bases determining the superiority or inferiority of the hybrid (for a recent review, see Burke and Arnold 2001).

5.3 The influence of parasitism on the future of hybridization

All living organisms are affected by parasites, and all parasites impose a cost on their host. Indeed, it is now commonly recognized that even 'silent' parasites or those with low pathogenicity have a negative impact on the physiological state of the host, and therefore on host fitness. In addition, parasites, by continually evolving, impose on their host populations a constant, dynamic, and ever-changing selective pressure. On the larger evolutionary scale this implies that parasites represent a major selective force which deeply influences the future of populations, and beyond that the life-history strategies of host taxa.

It is therefore assumed that parasites represent biotic factors which will favour or oppose hybridization. For example, one may assume that the propagation of parasites depends partially on ecological parameters, and that these parameters are different according to the environment(s) of the hybrid zone and the optimal environment(s) for the parental taxa. Conversely, all epidemiological and environmental parameters being equivalent, the intrinsic capacity to resist parasites may differ between individuals of different parental type and individuals with a hybrid genotype. As an exogenous environmental factor or an endogenous factor affecting genomic hybridization, parasites, by locally regulating competition between hybrids and parent taxa, can help us to understand the nature of a hybridization and its effect on the evolution of the parent taxa.

Within this framework, parasitological studies have principally been conducted on plant models and only rarely on animal models. After a short summary of the data and concepts from the plant kingdom, we will concentrate our discussion on the interactions between parasites and animal hybrids.

5.3.1 Parasitism in plant hybrids: hybridization favouring biodiversity

In plant hybrid zones, the consequences of selective pressures exerted by herbivores as a whole (i.e. all the members of the same community who feed on plants) have been analysed. In addition to feeding, parasites *sensu stricto* multiply, develop, and live on and/or in plants in a 'lasting interactive' manner. The first analyses were descriptive, comparing the herbivore distributions within each existing population (parental and hybrid). The distributions observed were different from one hybrid zone to another, and within the same hybrid zone from one herbivore to another. The results indicated a diversity of hybrid responses which reflected the following possible hypotheses (Fritz *et al.* 1994), in decreasing order of frequency:

• higher susceptibility in the hybrid than in the parents;
• higher resistance in the hybrid than in the parents;
• intermediate susceptibility in the hybrid;
• the same resistance (or susceptibility) in the hybrid as in one parental type;
• the same resistance (or susceptibility) whatever the host type.

Studies of the consequences of plant hybridization for several herbivores within the same zone gave diverse, sometimes contradictory, results (Straus 1994; Fritz *et al.* 1996, 1999). In modifying the

distribution of a single element of the community (increased or decreased presence of that herbivore), hybridization may have indirect consequences for local biodiversity (Dickson and Whitham 1996; Waltz and Whitham 1997). In addition, hybrid plants can be more resistant, or conversely more susceptible, to other herbivores, without any visible consequences for the community (Whitham *et al.* 1999).

The genetic and/or environmental origin of the differences in herbivore distribution observed between hybrid and parental populations has been amply studied and commented upon (for reviews see Fritz 1999; Fritz *et al.* 1999). For example, a disruption in co-adapted genes was mentioned as the origin of the susceptibility in certain models without ever having been satisfactorily demonstrated (Fritz 1999). More generally, it is quite clear that the mechanisms behind the origin of the different susceptibilities of hybrid plants can be quite different according to several criteria:

• the number of genes involved in the resistance of each parental type;
• the expression of this resistance (depending on dose, or threshold linked);
• the hybrid classes considered (F_1, F_2, backcrossing); and
• more generally, the genetic structure of the hybrid zone (hybrid types and genetic variability).

Now that we considered the question of an exogenous or endogenous origin for herbivore distributions in plant hybrid zones, we will review the consequences of the hybridization for herbivore evolution and for the interactions of herbivores with the 'hosts' (for review see Fritz 1999; Fritz *et al.* 1999; Moulia 1999). For example, hybrid zones in which hybrids are more affected by herbivores than the parental taxa are considered either as traps, limiting the possibility of adaptation of the herbivores to hosts with a more resistant genome and, therefore, limiting their proliferation outside the zone itself (the 'hybrid-as-sink hypothesis'; Whitham 1989; Whitham *et al.* 1994) or as refuges, favouring the maintenance of biodiversity (at several trophic levels) (Whitham *et al.* 1994; Dickson and Whitham 1996; Fritz *et al.* 1999).

Finally, certain plant hybrid zones, in which herbivore distributions are intermediate between the resistance of one taxon and the susceptibility of the other, have been considered as genetic bridges favouring the adaptation of herbivores to the new host genome, hence increasing their host range (the 'hybrid bridge hypothesis'; Floate and Whitham 1993).

5.3.2 Parasitism in animal hybrids: a regulator limiting hybridization

5.3.2.1 An underestimated evolutionary factor

In 1999, Moulia recorded six animal hybridization models involving very diverse hosts and parasites: Cyprinidea/helminth (Dupont and Crevelli 1988),[1] duck/protozoan (Mason and Clark 1990), mouse/intestinal helminth (Moulia *et al.* 1991), mussel/trematode (Coustau *et al.* 1991), Cyprinidea/Monogenea (Le Brun *et al.* 1992), and pocket gopher/Mallophaga (Heaney and Timm 1985). The recently described *Daphnia*/protozoan system provides a seventh example studied under natural conditions (Wolinska *et al.* 2004). And finally, the case currently under study of the parasites of water frog hybrids (a **hybridogenetic process**) completes our existing data. Nevertheless, available detailed data on the relationship between animal hybrids and parasite infestations remain limited, for two main reasons (Moulia 1999):

1. First, parasitism remains underestimated as a source of evolutionary pressure in the animal kingdom. Consequently, parasite diversity and the diversity of the pressures influencing a given host are little known and quantified in terms of parasite richness and biogeography, and are therefore not well identified by observers.

2. Secondly, the development of the concept of hybridization in the animal kingdom was very influenced by Mayr's vision (Mayr 1963), according to which it is a rare event resulting from secondary contact and part of the ineluctable process leading to total reproductive isolation of the two parental taxa. It is only recently been recognized

that hybridization between animal taxa occurs relatively often (Barton and Hewitt 1985), that it is important in solving certain speciation mechanisms (Hewitt 1988; Harrison 1993; Orr 1995), and that it has a role in gene exchange.

5.3.2.2 Studied models

As with plants (see Section 5.3.1), the most frequent types of parasite distribution in the case of the studied hybridizations are:

• increased susceptibility of the hybrids compared with their parents: this occurred in four of the seven models, including the mouse/helminth case which will be discussed later;
• intermediate susceptibility of the hybrids compared with their parents: this was observed in two of the seven cases, including mussels and their trematode parasites;
• a slightly (but significantly) higher resistance of the hybrids in the hybridogenetic complex of water frogs.

In contrast to what was observed in plant hybrids, environmental conditions do not seem to be behind the modification of host–parasite relationships in animal hybrid populations. Moreover, behavioural traits seem to be the major biological component of host–parasite relationships in animals. Indeed, the spread and transmission of parasites, and consequently their distribution in the population, show a strong dependence on the behaviour of the host. Finally, the features involved in the host–parasite relationship that are affected by the hybridization seem to be different (or are affected to a lesser degree) in animals and in plants (Moulia 1999).

In spite of their small number, studies of animal hybrid zones are the only ones to consider the consequences of the modification of parasite distributions on the hybridization itself (Moulia 1999). We will now present the three most significant models demonstrating that point of view:

1. Model 1: The first model is the trematode *Prosorhynchus squamatus* in the area of hybridization of the mussels *Mytilus edulis* and *Mytilus galloprovincialis* on the French Atlantic coast (Coustau *et al.* 1991). This trematode is a parasite with a **heteroxenous** cycle, having three successive hosts, the

---

[1] In this case, the hybrids are sterile F_1! Therefore, this is an 'extreme' case where we can consider that the hybridization is non-existent and that the taxa are in a total reproductive isolation. This needs to be considered with the other peculiarity of this model: the F_1 hybrids are not resistant (in contrast to most data on the F_1 phenotype).

mussel being its first intermediate host. It multiplies asexually and produces a large number of infective larvae in each infested mollusc. The difference in susceptibility between the taxa *edulis* and *galloprovincialis* is clear: *M. edulis* is castrated by the parasite while it does not succeed in infecting *M. galloprovincialis*. Studies of the presence of the parasite in the sympatric and hybrid zone of the two mussel taxa reveal that only individual *M. edulis* and hybrids with a genome mostly of the *edulis* type (we are dealing here with complex hybrids) are infested. Therefore, the parasite, by reducing to nil the reproductive fitness of *M. edulis*, can become a factor in favour of the taxon *M. galloprovincialis* in endemic areas where the two parental taxa hybridize.

2. Model 2: The second example involves helminth parasites in the hybrid zone between the two European mice *Mus musculus musculus* and *Mus musculus domesticus*. This model, the subject of 15 years of study, represents a fine example of the limitation of gene exchange between host taxa helped by parasites. The mouse, a mammal whose physiology, immunology, genetics, and population biology are well documented, is a laboratory animal *par excellence*, allowing many experimental approaches. Therefore, the descriptive analysis resulting from field observations was completed by studies in standardized and controlled conditions.

3. Model 3: The third example relates to the helminth parasite of water frog hybrids. The specific hybridization of the hosts (hybridogenesis) is associated with a relatively rare case in the animal kingdom (see above): resistance of the hybrid to certain helminths.

We will develop those two latter examples in the rest of this chapter.

5.4 The susceptibility of hybrid mice to helminths

5.4.1 The *musculus/domesticus* hybrid zone

The two European subspecies of the complex *Mus musculus*, *M. m. domesticus* (in the west) and *M. m. musculus* (in the east) hybridize in a zone

stretching from Denmark to Bulgaria (Fig. 5.2). This zone could be the result of secondary contact between the two recently differentiated taxa (Boursot *et al.* 1993).

5.4.1.1 *Characteristics of the hybrid zone*
Some slight morphological variations (Zimmerman 1949) and molecular diagnostic makers allow us to refine our understanding of this hybridization (Boursot *et al.* 1993). We are clearly dealing with complex hybrids arising from numerous recombinations. The variations in the frequency of **diagnostic markers** throughout the hybrid zone are defined by clines. The transition from one taxon to another happens rapidly (over about 20 km in most cases; see Raufaste *et al.* 2005). This was noted in different parts of the zone. The convergence of those observations suggests counter-selection within the hybrid zone. A major question emerges: what are the factors responsible for the small genetic flow and for the stability of the zone?

5.4.1.2 *The factors involved in the stability of the hybrid zone*
Considering exogenous (environmental) factors, the adaptation of *M. m. musculus* to a continental climate could explain the stability of the hybrid population in the area of transition between oceanic and continental climates (Klein *et al.* 1987). Indeed, the hybrid zone coincides in its southern part with the 0°C isotherm for January. Specific adaptations to a pattern of temperature for each of the taxa concerned could explain the location of the zone (Boursot *et al.* 1984). However, even if they contribute to the actual geographical location of the zone, those factors do not satisfactorily explain its rather narrow spread and its stability. We must turn to another hypothesis favouring the role of endogenous selective factors.

According to this hypothesis the hybrid zone is considered to be a zone of tension (Hewitt 1988): the clines prove the existence of incompatibilities between the two different genomes. And indeed, several indications, like the abrupt clines of markers related to sex chromosomes and the lower fertility of the F_1 generation from experimental *musculus/domesticus* crosses, confirm the existence

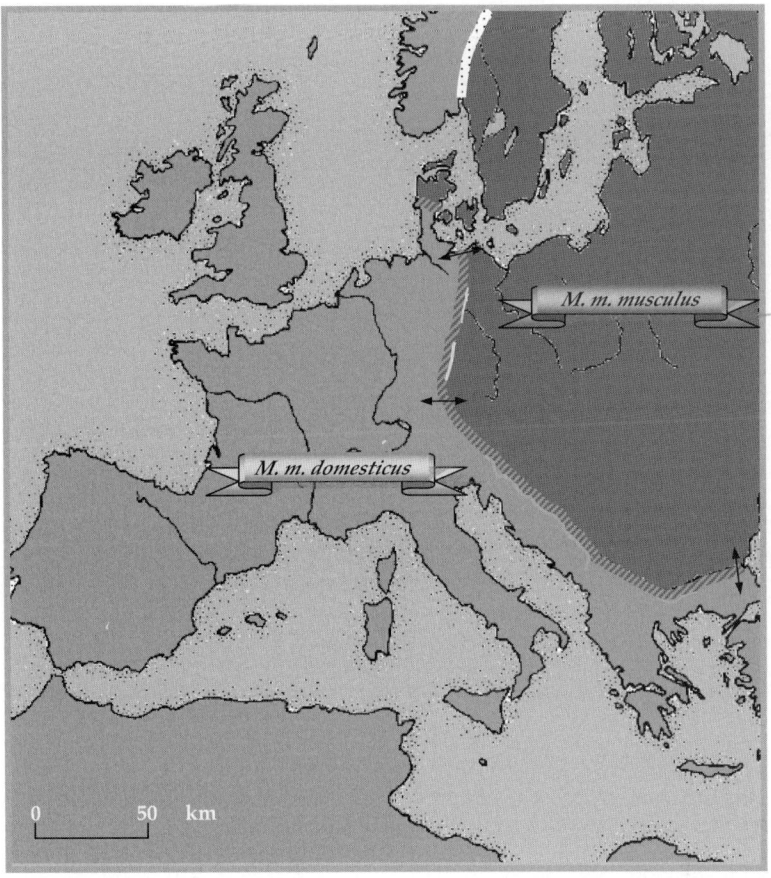

Figure 5.2 The hybrid zone between *Mus musculus domesticus* and *Mus musculus musculus*. The hybrid zone is represented by the striped area. In the dotted area continuity of the hybrid zone is unclear.

of hybrid counter-selection (and therefore support the hypothesis of a tension zone). In addition to this putative **post-zygotic** isolation, olfactory discrimination between the two taxa and a mating preference in the individual *musculus* for their conspecifics (Smadja and Ganem 2002) could form **pre-zygotic** barriers detrimental to gene flow. But some other indications, while not threatening this fundamental counter-selection theory, suggest that the hybrids exhibit better performances for certain traits, such as a reduced **fluctuant asymmetry** (Alibert *et al.* 1994). It is now thought that the outcome of all these factors corresponds to a relatively small constraint on the hybrids. The expected dysgenesis could even concern only a few genes (Alibert *et al.* 1994), between 43 and 120 (Raufaste *et al.* 2005).

5.4.2 The 'wormy' mice of the hybrid zone

Two studies compared the **infra-population**s of helminths (hepatic and intestinal **cestodes**, intestinal nematodes) in the digestive tracts of mice from the two parental taxa and of mice from the hybrid zone (Sage *et al.* 1986; Moulia *et al.* 1991). Simultaneously, an introgression index, characterizing each of the studied localities, was determined using four to ten **allozymic** loci, known diagnostics for the two taxa. The researchers observed the phenomenon known as 'wormy mice', i.e. heavier **parasite loads** in the hybrids than in the parents.

In accordance with the theories on the selective factors influencing the hybrid zone, the difference between the parasite loads in the parents and

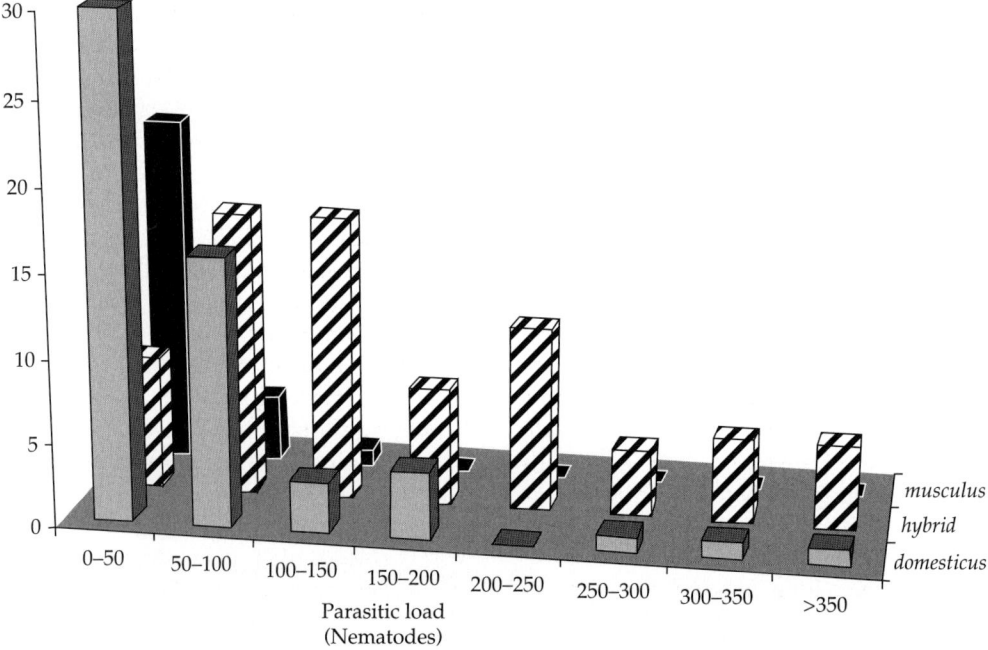

Figure 5.3 Distributions of parasite load in mice *musculus*, *domesticus* and *hybrids* with recombined genomes. Modified after Moulia *et al.* (1993).

hybrids could, in theory, come from two different sources:

1. An exogenous origin: the ecological factors specific to the hybrid zone could favour transmission of parasites (climate favouring infestation, the presence of other hosts necessary to the cycle, higher density, etc.). In this case, the fitness of the hybrids would not be a direct factor and the higher rate of parasitism would not indicate a genomic incompatibility between the two subspecies.
2. An endogenous origin: the high parasite load could mean a reduced resistance of the hybrids to parasites. In this case, the simplest hypothesis would be that this susceptibility results from a disruption of co-adapted genes.

In order to identify the causes of the hybrid susceptibility, Moulia's team carried out an experimental life cycle for a very abundant parasite in the hybrid zone: the oxyurid **nematode** *Aspiculuris tetraptera* (Moulia *et al.* 1993). The team effected controlled infestations of mice from the hybrid zone lineage and from lineages of the two parental taxa from areas outside the hybrid zone. As in natural

population studies, the distributions of parasite loads within the samples of hybrid mice were quite different from that in the samples of *musculus* and *domesticus* (see the shift towards higher parasite loads in Fig. 5.3). Therefore, the origin of this difference in parasite load is not environmental but intrinsic. Over-infestation of the hybrids is the consequence of their higher susceptibility, itself probably resulting from an immunodeficiency due to the incompatibility of the relevant murine genomes.

5.4.3 The nature and origin of hybrid susceptibility

Quantification of the true impact of parasitism on the mouse hybrid zone raises two fundamental questions:

1. Does hybrid susceptibility concern only helminths and other parasites of the digestive tract or is it generalized to all parasites?
2. Is hybrid susceptibility the direct expression of the recombination of low-compatibility alleles

within the multigenic immune system or is it the result of a more complex process?

In an attempt to answer the first question, an experimental cycle was carried out with two non-digestive-tract protozoan parasites (Derothe *et al.* 1999, 2001). Following an experimental protocol similar to the one described for *A. tetraptera*, i.e. infesting mice from the hybrid zone lineage and mice from the parental lineages with each of the protozoa, the researchers demonstrated that the hybrids were susceptible to parasites other than the digestive tract helminths, but also that they are not susceptible to all parasites. The hybrids could be susceptible to parasites with similar biology and/or life cycles and under the control of the same genetic system. For example, chronic parasitic infection (by recurrent re-infestation or by multiplication in the host) combined with widespread occurrence of the relevant parasites within the area of distribution of the host, would indicate that the same type of immune genes with perfect functionality would be frequently selected within each genome.

In answer to the second question, i.e. testing the direct implication of recombination for the susceptibility of hybrids, Derothe and colleagues performed experimental crosses of *musculus* and *domesticus* (intertaxa crosses) and control intrataxa crosses. Indeed, we would expect a disruption in the co-adapted genes only in F_2 individuals and further generations, as disruption only occurs when the recombination happens in F_1 gametes. The researchers compared the parasitic loads resulting from infestation with *A. tetraptera* in each category of individuals (parents, F_1, F_2, F_3, and F_4).

All individuals progeny from intrataxa crosses turned out to be more resistant than the parents, while none of the intertaxa hybrid progeny presented an increased susceptibility compared with the parents. However, the F_2 generation showed a higher level of susceptibility than F_1 (they were very resistant), while F_3 and F_4 presented a higher resistance than the parents (similar to F_1). The resistance generally observed seems to result principally from the phenomenon of **heterosis** expected when crossing different genetic pools with few variations. But the differences between F_2 intra- and interspecific crosses point to the presence of genomic incompatibilities between the subspecies. Those incompatibilities are minimal (no high susceptibility differences but instead a simple return to the standard parental level) and instead of increasing in F_3 and F_4 as expected under the hypothesis of the direct effect of recombination, they disappear during an unidentified selective process. This lack of increased susceptibility during the process of recombination strongly suggests that the disruption of co-adaptation does not directly affect the resistance (Derothe *et al.* 2004).

From all those results, what hypothesis emerges to explain the origin of the phenomenon? The hypothesis starts from the postulate that the parasitic loads obtained after experimental infestation are mostly an indication of the immune capacity of the individuals to give a 'proper' response (efficient regulation of the internal parasite population) or an 'inadequate' response (inefficient regulation). However, an immune response is not simply the expression of the present alleles on the immune locus. It depends on multiple factors which can favour or disfavour the successive steps of the response, starting from the presentation of antigens up to the production of cytokines and the activation of the effectors (Porcherie *et al.* 2003). An individual could possess the alleles to trigger the 'proper' response and resistance to a specific pathogen but could be in a physiological state such that the proper signals are not produced at the right time, resulting in an inadequate response. A physiological state preventing an efficient response could be temporary (e.g. gestation, an emotional state, stress, infection, nutritional deficiency) or constitutional (e.g. too low or too high a basic hormonal level). In this latter case, genetic determinism is probably involved, but we are dealing with a pleiotropic effect instead of direct action of the genes of the immune system.

Applying those data to the hybrid mice, we arrive at the following hypothesis: during the formation of the hybrid zone, a disruption in co-adaptation occurred in the hybrid genotypes. These would have performed less well than the parental genotypes for one or the other of the aspects of their biology but without being affected in their capacity to resist parasites. Within this context, one can suppose that one or several rare or new (arising from mutations) alleles partially compensated for the negative consequences of the disruption in co-adapted genes for a heavily affected function. Therefore, those alleles would have been selected and would spread in the hybrid zone. On the one hand those new alleles would greatly favour the global fitness of the hybrid, but on the other hand they could have an **antagonistic pleiotropic effect** on the response to parasites by preventing some methods of response. Therefore, the fitness of the hybrids could increase significantly for the primary selection factor and at the same time decrease because of their susceptibility to some parasites; the combination of those two effects would remain positive compared with the initial fitness (Fig. 5.4).

5.5 The resistance of water frog hybrids to helminths

5.5.1 The characteristics of water frog hybrids

In the 1960s Berger discovered that the hybridization between two taxa previously considered to be two subspecies of water frog, *Rana esculenta lessonae* and *Rana esculenta ridibunda*, did not follow Mendel's genetic rules (Berger 1977). Indeed, the **retro-crossing** of hybrid females with *lessonae* parents results only in hybrids, while it is expected that 50% of the progeny from such crosses should be of the *lessonae* parental form. Inversely, the retro-crossing of the hybrids with the other parental taxon produces 100% of the *ridibunda* parental form instead of the expected 50%. Obviously the hybrids only produce

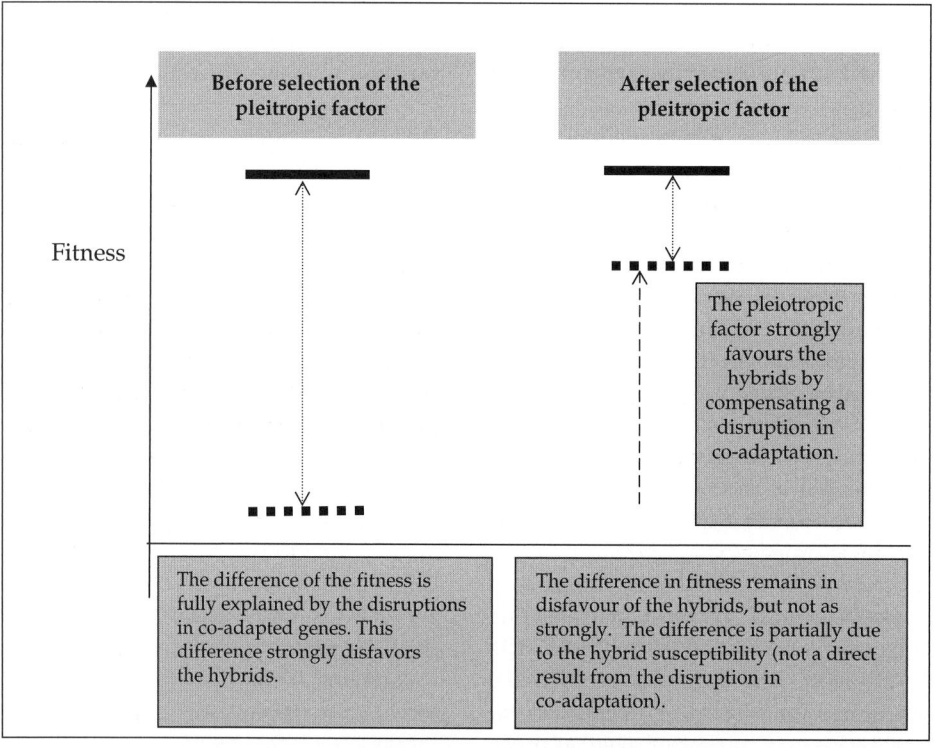

Figure 5.4 Pleiotropic factor, parasitism, and counter-selection of *musculus/domesticus* hybrids. The figure attempts to summarize the implications of the selection of a pleiotropic factor in the hybrid zone. The selection of this factor indicates that it gives an advantage to the hybrids while its pleiotropic effect is clearly a disadvantage, increasing the susceptibility of hybrids to certain parasites. Similar to the effect of medical drugs, the pleiotropic factor would cure the initial pathology but would have negative side-effects. The only explanation for its selection is that it is the 'lesser of two evils'—the side-effects are less damaging than the initial pathology. However, those side-effects render the hybrids a lower performance than their parents. Therefore, parasitism can be considered to be element of the counter-selection which currently seems to influence the hybrids.

the *ridibunda*-type gametes and the *lessonae* genome seems to be excluded from the germ line. This was observed with guppies from the genus *Poeciliopsis* and was described as the phenomenon of hybridogenesis (Schultz 1969, Tunner 1974; Graf and Polls-Pelaz 1989). Not all the mechanisms that support hybridogenesis are understood as yet. We assume that one of the parental genomes (in the case of the water frogs, the *ridibunda* genome) bears genes causing meiotic distortion (Joly 2001). Those genes only affect the germ line, where they cause the degeneration and exclusion of the *lessonae* genome which is physically eliminated from the germ cells in exocytotic vesicles (Ogielska 1994). During reproduction, the hybrids only transmit the *ridibunda* genome they received from one of their parents (usually from the mother). Therefore, in the populations arising from the *lessonae*–*esculenta* mix, this genome (in fact a **hemigenome**) is never recombined and forms what we commonly call a **hemiclone**. The mating of hybrid females with *lessonae* males (those matings occurred at a much higher frequency than expected by chance; Lengagne *et al.* 2006) result in hybrids, while the matings between hybrids mostly result in non-viable offspring. We suppose that the low viability of such descendants is due to the multiplicity of deleterious mutations in homozygous positions in the *ridibunda* genome resulting from those matings (Vorburger 2001).

5.5.2 Hybridogenesis in water frogs

This rare process of hybridization is known in two water frog complexes in Europe, both involving the *ridibunda* genome (bearing the meiotic **distorting genes**). The complex *esculenta* (*lessonae* × *esculenta* or LE) inhabits western Europe (the northern part of France, the Benelux, Germany, Switzerland) while the complex *grafi* (*perezi* × *grafi*) is found in the south of France and the Iberian Peninsula. For each complex, the hybrids show good demographic success. In several zones they even do better than the parental species, proving a higher fitness. Biologically, the hybrids correspond to the F_1 generation of a primary hybridization as the two parental genomes are not recombined and are as heterozygous as possible. Nevertheless they present some divergence from a primary hybridization

in the fact that one of the parental genomes corresponds to a clonal transmission and therefore may have either accumulated some deleterious mutations or have been selected because of the environment (interclonal selection). The low viability of the progeny of crosses between hybrids supports the first possibility, while the small number of hemiclonal lines in the natural populations favours the second (Colon 2004).

The success of these hybrids has aroused much interest because, as we saw, hybridization generally results in a lower fitness of the hybrids. Research has concentrated around two hypotheses. The first supposes that the large heterozygosity gives the hybrids a high **tolerance** to fluctuating environments: this is the 'general-purpose genotype' hypothesis. The second hypothesis considers that each hybrid line takes advantage of ecological conditions intermediate between the optima for the two parental taxa. As part of the genome is of clonal origin, we suppose that the diverse hybrid lines occupy different and well-defined niches: this is the 'frozen niche variation' hypothesis (Lynch 1984; Moore 1984).

In order to verify the validity of those hypotheses, we evaluated the different physiological parameters of the hybrids and the parents. The hybrid tadpoles show better performance than the parental species with regard to the variables of density and predation (Semlitsch 1993; Semlitsch and Reyer 1992). Similarly, juvenile hybrids resist hypoxia better than the parental taxa (Tunner and Nopp 1979). Meanwhile, the hybrid tadpoles show a performance intermediate between that of the two parental taxa when in **hypoxic** situations, in both laboratory and natural conditions (Plénet *et al.* 2000, 2005). The first results indicate a genotype with an important tolerance while the latter results indicate an ecological optimum in an intermediary niche.

The last selective factor to be analysed in this case hybridization is parasitism.

5.5.3 Hybrid resistance to helminths

Comparisons of parasite loads between taxa were conducted on the adults (Joly *et al.* 2007; Moyen *et al.*, submitted). The coexistence of the two taxa in the same habitats and microhabitats makes this comparison particularly pertinent as the two taxa are exposed to similar infestation pressures. We concentrated our attention on two helminth parasites of the lungs: the nematode *Rhabdias bufonis* and the trematode *Haplometra cylindracea*. Because these two parasites feed on blood in the lungs and, more specifically, because of the lesions they cause, we can assume that they have a strong impact on the host organisms.

In the case of *Rhabdias bufonis*, the frog hybrids are generally less infested than the parental *lessonae* frogs; while in the case of *Haplometra cylindracea* the infestations are similar but the relationships between the host and the parasite are differently expressed. Indeed, hybrids generally have a parasite load that is higher than that of the parent species. But if we consider the age of the individuals, we notice that parasite load increases with age in the hybrids while it decreases in the parent populations. On the one hand the accumulation of parasites with age suggests a tolerance phenomenon, but on the other hand the decreased parasite load suggests a phenomenon of selection, as infested individuals have poorer survival.

In summary, these analyses of the variations in parasite load within the hybrid complex demonstrate that the hybrids have a better resistance to lung parasites (a simultaneous analysis of the resistance to intestinal helminth parasites showed no differences between the hybrid and the parental populations). We could conclude that the hybridization of the genomes implies a lower susceptibility to some parasites. But we need to consider the related influence of two factors on the level of infestation. First, the hybrid individuals are larger than the parents and this bigger size could influence the mechanism of infestation (by increasing its frequency). Everything being equal, we would expect a higher parasite load in the hybrids because the volume of the organs to be colonized (lungs, intestine) is larger. But instead we observe the opposite, confirming the hypothesis of a lower hybrid susceptibility. Then, secondly, we need to take into account that the hybrids arising from a hybridogenetic process are not exactly the same as first-generation hybrids issuing from a primary hybridization. Indeed, the hemigenome *ridibunda* involved in hybridogenetic complexes is a hemiclone transmitted from generation to generation after distant primary hybridization between parental species, dating from the last glaciation and localized in the Greco-Anatolian glacial refuge (Hotz *et al.* 1997). The colonization of western Europe by those hybrid lines probably resulted in the selection of those hemiclones producing the best performing hybrid phenotypes (see Box 5.1). When considering the genetic diversity in natural populations, it is indeed surprising to identify only a small number of hemiclones (Semlitsch *et al.* 1997; Colon 2004). Even if they are heterozygous for all loci, the hybrid individuals are different from an F_1 generation resulting from parental individuals selected randomly from a population. Therefore, it is possible that the *ridibunda* part of the hybrid genome has been selected specifically for its better resistance to parasites when it is interacting with a *lessonae* genome (interclonal selection of co-adapted genomes). Meanwhile, the green frog hybrid complex remains a remarkable example in the animal kingdom of the increased resistance of hybrids to pathogens.

5.6 Conclusion and perspectives

The hybridization process is like a window allowing us to explore the interactions between a genome and its environment. This is particularly so in the case of parasitism, which constitutes an important selective pressure in the natural world. The variations in the functional organization of the genome and the interactions with the environment provide a complex conceptual framework within which contradictory hypotheses have evolved, and where most of them are validated. The two examples we discussed above are good illustrations of this complexity. In the mice, whatever the origin of the phenomenon, the hybrids can bear higher parasite loads than the parents, and those loads have a cost for their hosts. According to the antagonist **pleiotropy** hypothesis, parasites represent the 'lesser of two evils' for the hybrids. Nevertheless, they are a source of selective pressure which contributes to a limitation of gene flow between the two taxa. Today, parasites could be a substantial element maintaining this tension zone. On the other hand, in water frogs, the lesser susceptibility of the

Box 5.1 Hybridogenesis in water frogs

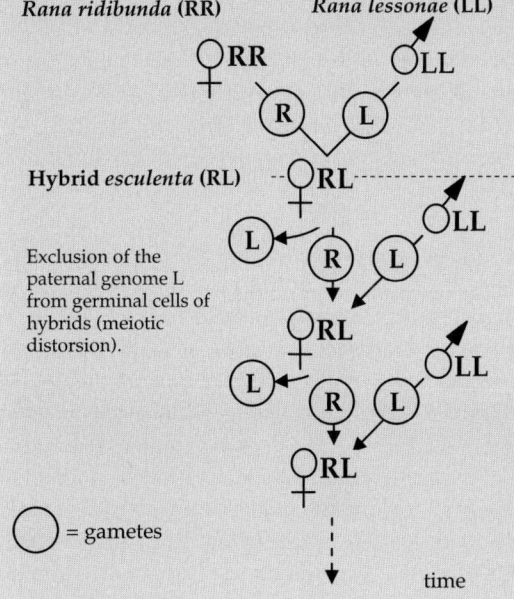

Rana ridibunda (RR)   *Rana lessonae* (LL)

Hybrid *esculenta* (RL)

Exclusion of the
paternal genome L
from germinal cells of
hybrids (meiotic
distorsion).

◯ = gametes

time

Primary hybridization between a *ridibunda*
individual (RR) and a *lessonae* **one (LL)**

Primary hybridizations that are responsible from hybridization
complexes in western Europe probably occurred in eastern
Europe during the last glaciation.

Because of the pre-meiotic exclusion of one parental genome,
the presence of hybrids mainly depends on backcrossing
between hybrids (usually females) and the parental
species whose genome is excluded (sexual parasitism).

The distorting hemigenome (ridibunda) can spread
outside the area originally occupied by the parental species.
It has then been able to invade western Europe where
this species was not present.

The clonal transmission of these hemigenomes leads to
the accumulation of deleterious mutations. The fertility of
crossings between hybrids slowly decreases. Because there is
no recombination between the two genomes, hybrids have a
level of heterozygosity that is maximal.

How does hybridogenesis work?
Here we show a simplified scheme of the L–E complex (lessonae–esculenta). We do not show
crossings between hybrids that regain the parental form RR because descendants usually
have a low viability. Similarly, we do not consider crossings between hybrid males and parental
females since they do not occur, usually for ethological reasons.

hybrids to parasites probably contributes to their
success.

These examples are perfect illustrations of what
studies of parasitism in hybrid zones can bring to
biology and evolutionary parasitology. Important
modelling work is now necessary in order to allow
us to make testable predictions. Meanwhile, initial
results are promoting further study of parasitism
in situations of natural hybridization between animal taxa and consideration of the role of parasitism
in the phenomenon of speciation. The hypotheses
constructed from plant models concerning effects
on ecosystem evolution and biodiversity have
encouraged work in animal models which it is
hoped will open up new perspectives in this exciting field.

Important points

• Hybridization may allow us to understand the
mechanisms favouring reproductive isolation
between taxa (and therefore speciation) and the
emergence of novelties at the level of alleles, the
genome, and taxa.
• Gene exchanges between host taxa through
hybridization can be favoured or disfavoured by
parasitism. Therefore, parasites can potentially
determine the outcome of the mixing of host
genomes and subsequent possible innovations.

Questions for discussion

• Does hybridization create novelty? It does seem
indeed to allow the emergence of rare, even new,

alleles, but how does this happen? Is it the result of a high rate of mutation? Of intragenetic recombinations? The novelties would be selected if their effect were to reduce significantly counter-selection on the hybrids.

• Can the parasites of parental host species adapt to the hybrids and diverge? The adaptation to new host genomes, if associated to a mechanism limiting the gene flow between different parasite populations, may lead to a divergence of those populations, even to speciation.

• Can the parasite taxa hybridize a host hybridization further? If the parasites of the two host taxa are divergent but without reproductive isolation, the two gene pools could mix within a 'hybrid' parasite population. Unfortunately, very little information is available on this subject even if the epidemiological incidence is in theory important: new, more virulent, genotypes could emerge.

Further reading

• Burke, J.M., Arnold, M.L. (2001). Genetics and the fitness of hybrids. *Annual Review of Genetics* **35**: 31–52. A review with a self-explanatory title.

• Derothe, J.M., Porcherie, A., Perriat-Sanguinet, M., Loubes, C., Moulia, C. (2004). Recombination does not generate pinworm susceptibility during experimental crosses between two mouse subspecies. *Parasitology Research* **93**: 356–363. Discusses the susceptibility of hybrid mice and the hypothesis of antagonist pleiotropy.

• Moulia, C. (1999). Parasitism of plant and animal hybrids—are facts and fates the same? *Ecology* **80**: 392–406. A review of parasitism in animal hybrid zones.

CHAPTER 6

Parasitism and regulation of the host population

Serge Morand and Julie Deter

6.1 Introduction

Parasites are found in all living organisms. Parasite diversity is reflected in the number of parasite species (probably half of all living species), but also in terms of their life cycles (direct vs. indirect) or their size (from a few micrometres to metres). As a result of their impact on host reproduction, growth, and survival, parasites can be important agents regulating populations, communities, and ecosystems. The aim of this chapter is to illustrate, using several simple theoretical approaches, the role of parasites in the regulation of host populations.

It is first necessary to introduce some basic facts. First, we describe the heterogeneity in parasite distribution among host populations or communities. Secondly, we consider parasite **virulence** as a measure of the impact of parasites on hosts. The interaction between virulence and aggregation allows us to estimate the impact of parasite regulation (destabilization) on host populations. After presenting real examples illustrating the existence of regulatory parasites, we introduce the use of multi-host-parasite models. Finally, we consider host–parasite systems as co-evolutionary systems; host defences (behaviour, immunity) evolve in response to the evolution of parasite life-history traits (**transmission**, virulence), and vice versa. In the context of global change (global warming, habitat fragmentation, decline of biodiversity), it is more important than ever that account is taken of parasites when considering health (disease emergence) and conservation strategies.

6.2 General principles

More than 10% of the **metazoan** species that have been described are parasites (Poulin and Morand 2004). Including micro-organisms, parasites are estimated to represent more than half of all living species (de Meeûs and Renaud 2002), suggesting than no species is free from infection. But not all individuals in a host population are equally infected. Actually, host–parasite relationships are characterized by both variability (between individuals and host species) and asymmetry (each host can be parasitized by several parasites while a parasite can infect only one or a few hosts) of infection. A few individuals in a single host population generally harbour the majority of parasites (either individuals of one parasite species and/or individuals of several parasite species). For hosts, this variability comes from differences in the risk of infection and the negative effects of parasitism, due to genetic, environmental, and physiological factors. For example, the sex of the host can influence the risk of infection and the impact of the infection on individual survival or reproduction. In mammals, males are often more susceptible. **Sexual dimorphism** in size, with males being larger, is usually observed, and is often associated with a mortality bias in males. Males of species that are highly dimorphic are often more parasitized than males of less sexually dimorphic species (Fig. 6.1) (see also Chapter 3).

At an interspecific level, another important variability is also observed: a few host species harbour high parasite diversity (Fig. 6.2). This can be explained by differences in the geographical

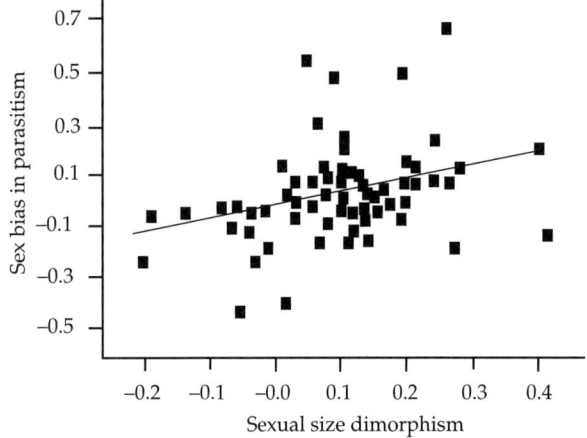

Figure 6.1 Mammals and parasitic infections: difference in susceptibility due to sex. An infection bias in males is associated with sexual dimorphism in favour of males. Sexual bias in infection is measured by the ratio mean infection of males/mean infection of females; sexual dimorphism is measured by the ratio male weight/female weight. After Moore and Wilson (2002).

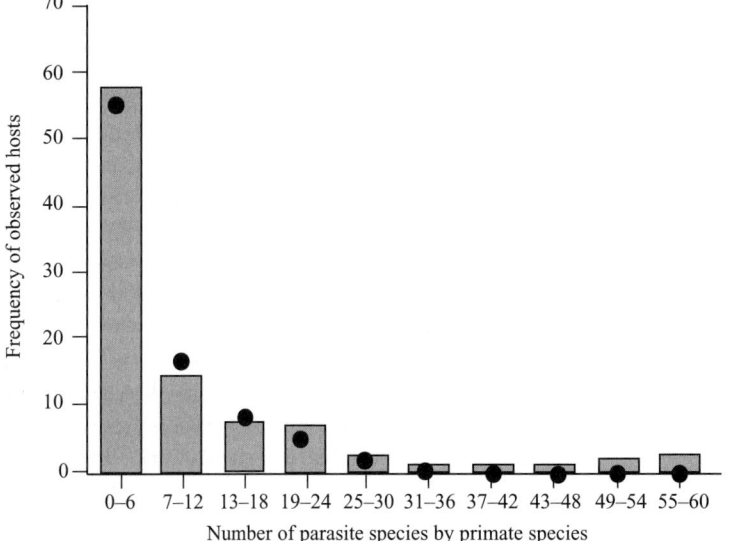

Figure 6.2 Distribution of the number of parasite species among primates with adjustment for the negative binomial distribution (black circles). After Nunn et al. (2003).

distribution of species. We suspect that this inequality in parasite infection may have important consequences for regulation and stability of the host population.

The relationships between hosts and parasites are also characterized by a strong asymmetry. Generally, a parasite interacts with a small number of hosts. This number depends on the complexity of the parasite life cycle (the presence or not of **intermediate hosts**) and on the specificity of the parasite (the number of host species that a parasite can infest).

A host (definitive or intermediate) generally encounters many parasite species. Consequently, this host has to adapt and invest energy in various immune responses depending on the diversity of parasites and pathogens in its environment (see Chapter 1). Most parasites are specialized; they depend entirely on their hosts for their nutrition and reproduction. These parasites experience strong pressures from their hosts and have to continually adapt. We can therefore suppose that the evolution of parasite life-history traits (size, age of sexual maturity, fecundity) is constrained by host traits, ecology, and defence mechanisms (see Chapter 2).

Parasites are able to evade, modulate, and manipulate host defence mechanisms (involving immune defences) and behaviour (see Chapter 4). In turn, these complex mechanisms experience strong pressure from the evolutionary responses of host defences.

From this introduction, we can conclude that the regulation of host–parasite interactions has to be studied at different scales: from individuals to populations and communities. In order to better manage the current problems of biological invasions and emerging diseases, a good understanding of the ecology and evolution of host–parasite interactions is thus necessary. These are central objectives in the fields of **biological control** (Chapter 7), conservation biology (Chapter 9), and health ecology (Chapter 8).

6.3 Parasite diversity: consequences for regulation

Parasites are divided into two categories depending on their life-history traits and duration of infection: microparasites and macroparasites (Table 6.1). Microparasites are represented by viruses, bacteria, and protozoa; they are associated with diseases with a direct or indirect (intermediate hosts, **vectors**) agent of transmission. Because of the very small size of these parasites, we cannot generally count their number in a host. However, recently developed methods such as the real-time polymerase chain reaction (RT-PCR) now allow us to count these parasites, and will no doubt be improved in the future (see the Appendix on Methods). The modelling of microparasite

(pathogen) transmission involves a subdivision of host populations into three parts: susceptible, infected–infectious, and recovered hosts (the SIR model; see below).

Macroparasites include metazoans such as helminths (cestodes, monogeneans, trematodes, nematodes, acanthocephalans) and arthropods (ticks, fleas, lice, mites, dipterans). These parasites can live and reproduce on the external surface of the host (ectoparasites), in internal cavities such as the stomach, lungs, and intestines (mesoparasites), or inside the organism (blood or lymph) or cells (intraparasites). Macroparasites present simple or complex life cycles (with a vector and intermediate hosts), and sometimes complex migrations inside the host.

Numerous modes of transmission exist for parasites: by ingestion, bites, transcutaneous penetration, sexual, trans-ovarian transmission, etc. Note that many parasites exploit host behaviour for their transmission. For example, mesoparasites (cestodes, nematodes) can be transmitted via grooming, a defence mechanism against ectoparasites. Others (acanthocephalans, trematodes) can manipulate host behaviour and increase their transmission efficiency (see Chapter 4).

The size, mode of transmission, location, feeding (taking of blood or substances), and injuries caused by parasites have consequences for host life-history traits—survival, fecundity, and even the ability to perform skilled tasks. These consequences can be measured at the level of the host population either by a regulatory effect (a decrease in host population density), or by a destabilization effect (inducing cyclical fluctuations in host–parasite dynamics). These effects depend on the distribution of parasites among host individuals.

Table 6.1 Differences between micro- and macroparasites

	Microparasites	Macroparasites
Generation time	very high compared to host generation time	Similar to host generation time
Number of infectious forms and eggs	Very high	High to very high
Immunity	Usually specific	Usually non-specific
		Often absent in epidemiological models
Epidemiological parameters measured	Prevalence	Mean abundance (mean number of parasites per individual host, including uninfected ones)

6.4 Parasite aggregations: causes and consequences

Macroparasite infections are characterized by an aggregated distribution of parasites among host populations. In human communities, 20% of individuals generally harbour 80% of the parasitic helminths. Such patterns are observed in many other animals. A small number of host individuals are responsible for most of the transmission by allowing parasites to persist (Anderson and May 1985; Woolhouse *et al.* 1997). Parasite aggregation is estimated by the ratio between position (mean **abundance**) and dispersion (variance) of parasite distribution among the host population (see Box 6.1).

Although aggregation has been less studied, it was recently suggested to be important for

---

Box 6.1 Parasite aggregation and population quantification

Mean abundance, i.e. the mean number of parasites per host individual, is not sufficient to inform us about parasite distribution, which is rarely regular or random. Most of the individual parasites are harboured by a small number of individual hosts: parasites are aggregated within their host population. Several measurements of aggregation have been proposed: the ratio of variance to parasite abundance, the coefficient *k* of the negative binomial distribution, the power law or Taylor's law linking variance in abundance to abundance and Ives' coefficient (see Box 6.3) applied to intra- and interspecific aggregations.

The variance/abundance ratio

A first approach consists of estimating the ratio of variance of parasite abundance Var(M) or s^2 to mean parasite abundance M (mean number of parasites per individual host): Var(M)/M = 1 corresponds to a Poissonian distribution, Var(M)/M < 1 corresponds to an under-dispersed distribution, while Var(M)/M >> 1 corresponds to an over-dispersed or an aggregated distribution. The ratio of variance/mean is used to appreciate the regulatory effect of a parasite on the host populations (Anderson and Gordon, 1982).

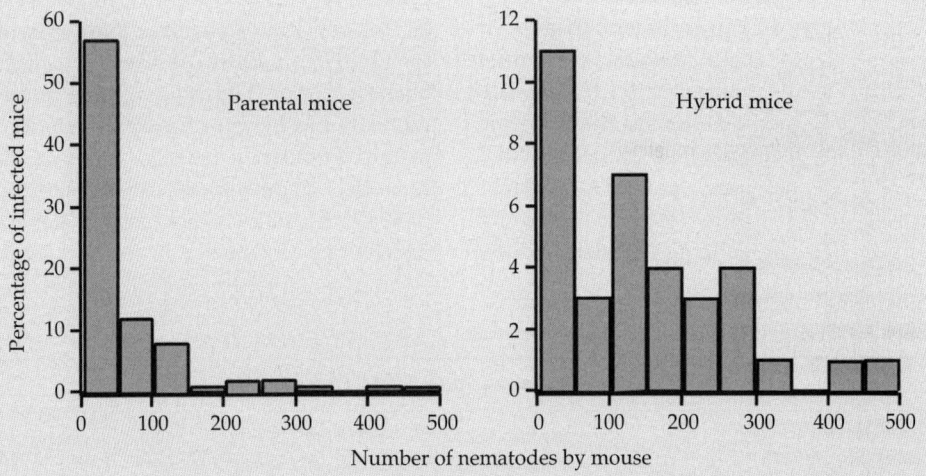

Box 6.1a Distribution of the number of nematodes (*Aspiculuris tetraptera* and *Syphacia obvelata*) per mouse in parental mice (*Mus musculus* and *Mus domesticus*) and their hybrids. After Moulia *et al.* (1991).

continues

Box 6.1 continued

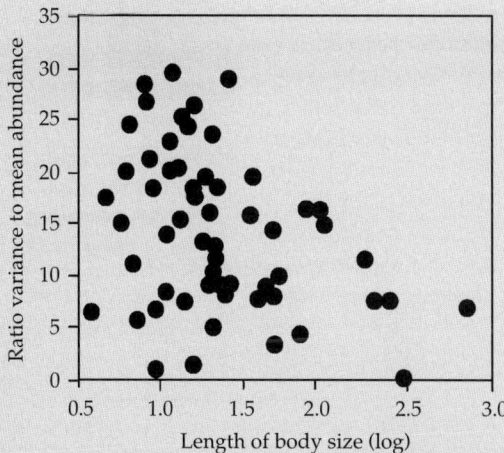

Box 6.1b Relationship between the variance/mean abundance ratio and the body volume of nematodes parasitizing vertebrates (in mm³) (Poulin and Morand, 2000). The diagram has a triangular pattern suggesting that small parasite species show a more important and variable aggregation than large species.

The parameter of the negative binomial distribution, k

In numerous examples, the best adjustment to a statistical distribution is made with the negative binomial law. This law can be defined with two elements: the mean (here, mean abundance M) and a dispersal parameter k.

Negative binomial distribution is defined by

$$s^2 = M + M^2/k.$$

The parameter k, or parameter of the negative binomial law, describes the best adjusted distribution shape for empirical data. k is relatively easy to estimate with empirical data.

An estimation based on the sample size (n) is given by

$$k = (M^2 - s^2)/(s^2 - M).$$

An estimation of this parameter k is obtained by minimizing the likelihood L (Hilborn and Mangel 1997):

$$L(Y\ m,k) = -\sum_{i=1}^{n} \ln\left(\frac{\Gamma(k+Y_i)}{\Gamma(k)+Y_i!} \left(\frac{k}{k+M} \right)^k \left(\frac{M}{M+k} \right)^{Y_i} \right)$$

where Y is the vector of the number n of parasites observed for each host individual, M is the abundance, and k the aggregation coefficient.

When k tends to 0, the aggregation becomes more important. On the contrary, when k is big (>20), the parasite population tends to a Poissonian distribution.

More recently, the Weibull distribution has been proposed as an alternative to the negative binomial law. In certain cases, like gastrointestinal nematodes found in small ruminants, the Weibull distribution seems to be more appropriate for describing macroparasite distribution (Gaba *et al.* 2005).

Discrepancy index

This index has been proposed by Poulin (1993) to quantify the aggregation as a departure from a hypothetical distribution for which all hosts are used in an identical way, and all parasites are partitioned in similar infrapopulation sizes. This corresponds to a regular distribution:

$$D = 1 - \frac{2\sum_{i=1}^{n}\left(\sum_{j=1}^{j} x_i \right)}{xn(n+1)}$$

where x is the number of parasites in a host j (hosts are ranged from the least to the most infected), and n the number of hosts in the sample. D varies from 0, corresponding to an absence of aggregation, to 1, corresponding to the maximum aggregation where all parasites are harboured by a single host. This index D is useful for comparing parasite populations that differ in prevalence or abundance.

Interpopulation aggregation: the power law or Taylor's law applied to parasites

The mean abundance M is linked to its variance Var(M) by a power law or Taylor's law (Taylor, 1961):

$$\log(V) = b \log(M) + \log(a)$$

with the parameters a (the intercept between the line and the abscissa) and b (the slope of the line corresponding to the exponant of the power law). The coefficient b represents an aggregation index. This relationship can be applied to all types of macroparasites (Shaw and Dobson, 1995; Morand and Guégan, 2000; Krasnov *et al.*, 2005).

continues

Box 6.1 continued

Note also that k is linked to the parameters a and b of Taylor's law:

$$1/k = aM^{(b-2)} - (1/M).$$

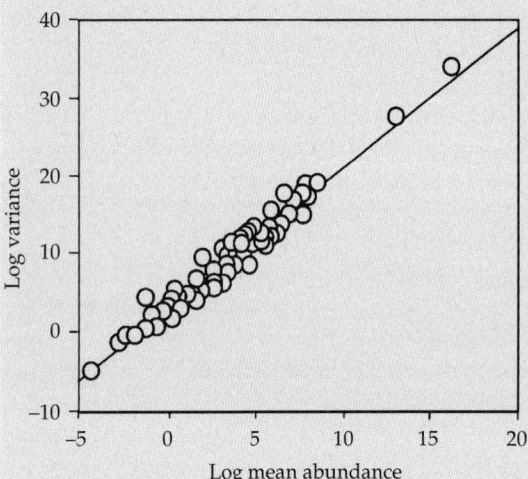

Box 6.1c Another approach consists of estimating the allometry of Var(M) as a function of M. The regression line in log scale indicates a relationship between log variance and log abundance: Var(M) = $1.79M + 3.09$ in nematodes parasitizing vertebrates (Poulin and Morand, 2000).

Relationship between indices

The most used indices are the ratio 'variance/mean abundance' and k. However, the relationships between both of these indices are not simple. Scott (1987) preferred the ratio 'variance/mean abundance variance' to measure parasite aggregation and compare parasite distribution patterns. Actually, k is not independent of mean abundance (see the preceding formula).

References

Anderson, R.M., Gordon, D.M. (1982). Processes influencing the distribution of parasite numbers within host populations with special emphasis on parasite-induced host mortalities. *Parasitology* **85**: 373–398.

Gaba, S., Ginot, V., Cabaret, J. (2005). Modelling macroparasite aggregation using a nematode–sheep system: the Weibull distribution as an alternative to the negative binomial distribution? *Parasitology* **131**: 393–401.

Hilborn, R., Mangel, M. (1997). *The ecological detective: confronting models with data*. Princeton University Press, Princeton, NJ.

Krasnov, B.R., Stanko, M., Miklisova, D., Morand, S. (2005). Distribution of fleas (Siphonaptera) among small mammals: mean abundance predicts prevalence via simple epidemiological model. *International Journal for Parasitology* **35**: 1097–1101.

Morand, S., Guégan, J.-F. (2000). Abundance and distribution of parasitic nematodes: ecological specialisation, phylogenetic constraints or simply epidemiology? *Oikos* **55**: 563–573.

Moulia, C., Aussel, J.-P., Bonhomme, F., Boursot. P., Nielsen, J.T., Renaud, F. (1991). Wormy mice in a hybrid zone: a genetic control of susceptibility to parasite infection. *Journal of Evolutionary Biology* **4**: 679–687.

Poulin, R. (1993). The disparity between observed and uniform distributions—a new look at parasite aggregation. *International Journal for Parasitology* **23**: 937–944.

Poulin, R., Morand, S. (2000). Parasite body size and interspecific variation in levels of aggregation among nematodes. *Journal of Parasitology* **86**: 642–647.

Scott, M.E. (1987) Temporal changes in aggregation: a laboratory study. *Parasitology* **94**: 583–595.

Shaw, D.J., Dobson, A.P. (1995). Patterns of macroparasite abundance and aggregation in wildlife populations: a quantitative review. *Parasitology* **111**: S111–S133.

Taylor, L.R. (1961) Aggregation, variance and the mean. *Nature* **189**: 732–735.

microparasites. The number of microparasites inoculated in a first infection would not be without consequences for microparasite dynamics or the level of the induced pathology.

Different heterogeneities are at the origin of aggregation: variability in host exposure and susceptibility to parasites. Other factors (extrinsic and intrinsic) can increase or decrease this heterogeneity: multiple infections, immune interactions, genetics, or behaviour (Anderson and May 1985; Grenfell *et al.* 1995; Quinnell *et al.* 1995; Wilson *et al.* 2001; Rosa and Pugliese 2002).

Aggregation is a universal characteristic of parasites (Crofton 1971; Shaw and Dobson 1995; Galvani

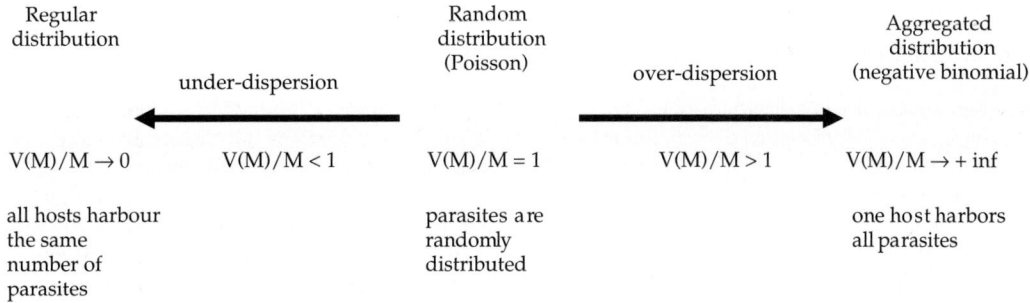

Figure 6.3 Mode of distribution of parasites among a host population—regular, random, and aggregated—with identified causes for the origin of these distributions. The type of distribution can be characterized by the ratio between variance of abundance V(M) and abundance M of parasites (see Box 6.1). After Anderson and Gordon (1982).

2003a) (Fig. 6.3). The processes responsible for parasite aggregation include:

- reproduction of the parasite inside the host;
- heterogeneity in susceptibility of the host to the infection (e.g. differences in behaviour, genetics, ecology);
- heterogeneity in host exposure depending on the distribution of infectious stages in time and space.

Aggregation can decrease:
- when host mortality is a density-dependent function varying with the number of parasites;
- when mortality increases or fecundity decreases in a density-dependent fashion;
- when immunity protects against re-infection— but an immune reaction that affects parasite fecundity has few effects on the level of aggregation.

Aggregation is a key parameter in host–parasite dynamics (Anderson and May 1978; May and Anderson 1978). A large aggregation of parasites corresponds to situations in which a small number of hosts harbour most of the parasites. These hosts will exert regulatory effects— mortality or **morbidity** induced by the parasites. In response, those numerous parasites harboured by a few hosts will encounter most of the regulatory processes, such as density-dependent effects,

parasite-induced host mortality, or the effects of immunity.

6.5 Interspecific aggregation and co-existence between parasites

Parasite communities constitute a mixture of parasite species resulting from differential exposure and transmission factors depending on the host and parasites, and from interactions between the different parasite species. Holmes and Price (1986) have suggested that interactions between parasites in a host population should be stronger between the most common and abundant species of parasites (called *core species*) than between rare species (called *satellite species*). Numerous studies of parasite communities have shown that negative interactions are rare. In most cases, a positive interaction or a lack of interaction is observed, even if most parasite communities do not seem to be random assemblages (Box 6.2).

However, the detection of competition between parasite species is difficult to demonstrate on the basis of correlative studies alone, and there have been few experiments involving controlled infections.

Theoretical arguments suggest that parasite species are rarely in direct competition with each other,

Box 6.2 Nested pattern in parasite species assemblages

Nestedness corresponds to a simple organization of communities where the poorest species sub-assemblage is a perfect subset of a richer assemblage. A nested pattern corresponds to a non-random sampling type of the considered set of species. This is largely observed in free-living species occupying insular environments or fragmented habitats (Atmar and Patterson, 1993), and in parasite species (Guégan and Hugueny, 1994; Poulin and Guégan, 2000; Šimková et al., 2003).

However, perfectly nested patterns are rare (A in figure) and generally we observe imperfectly nested patterns with missing species or outliers (B). A perfectly random pattern would resemble a chessboard (C) where black squares represent the presence of a species (parasite) in a given locality (host) and white squares, the absence. An anti-nested pattern is represented in (D).

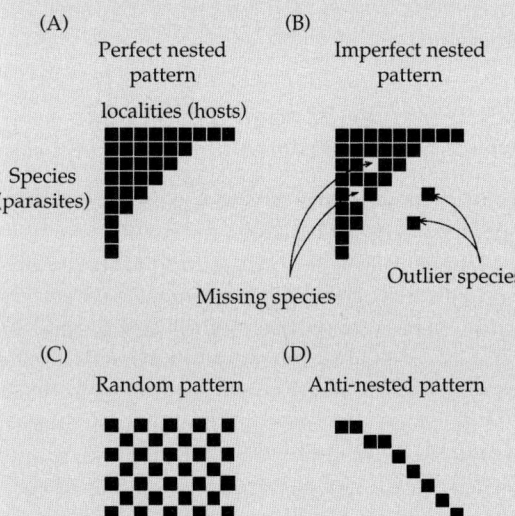

Box 6.2a Community assemblage patterns. Black squares correspond to the presence of a species (columns) in localities (rows). After Šimková et al. (2003).

Atmar and Patterson (1993) proposed a direct measure of the order of a matrix (entropy) by the presence or absence of species among the different habitats. The measure of entropy T would correspond to the temperature disorder in the matrix. A perfectly nested matrix corresponds to a temperature of 0°C,

and a random matrix corresponds to a temperature of 100°C. The significance of this temperature is tested by simulations (Atmar and Patterson,1995).

Using simple statistics, Goüy de Bellocq et al. (2003) showed that a nested pattern for the presence or absence of helminths in different populations of a European rodent, the wood mouse, could be explained by parasite life-history traits and by the specific environment of each rodent population.

Host populations are ordered from left to right (see figure), with host populations harbouring the most parasites on the left. Parasite species are ordered from the top to the base, with parasite species occupying the majority of localities at the top of the matrix. The application of Atmar and Patterson's measure shows a temperature significantly colder than the mean temperature calculated by simulations. This demonstrates the nested character of this matrix for presence–absence of parasites in different host populations.

Parasite species are ordered following their life-history traits (direct or indirect cycles, specificity) responsible for strong colonization and/or weak extinction rates. Parasites located at the basis of the matrix are parasites with complex life cycles involving the presence of intermediate hosts in the environment.

Host populations live in different localities with contrasting ecological conditions. Localities on the left of the matrix are characterized by low extinction rates or a large area (continent), while localities on the right are characterized by high extinction rates or small areas (islands). Biodiversity indices such as total host richness are correlated with the locality order.

Epidemiological factors can be established on the basis of nested patterns. The epidemiological processes of colonization and extinction act on each parasite species (Morand et al., 2002; Guégan et al., 2004).

References

Atmar, W., Patterson, B.D. (1993). The measure of order and disorder in the distribution of species in fragmented habitat. Oecologia **96**: 373–382.

Atmar, W., Patterson, B.D. (1995). The nestedness temperature calculator. A visual BASIC program, including 294 presence–absence matrices. AICS Research, Inc., University Park, NM and the Field Museum, Chicago.

Guégan, J-F., Hugueny, B. (1994). A nested parasite species subset pattern in tropical fish: host as major determinant of parasite infracommunity structure. Oecologia **100**: 184–189.

continues

Box 6.2 continued

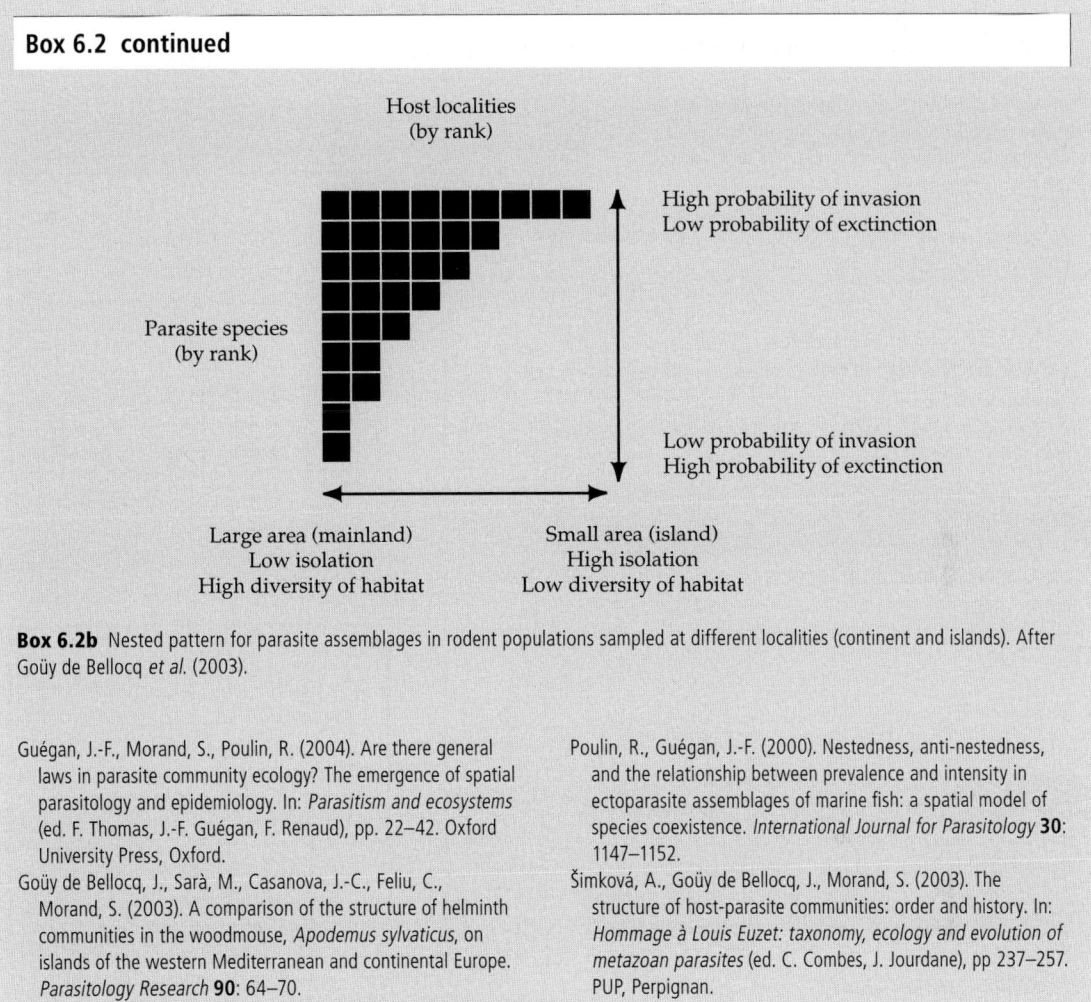

Box 6.2b Nested pattern for parasite assemblages in rodent populations sampled at different localities (continent and islands). After Goüy de Bellocq *et al.* (2003).

Guégan, J.-F., Morand, S., Poulin, R. (2004). Are there general laws in parasite community ecology? The emergence of spatial parasitology and epidemiology. In: *Parasitism and ecosystems* (ed. F. Thomas, J.-F. Guégan, F. Renaud), pp. 22–42. Oxford University Press, Oxford.

Goüy de Bellocq, J., Sarà, M., Casanova, J.-C., Feliu, C., Morand, S. (2003). A comparison of the structure of helminth communities in the woodmouse, *Apodemus sylvaticus*, on islands of the western Mediterranean and continental Europe. *Parasitology Research* **90**: 64–70.

Morand, S., Rohde, K., Hayward, C. (2002). Order in ectoparasite communities of marine fish is explained by epidemiological processes. *Parasitology* **124**: S57–S63.

Poulin, R., Guégan, J.-F. (2000). Nestedness, anti-nestedness, and the relationship between prevalence and intensity in ectoparasite assemblages of marine fish: a spatial model of species coexistence. *International Journal for Parasitology* **30**: 1147–1152.

Šimková, A., Goüy de Bellocq, J., Morand, S. (2003). The structure of host-parasite communities: order and history. In: *Hommage à Louis Euzet: taxonomy, ecology and evolution of metazoan parasites* (ed. C. Combes, J. Jourdane), pp 237–257. PUP, Perpignan.

and this seems to be confirmed by empirical studies. One of these arguments considers that parasite communities are not saturated, that numerous empty niches exist inside hosts, and thus, that communities are not structured by competition (Rohde's hypothesis; Rohde 2005). Another theoretical argument, based on mathematical models, suggests that regulation between species occurs via regulation of the host population ('all against one, one against all') and not via direct competition for host exploitation (cf. Dobson and Roberts 1994).

Species of parasites and individuals from a population of parasites are aggregated among their host populations. A host that is highly infected with one species of parasite is often highly infected by several other parasite species as well (covariance in infections). The mechanisms of intra- and interspecific aggregation would be similar. Comparisons between levels of intra- and interspecific aggregations would permit us to explain the coexistence of individuals and species of parasites (Box 6.3).

Box 6.3 Intra- and interspecific aggregation

Ives (1991) proposed an intraspecific aggregation index J, which represents the proportional increase in the number of parasites harboured by a given host, relative to a random distribution:

$$J_1 = \frac{\sum_{i=1}^{p} \frac{n_{1i}(n_{1i}-1)}{m_1} - m_1}{m_1} = \frac{\frac{V_1}{m_1}-1}{m_1}$$

where $n_1 i$, m_1, and V_1 are the number of parasites in a host i (p is the number of hosts), mean abundance, and its variance for the species of parasite, respectively.

A value of $J = 0$ indicates that individuals are randomly distributed, while a value of $J = 0.5$ indicates a 50% increase of the number of individual parasites in a given host, compared with a random distribution.

Ives also proposed a measure for the interspecific aggregation C, which quantifies the proportional increase in the number of parasite species harboured by a given host, compared with a random distribution:

$$C_{12} = \frac{\sum_{i=1}^{p} \frac{n_{1i} n_{2i}}{m_1 P} - m_2}{m_2} = \frac{Cov_{12}}{m_1 m_2}$$

where $n_1 i$ and $n_2 i$ are the number of individuals of parasite species 1 and 2 in host i, m_1 and m_2 are the mean abundance of parasite species 1 and 2 in the host, P is the number of hosts, and Cov is the covariance between pairs of parasite species. When $C > 0$ both of the species are positively associated, when $C < 0$ they are negatively associated.

These indices have been developed and used for parasites by Boulinier *et al.* (1996). Morand *et al.* (1999) demonstrated that almost all fish ectoparasite communities studied show mean values of $J > 0$. This indicates that individuals are aggregated in their hosts (figure a). Among these 36 communities, 15 show negative associations between parasite species with a mean of $C < 0$, and 21 communities show positive associations with a mean of $C > 0$ (figure, b).

References

Boulinier, T., Ives, A.R., Danchin, E. (1996). Measuring aggregation of parasites at different host population levels. *Parasitology* **112**: 581–587.

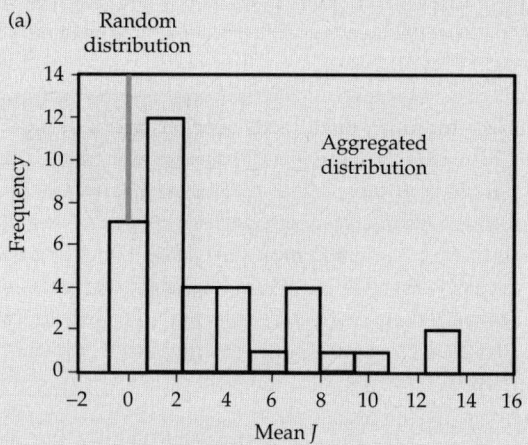

(a)

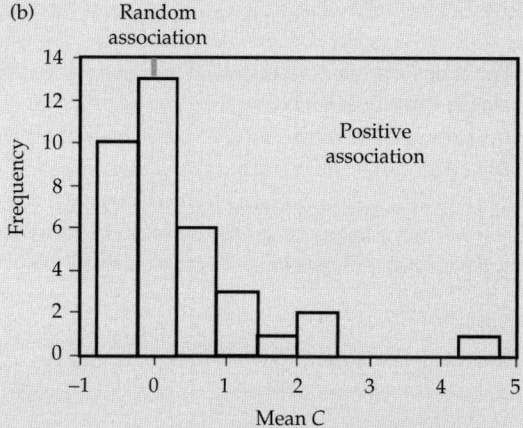

(b)

Box 6.3 (a) Distribution of mean intraspecific aggregations J in fish ectoparasite communities. (b) Distribution of mean interspecific aggregations C in fish ectoparasite communities (Morand *et al.* 1999).

Ives, A.R. (1991). Aggregation and coexistence in a carrion fly community. *Ecological Monographs* **61**: 75–94.

Morand, S., Poulin, R., Rohde, K., Hayward, C. (1999). Aggregation and species coexistence of ectoparasites of marine fishes. *International Journal for Parasitology* **29**: 663–672.

Laboratory studies have shown the importance of immunity in concomitant infections with microparasites. In certain cases, immunity can increase the heterogeneity of infection. For example, the immunosuppressive role of certain parasites (e.g. HIV), favours the emergence of many diseases caused by opportunistic pathogens. In humans, cross-immunity has already been described between helminths and malarial agents or between helminths and HIV (Nacher 2002; Wolday *et al.* 2002).

In rabbits, *myxoma virus* is another example. This virus suppresses the immunity of an infected host. Actually, the virus leads to an important depression of the immune response against macroparasites. Boag *et al.* (2001) showed that infected rabbits harbour a higher mean number of individuals and species of macroparasites (Fig. 6.4).

Cross-immunity can also be the cause of the negative interactions observed between two species of parasites, suggesting direct competition between them. For example, infections of bovines with dipterans of the family Oestridae (the larvae of which infest the nasal cavity), enhance the production of eosinophil cells. The increase in these immune cells affects strongylid nematodes living in the digestive tract, increasing their mortality and decreasing their fecundity. The negative interaction between both these species is not direct, as they live in different organs, but occurs via the immune response (Dorchies *et al.* 1997).

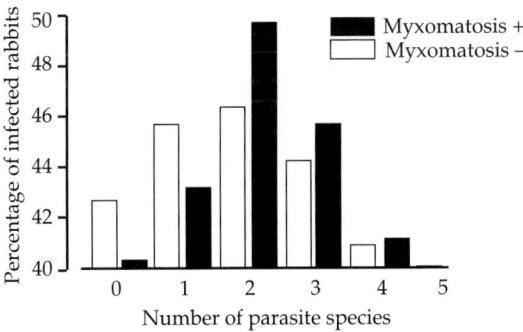

Figure 6.4 Immunosuppressive effect of *myxoma virus* on parasite richness among a rabbit population (Boag *et al.* 2001). Myxomatosis + = seropositive and infected animals; myxomatosis− = seronegative or non-infected animals. After Cattadori *et al.* (2006).

These examples show the importance of interactions between parasite species. These are complex interactions, which are rarely taken into account in the transmission and regulation of host–parasite systems.

6.6 Measuring regulation using population dynamics modelling

Regulation is defined as the tendency for a population to decrease when it is above a certain threshold, and to increase when it is below this threshold. Generally, this threshold corresponds to the **biotic capacity**, i.e. the maximum number of individuals that an environment can support. Parasites are supposed to regulate their host populations by diminishing the demographic equilibrium threshold below the biotic capacity. One way to quantify the regulation of host demography by parasites is to model the dynamics of both these populations. Mathematical models have been developed in response to multiple objectives, such as:

• the prediction of growth (epidemics) and conditions of persistence (endemism) of diseases in host populations;
• estimation of the potential to control (regulation) a host population by a given parasite (Anderson and May 1978; May and Anderson 1978).

Population dynamics are generally described using systems of differential equations, describing the observed variables as a function of time. In the case of microparasites (viruses, bacteria, protozoa) for which we generally estimate the infected stage, different classes of hosts are considered: susceptible, infectious, and recovered (the SIR model, see below). In the case of macroparasites, we can estimate the number of parasites (**intensity**, abundance). We model parasite population dynamics by incorporating an essential parameter reflecting parasite aggregation: k (negative binomial law, Box 6.1).

6.6.1 The SIR model

SIR models are used for modelling microparasite infection dynamics. The classical model considers three compartments: Susceptible, Infected and

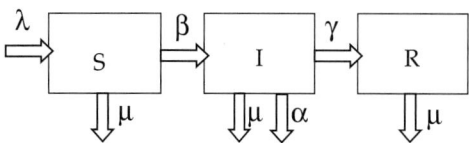

Figure 6.5 Basic SIR (Susceptible–Infected–Recovered) model: λ, arrival of new non-infected hosts; β, infection rate; μ, host death rate; γ, recovery rate; α, virulence.

infectious, and Recovered (Fig. 6.5). Two basic models exist, taking into account a density- or a frequency-dependent transmission.

Density-dependent transmission (βI):

$$dS/dt = a - \mu S - \beta IS$$
$$dI/dt = \beta IS - (\gamma + \mu)I$$
$$dR/dt = \gamma I - \mu R$$

with S the density of susceptible hosts, I the density of infected and infectious hosts, and R the density of recovered hosts; the total population is $N = S + I + R$, a is the arrival of new non-infected hosts, β is the infection rate, μ is the host death rate, and γ is the recovery rate.

Frequency-dependent transmission (βI/N):

$$dS/dt = a - \mu S - \frac{\beta I}{N}S$$
$$dI/dt = \frac{\beta I}{N}S - (\gamma + \mu)I$$
$$dR/dt = \gamma I - \mu R.$$

In the case of a parasite-induced mortality (**pathogenicity** or virulence), we can modify the equation concerning I by:

$$dI/dt = \beta IS - (\gamma + \mu + \alpha)I$$

where α is the pathogen-induced mortality rate, or the virulence. We note that the **prevalence** of infection can be expressed as:

$$\text{prevalence}(\%) = \frac{I}{S + I + R} \times 100.$$

6.6.2 The macroparasite model

The basic host–macroparasite model is a little more complicated and has three parts—host H, free

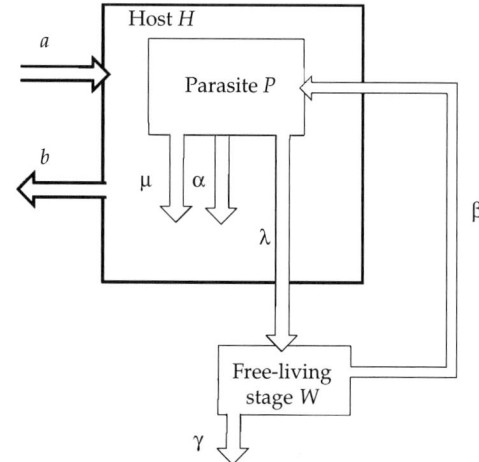

Figure 6.6 Macroparasite model: a, host birth rate; b, host death rate; α, host mortality induced by parasites (P); β, transmission rate; λ, parasite (P) fecundity; μ, intrinsic mortality of adult parasites (P); γ, intrinsic mortality of free stages (W) of the parasite; k, parameter for the negative binomial distribution describing how adult parasites (P) are dispersed among the host population (H).

stages W, and adult parasite P harboured by the hosts (Fig. 6.6):

$$dH/dt = aH - bH - \alpha P$$
$$dW/dt = \lambda P - \gamma W - \beta WH$$
$$dP/dt = \beta WH - (\mu + b + \alpha)P - \alpha\left(\frac{P^2}{H}\frac{k+1}{k}\right)$$

where a is the host birth rate, b the host death rate, α the parasite-induced death rate, β the transmission rate of free stages W to host H, λ the parasite fecundity, μ the intrinsic mortality of adult parasites P, γ the intrinsic mortality of parasite free stages W, and k the parameter for the negative binomial distribution describing how adult parasites are dispersed among the host population.

From this basic epidemiological model, we can link prevalence P to the abundance M, at a time (t) in the infection dynamics, by the following equation (Anderson and May 1985):

$$P(t) = 1 - [1 + M(t)/k]^{-k}$$

where k is the parameter of the negative binomial (Box 6.1).

6.6.3 Sexually transmitted pathogens: the STD model

This is a particular case of frequency-dependent transmission, i.e. transmission depending on the proportion of susceptible individuals compared with the total number of individuals, in a microparasite model:

$$dS/dt = a - \mu S - \frac{c\beta IS}{I+S}$$
$$dI/dt = \frac{c\beta IS}{I+S} - (\alpha + \mu)I$$

where c is the number of couplings between individuals, β is the transmission rate, and α is the host mortality induced by the microparasite.

6.6.4 Microparasites transmitted by a biting vector

This type of model is used for viral pathogens, bacteria, or protozoa transmitted by arthropods like mosquitoes, ticks, or fleas (e.g. as in the human diseases of encephalitis, West Nile fever, Lyme disease, etc.).

The basic model used for the transmission of the agent of malaria is described by two equations (Anderson and May 1991):

$$dX/dt = ab\frac{M}{N}(1-X) - rX$$
$$dY/dt = aX(1-Y) - \mu Y$$

where X is the proportion of infected humans, Y is the proportion of infected female vectors, N is the size of the human population, M is the size of the female mosquito population, a is the biting rate, b is the proportion of bites per infected female causing an infection in humans, r is the recovery rate for humans, and μ is the mosquito death rate.

6.6.5 Basic reproduction rate, R_0

The epidemiological models presented above permit to us to derive, via the basic reproduction rate R_0, an estimate of the capacity for a parasite to invade the host population. This measure can be defined by the number of secondary infections produced after the introduction of an infected host,

or a mature parasite, into a susceptible population. This quantity can be estimated at the beginning of the infection when the density-dependent processes do not act, assuming:

$$\frac{1}{I}dI/dt > 0$$

in the case of the SIR model or

$$\frac{1}{P}dP/dt > 0$$

in the case of a macroparasite model, and considering the host population at equilibrium, i.e. at the biotic capacity. R_0 can also be defined as the product of transmissions at all stages (**free-living**, larvae in the intermediate host, adult in the definitive host, etc.) divided by the product of mortality at all stages (free stages, in the intermediate host, in the definitive host, etc.). Different expressions for R_0 are given in the Table 6.2.

6.6.6 Host transmission threshold

The basic reproduction rate R_0 permits us to define the minimum host threshold necessary for introduction and persistence of a parasite in the host population. This threshold is obtained for $R_0 = 1$. In the case of the SIR model, the host transmission threshold is N_0:

$$N_0 = \frac{\lambda + \mu + \alpha}{\beta}.$$

For a direct cycle macroparasite, this threshold H_t is:

$$H_t = \frac{(\gamma + \beta H)(\mu + b + \alpha)}{\lambda \beta}.$$

6.6.7 Infection peaks, age, and immunity: parasite regulation by immunity

When a host acquires an immune response after a parasite infection, the interaction between host and parasite is not simply dependent on the current abundance of parasites, but also on the entire history of exposure to this parasite. This interaction also depends also on the competency and maturity of the immune response to secondary infections.

Table 6.2 Examples of formulae for R_0 for simple models with micro- or macroparasites

Transmission	R_0	Parameters
Microparasites		
SIR density-dependent transmission	$R_0 = \dfrac{\beta N}{\lambda + \mu}$	N, host population size; β, transmission rate; μ, host mortality; γ, recovery rate; λ, the parasite fecundity
SIR frequency-dependent transmission	$R_0 = \dfrac{\beta}{\lambda + \mu}$	Definitions as above
SIR density-dependent transmission and pathogen virulence	$R_0 = \dfrac{\beta N}{\lambda + \mu + \alpha}$	α, host mortality induced by the pathogen, or virulence. Other definitions as above
STD: sexually transmitted pathogen	$R_0 = \dfrac{c\beta}{\alpha + \mu}$	c, . Other definitions as above; c, the number of couplings
Pathogen transmitted by vectors	$R_0 = \dfrac{a^2 \beta M p^n}{N r(-\ln p)}$	M, size of vector population; N, size of host population size; a, vector biting rate; r, duration of the infective period; p, vector survival (per day); β, transmission rate
Macroparasites Direct cycle	$R_0 = \dfrac{\lambda \beta H}{(\gamma + \beta H)(\mu + b + \alpha)}$	H, host population size without any parasite; b, host death rate; α, mortality induced by the parasite; β, transmission rate; λ, parasite fecundity; μ, intrinsic mortality of parasites; γ, intrinsic mortality of parasite free stages
Indirect cycle	$R_0 = \dfrac{\lambda \beta_1 \beta_2 H_1 H_2}{(\gamma + \beta_1 H_1)(\mu_1 + b_1 + \beta_2 H_2)(\mu_2 + b_2 + \alpha)}$	H_1, H_2, definitive and intermediate host population size; b_1 and b_2, host death rate in H_1 and H_2; β_1, transmission rate for free stages of parasites to H_1; β_2, transmission rate from H_1 to H_2; μ_1 and μ_2, parasite intrinsic death rate in H_1 and H_2

SIR [model], Susceptible, Infected and infectious, and Recovered; STD, sexually transmitted disease.

Mathematical models suggest that acquired immunity can be modelled as cumulative exposure to the parasite as a function of host age (Anderson and May 1985; Woolhouse 1998).

The relationship between infection and age of individuals, in populations subject to constant rates of transmission, is given by the following equation:

$$P(a) = \frac{\lambda}{\lambda - v}[\exp(-va) - \exp(-\lambda a)]$$

where $P(a)$ is the prevalence at age a, λ is the infection rate (proportional to the transmission rate), and v is the recovery rate.

For high transmission rates, parasite intensity increases rapidly and reaches a peak in relatively young hosts. This peak intensity is followed by a decline caused by the activation of and the increase in the acquired immune response. When transmission rates are low, the intensity of parasitism

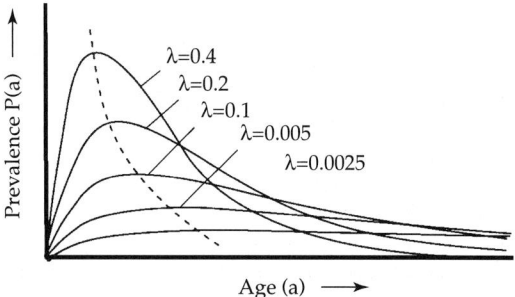

Figure 6.7 Predicted relationship between infection and the age of host individuals in populations with different infection rates λ. The maximum value of prevalence is a negative function of the age at which the maximum value is observed (dotted curve). After Woolhouse (1998).

increases more slowly and the peak is reached in older hosts. The parasite intensity is lower for this peak than for peaks obtained with higher transmission rates (Fig. 6.7). Thus the age at which the peak occurs depends on the force of infection. The epidemiological pattern, described as the 'peak shift' (Woolhouse 1998), predicts a negative link between the observed parasite intensity and the age at which the peak is observed.

6.7 Theoretical consequences for regulation of the host population

6.7.1 Regulation, virulence, and aggregation

Mathematical models demonstrate that parasites are able to regulate the populations of their hosts. This regulation is appreciated by the difference observed between the host population at the **biotic capacity** (without any parasites) and the

equilibrium size of the host population in the presence of parasites (Box 6.4).

Basic mathematical models show that moderately virulent parasites are better at regulating the host population (Fig. 6.8). Logically, parasites that are not, or are only weakly, virulent do not decrease, or decrease only slightly, the equilibrium size of the host population. Highly virulent parasites have a weak potential for regulation because they kill their hosts too rapidly and disappear before they can infect new hosts. Note also that highly virulent parasites are rare (low prevalence) and that the most prevalent parasites are weakly or moderately virulent. These latter parasites can potentially regulate their host populations. The most cited evidence supporting a **trade-off** between

---

Box 6.4 Measuring the impact of an infection in a natural population

When a host–parasite system is at equilibrium, it is possible, with the help of basic mathematical models, to estimate the depression, or regulation factor, of the host population by the parameter D. D is the value of the size of the population at equilibrium when the parasite is present, divided by the biotic capacity value. (McCallum and Dobson, 1995).

For a microparasite affecting fecundity and survival:

$$D = \left(\frac{\alpha + a(1-f_1)}{r} \right) \Phi$$

where α is the parasite-induced host death rate (virulence), r is the rate of growth of the host population ($r = a-b$), f_1 is the impact of the parasite on fecundity (1, no impact; 0, host sterilized by parasites), a is the birth rate of the uninfected host population, and Φ is the prevalence of the pathogen at equilibrium.

For macroparasites, we have the following equation:

$$D = \left(\frac{\alpha}{r} \right) M$$

where M is the mean parasite abundance and α is the increase in host mortality per parasite.

For a host infected by microparasites the pathogenicity α can be estimated as the inverse of the life expectancy of the infected host; for macroparasites, it is the degree of relationship between parasite intensity and the inverse of life expectancy. r can be estimated from host demographic data, but also by a simple allometric relation

knowing adult weight (in mammals the link between mortality and weight is expressed by a power law with an exponent of −0.25).

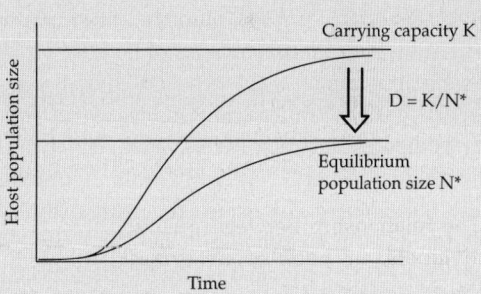

Box 6.4 Trajectory of host population dynamics with and without parasitism. The host population reaches equilibrium at the biotic capacity K, and reaches an equilibrium N^* in the presence of the parasite. The depression factor D is the ratio of K to N^*.

Reference

McCallum, H., Dobson, A. (1995). Detecting disease and parasite threats to endangered species and ecosystems. *Trends in Ecology and Evolution* **10**: 190–194.

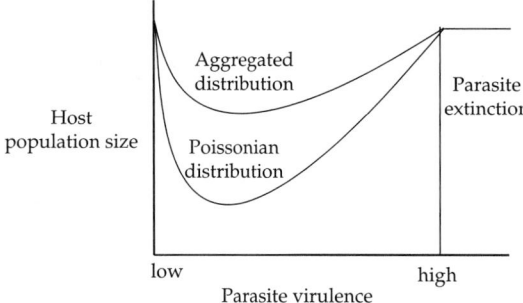

Figure 6.8 Effect of parasite virulence and aggregation on the size of the host population at equilibrium.

virulence and transmission and microparasite–host co-evolution comes from the use of *myxoma virus* to control European rabbit populations in Australia and Europe, resulting in more resistant rabbit populations and less virulent viruses (see Section 6.2) (Fenner and Cairns 1959; Fenner and Ratcliffe 1965; Anderson and May 1982). Within a relatively short time after the release of highly virulent *myxoma virus*, the viruses recovered from the remaining decimated and sometimes more resistant wild rabbit populations were less virulent and had lower rates of disease-induced mortality in control laboratory rabbits than the initial populations. However, the attenuation of *myxoma virus* in natural populations was less severe than in experimental studies (Fenner and Ratcliffe 1965). This is also evidence for a trade-off between virulence and transmission. Highly virulent forms of the virus were disadvantaged because they killed the rabbits too quickly and thus reduced the time available for them to be picked up by the insect vectors (mosquitoes or fleas) required for their infectious transmission. Viruses that were too attenuated were disadvantaged because they generated fewer skin lesions and were transmitted less. The *myxoma virus* story is particularly interesting because it was a single case with a eukaryotic host where the relationship between virulence and transmissibility inferred from the epidemiological data and models was independently tested and demonstrated experimentally (Fenner *et al.* 1956; Mead-Briggs and Vaughan 1975).

For macroparasites, the degree of aggregation plays an important role. Weakly aggregated

parasites present a higher regulation potential for similar virulence because they have an impact on a larger number of hosts (see Fig. 6.8).

However, basic theoretical models consider that the causes of mortality that can affect the hosts are independent. Total mortality affecting the hosts is the sum of deaths from all causes (additive mortality). However, observed mortalities are often the results of interactions or trade-offs. Numerous studies show that in many experimental or natural situations, the nutritional status of the host determines the severity of the infection. Host malnutrition can also inhibit or reduce the host immune response, leading to an increase in parasite intensity in the individual. Basic models do not take host behaviour into account, especially the behaviours of avoidance of contaminated habitats or infectious animals. These behavioural decisions can also be submitted to trade offs.

6.7.2 Destabilization of host–parasite dynamics

Mathematical models of host–parasite dynamics have identified three characteristics leading to instability (May and Anderson 1978; Anderson and May 1978):

• a strong reduction in host fecundity caused by the parasite, in comparison with the impact of the parasite on host survival (parasites affecting host fecundity lead to less destabilization than parasites affecting survival);
• delays in recruitment of infecting stages (a long development in the environment with quiescent stages or the existence of resting stages in hosts have destabilizing effects);
• a random or regular parasite distribution among the host population (strong aggregations of the parasite reduce the destabilizing effect).

6.8 The role of parasites in regulation of the host population: empirical evidence

Parasites modify host behaviour, energy requirements, reproductive capacity, and survival (Grenfell and Dobson 1995; Poulin 1998; Combes

Table 6.3 Examples showing the impact of parasites on wild vertebrates (modified from Tompkins and Begon 1999)

Host	Parasite	Impact
Birds		
Hirundo rustica	*Ornithonyssus bursa* (mite)	Fecundity
Hirundo pyrrhonota	*Oeciacus vicarious* (bug)	Fecundity
Sturnus vulgaris	*Dermanyssus gallinae* (mite)	Fecundity
	Ornithonyssus sylvarium (mite)	Fecundity
Parus major	*Ceratophyllus gallinae* (flea)	Fecundity
Delichon urbica	*Oeciacus hirundinis* (bug)	Fecundity
Margarops fuscatus	*Philinus deceptivus* (dipteran)	Fecundity
Progne subis	*Dermanyssus prognephilus* (mite)	Fecundity
Lagopus lagopus	*Trichostrongylus tenuis* (nematode)	Fecundity and/or survival
Mammals		
Gerbillus andersoni	*Synoternus cleopatrae* (flea)	Fecundity
Mus musculus	*Capillaria hepatica* (nematode)	Fecundity and/or survival
Lepus americanus	*Obeliscoides cuniculi* (nematode)	Survival
Ovis aries	*Teladorsagia circumcincta* (nematode)	Survival

2001). Theoretical studies (see below) have demonstrated that parasites are able to regulate or destabilize host populations. This is possible because parasites can reduce host survival and/or fecundity, as widely demonstrated by empirical studies (Table 6.3). However, to show that a parasite is able to reduce host fecundity or survival does not necessarily demonstrate a capacity for the regulation of host population dynamics.

Experimental studies remain rare, even more so in natural situations. The most famous studies to show that parasites are able to regulate the populations of vertebrate hosts were carried out in enclosures.

Scott (1987) conducted an experimental study showing the ability of a nematode, *Heligmosomoides polygyrus*, to regulate the abundance of its rodent host. An uninfected mouse population could reach a size 20 times larger than a population infected by this nematode. Using antihelminth treatments, the nematode was eliminated from the infected population, and this population then reached an abundance similar to that observed for the uninfected population.

Singleton and Spratt (1986) showed that the nematode, *Capillaria hepatica*, significantly reduced the fecundity and survival of domestic mice in the laboratory. This nematode does not have a free stage, and is directly transmitted by predation or cannibalism. Death of the host is thus necessary to ensure transmission. Using modelling, McCallum and Singleton (1989) demonstrated that the necessary host death has a destabilizing effect on host dynamics by inducing important fluctuations. The model suggests that the parasite is able to maintain host density below the biotic capacity obtained when the parasite is absent. However, experimental designs with wild domestic mice did not confirm this regulation efficiency (Singleton and Chambers 1996).

The impacts of parasitism on survival and fecundity have also been demonstrated in two natural ungulate populations: Soay sheep on St Kilda and Svalbard reindeer in Spitsbergen. On St Kilda, an island where predators and competitive herbivores are absent, sheep populations are unstable, exhibiting regular declines every 3–4 years (Clutton-Brock *et al.* 1997). Declines would be due to a density-dependent winter mortality associated with a lack of food and the additive effect of macroparasites (Clutton-Brock *et al.* 1991; Grenfell *et al.* 1992; Gulland and Fox 1992; Catchpole *et al.* 2000). Reduction of the parasite burden by using antihelminths, confirmed the additive role of parasites in certain age classes during sheep population declines (Gulland 1992). Svalbard reindeer populations on Spitsbergen fluctuate and show a covariation in their birth and death rates (Solberg *et al.* 2001). An important part of the fluctuation can be attributed to the non-random nature of climatic events. Modelling studies and experimental reduction of parasitism showed that intestinal helminths have an effect on reindeer body condition and female fecundity and growth rate (Albon *et al.* 2002; Stien *et al.* 2002).

In the case of lagomorphs, changes in prevalence and intensity are observed in the cyclic variation in hare populations. Parasites reduce the survival of arctic hare, in relation with predation (Murray 2002). Models show that predators and parasites can destabilize populations by inducing cycles (Ives and Murray 1997).

Finally, the most famous example is the red grouse and its intestinal nematode parasite *Trichostrongylus tenuis*. In Scotland, grouse populations show 4–10-year cycles attributed to this nematode (Haydon *et al.* 2002; Hudson *et al.* 2002). Mathematical and field studies have demonstrated that this weakly aggregated nematode reduces its host's fecundity. Both of these factors (aggregation and reduction of fecundity) destabilize host populations and induce population cycles (Dobson and Hudson 1992; Hudson *et al.* 1992). Antihelminth treatments reduced the rate of population decline and removed the crash (Fig. 6.9). These results agree with theoretical predictions (Hudson *et al.* 1998) that the nematode *T. tenuis* would be responsible for red grouse population cycles. The *T. tenuis*–red

grouse system remains a unique experimental study demonstrating the destabilizing role of parasites on wild host populations (see also criticisms from Lambin *et al.* 1999; Tompkins and Begon 1999; Turchin 2003).

6.9 The host community and regulation by parasites: a few examples of multi-host-parasite models

6.9.1 Invading a host community

Many micro- and macroparasites attack multiple host species, so their ability to invade a host community can depend on the composition of that community. A theoretical framework can help us to understand how a parasite can become established in a host community and regulate one or a few host populations. Holt *et al.* (2003) presented a graphical isocline framework for studying the establishment of disease in systems with two host species, based on the host species being treated as resources. The isocline approach provides a natural generalization to multi-host systems of two related concepts in disease ecology—the basic reproductive rate of a parasite (R_0, see Section 6.6), and the threshold host density. A qualitative isocline shape characterizes the threshold community configurations that permit the establishment of a parasite.

Species may be threatened by invaders via shared parasites. The graphical framework can be applied to the question of how shared parasites affect the coexistence of hosts. Depending on the relationship

(a)

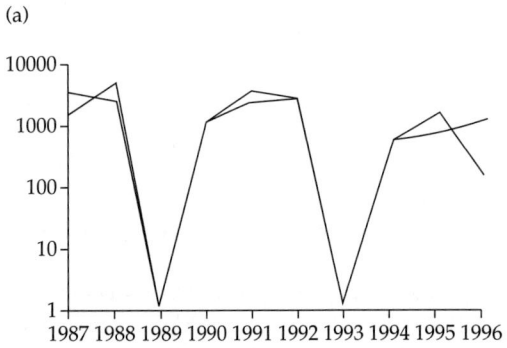

(b)

Figure 6.9 Red grouse dynamics at two control sites (a) and two sites (b) where birds were treated with antihelminths. The elimination of the parasite leads to the absence of a crash in the host population.

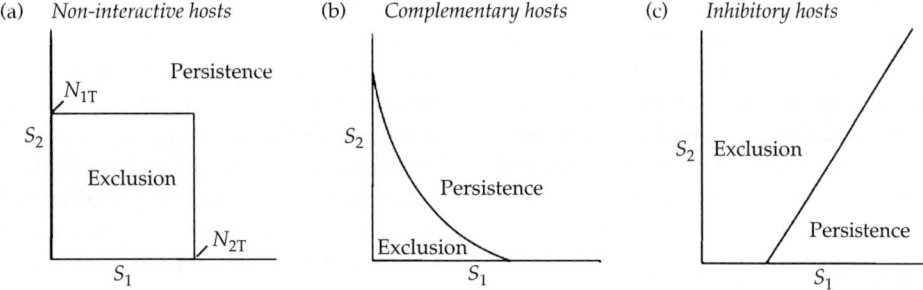

Figure 6.10 Potential isocline shapes for the establishment of parasites on two host species. (a) Non-interactive hosts: the existence of alternative hosts is irrelevant for parasite establishment. (b) Complementary hosts: strong cross-species infection. If infection occurs much more readily between than within host species, the isocline bows towards the origin. This might arise if there are strong mechanisms for spacing within species (e.g. territoriality), leading to more potential contacts across than within species. In this case, a mixture of host species more readily sustains the parasite than does either host alone. (c) Inhibitory hosts: one host cannot sustain the parasite on its own, at any density, and moreover diminishes the rate at which the other host becomes infected. This implies an isocline with positive slope. S, density of susceptible hosts; N, total number of hosts (infected + susceptible). After Holt *et al.* (2003).

between the host species and the parasite (non-interactive hosts, cross-species infections, etc.) different patterns can be observed (Fig. 6.10), with the persistence of the exclusion of one or both host species. For example, if a pathogen is highly virulent to host species i, the intercept of the isocline on the S_i axis (density of susceptible hosts) is likely to exceed K_i (the carrying capacity of host i) and the pathogen will not persist in a population of host i alone. If host species j is more tolerant of infection, then the isocline will intersect the S_j axis near the origin and the pathogen may persist in host j alone. Consequently, if host j is introduced and reaches a sufficiently high density, the pathogen can persist and the vulnerable host i will be exposed. Given sufficient interspecific transmission, host i will not persist in the presence of both host j and the pathogen (this case of shared parasitism is examined formally in Holt and Pickering (1985), Begon *et al.* (1992), and Greenman and Hudson (1997)).

6.9.2 Apparent competition

When several host species are involved in a host–parasite system, apparent competition can occur. Apparent competition is a process that results in a decrease in the population growth of two host species that do not compete for the same resource but do share the same natural enemy (parasite). Often, this leads to the exclusion of one of the host species. Recent theoretical and empirical work has shown that this ecological process (for prey–predator or host–parasite systems) is instrumental in structuring species assemblages (Tompkins *et al.* 2000; Rushton *et al.* 2000; Gilbert *et al.* 2001).

Gilbert *et al.* (2001) studied the grouse–hare–deer–tick–louping ill virus system of upland Britain. Each host differed in its interaction with the vector and pathogen. Grouse amplify virus only, deer amplify vector only, and hares amplify both. Grouse alone suffer high virus-induced mortality. An analytical model of the system was parameterized using empirical data from two wild populations with different community structures. It was shown that apparent competition may occur between mountain hares and grouse. The addition of a third host type increased the likelihood of disease persistence. This work highlights the importance of taking the entire host community of a parasite into account for understanding the host and disease dynamics.

A more well known model of apparent competition mediated by an infectious agent was developed by Rushton *et al.* (2000) to understand the decline of red squirrels in Norfolk. The model simulates the spread of parapoxvirus between two species of squirrels (resident red squirrels and invasive grey squirrels) in fragmented populations based on the dispersal of infected animals. The model predicted that parapoxvirus, like interspecific competition, could have led to the extinction of the red squirrel, which is more susceptible to the virus. These results have implications for conservation management of the red squirrel in the UK. Schemes in which animals are translocated or given supplementary feeding may enhance disease spread by bringing infected animals into contact with others.

6.10 Pathology and the evolution of virulence

In addition to aggregation, theoretical models and empirical observations demonstrate the importance of the pathology induced by the parasite (virulence) in the reduction of host fecundity and survival.

Many definitions exist for the term virulence, depending on the scientific field. In microbiology, virulence corresponds to the capacity of a pathogen

to infect and develop in a host. Generally, this virulence is associated with one or several virulence genes. In pathology, virulence is the effect of the pathogen linked to the infection. In evolutionary ecology, virulence is defined as the increase in host mortality induced by a parasite, or by the reduction in host **fitness** caused by a parasite.

Most of the epidemiological models predict an association between reduced host survival and increased parasite virulence. For example, an increased host mortality selects changes in parasite life-history traits (Morand and Poulin 2000), like faster parasite maturation.

Various factors linked to transmission or host and parasite ecology have a theoretical effect on the evolution of virulence (Table 6.4). Vertical transmission, from parents to descendants, would favour a reduction in virulence, whilst horizontal transmission would favour its increase. Several examples

Table 6.4 Factors affecting the evolution of virulence (after Galvani 2003b)

Factor	Evolutionary response
Transmission	
Horizontal/vertical	Increased virulence with horizontal transmission is more important than vertical transmission
Spatial structure	Decreased virulence
Indirect transmission	Increased virulence
Propagule production	Increased virulence
Host ecology	
Heterogeneity in host resistance	Decreased or increased virulence
Recognition and avoidance of infected host	Increased virulence
Number of sexual partners	Increased virulence for STD
Host population size	Increased virulence during epidemics but not during endemicity
Host life expectancy	Decreased virulence
Co-evolutionary dynamics	The system can evolve from a unique stable evolutionary equilibrium to a system with several stable equilibria
Parasite ecology	
Competition between strains	Increased virulence, with reduced effects as a function of degree of relatedness between strains
Parasite with density-dependent replication	Increased virulence if the relationship between parasite replication and reduction of host fitness is not linear
Sexual reproduction	Parasite sexuality, increased virulence. Host sexuality, decreased virulence
Medicine	
Vaccines reducing parasite growth and toxicity	Increased virulence
Vaccines reducing parasite transmission and host susceptibility	Decreased virulence
Use of antibiotics	Increased or decreased virulence

STD, sexually transmitted disease.

agree with this hypothesis, for example bird lice or nematodes associated with fig-tree-pollinating insects. An increased number of infective parasite stages increases virulence, while the longevity of the infective stage does not seem to have an effect ('the curse of the pharaoh' refers to the numerous deaths that occurred after Egyptian tomb openings). Heterogeneity of hosts with regard pathogen resistance can decrease or increase virulence depending on the type of heterogeneity. An increase in host life expectancy favours a reduction in virulence. Sexual reproduction of a host or parasite also leads to the evolutionary response of virulence.

Note that parasitism is an environmental pressure that differs from the other pressures. Actually, the evolutionary response of hosts modifies parasite pressure in turn. Host life-history traits and parasite virulence are thus determined by co-evolutionary links. The results are difficult to predict, simply because of the complexity of the processes affecting parasite infection rate and the environment of host and parasite (Koella and Restif 2001).

The force of infection of the parasite will determine the pressure exerted on the host's life-history traits, which in turn will determine the selection pressure on parasite virulence. The most cited evidence in support of microparasite–host co-evolution comes from the trial using *myxoma virus* to control European rabbit populations in Australia and Europe resulting in more resistant rabbit populations and less virulent viruses (see Section 6.7.1).

The quality of the environment can also influence population dynamics and host life-history traits. The quality of the environment can potentially modify the pressures imposed by parasites. Environments with a high productivity are predicted to be dominated by hosts that invest highly in resistance and by virulent parasites. Low-quality environments should favour less antagonistic or mutualistic interactions (Hochberg *et al.* 2000).

Finally, human intervention, especially the use of vaccines or antibiotics, considerably modifies the environment of parasites. The mode of action of a vaccine, reducing the growth or transmission of a parasite, can favour an evolutionary response from the parasite of either increased or a decreased virulence (Gandon *et al.* 2001). Note also that basic epidemiological mathematical models rarely take evolutionary mechanisms into account.

6.11 Parasitism and extinction

Microparasites are generally considered to be a bigger threat than macroparasites in conservation biology (Daszak and Cunningham 1999; Daszak *et al.* 2000; Cleaveland *et al.* 2001; see also Chapter 9). Most conservation textbooks refer to infectious diseases due to viral agents, microbes, or protozoa. Pathogens have been thought to be implicated in the extinction of numerous species (McCallum and Dobson 1995; Vitousek *et al.* 1997), such as several endemic Hawaiian birds (VanRiper *et al.* 1986) or the thylacine (Australian marsupial carnivore). In a recent review, Daszak *et al.* (2000) mentioned 19 microparasites, but no macroparasites, as important dangers for conservation or for human health (major **zoonotic** agents). The absence of macroparasites suggests that these parasites are less important and their monitoring presents little interest, except for ectoparasites such ticks and fleas because of their role as vectors for numerous micro-organisms

Parasitism can interfere with the critical size of a population, below which the population can go extinct. Theoretically, several mechanisms can produce extinction due to parasitism (Table 6.5):

1. Small population syndrome: small populations are characterized by low densities, which are reduced in the presence of a pathogen. Fluctuations below a persistence threshold can increase and proceed to extinction because of random environmental processes. The impact of parasites can also drive the host population close to the threshold for the **Allee effect** (the threshold below which the population inexorably declines), leading the host population into a spiral to extinction (Deredec and Courchamp 2003).

2. A reduced genetic variability: small populations are often characterized by a high degree of relatedness, with a reduced genetic variability, especially for genes involved in resistance against parasites such as major histocompatibility complex (MHC) genes.

3. The impact of density-dependent or vector-transmitted parasites: sexually or vector-transmitted

Table 6.5 Several examples of a proven or potential role for parasitism in extinction of the host population (modified, from Gog *et al.* 2002)

Mechanism	Host species	Pathogen	Impact	References
Small population syndrome	Thylacine (*Thylacinus cynocephalus*)	Virus	Probable extinction	Guiler 1961 (in McCallum and Dobson 1995)
Random	Golden toad (*Bufo periglenes*)	Virus	Probable extinction	Pounds *et al.* (1997)
	Black-footed ferret (*Mustela nigripes*)	Virus	Probable extinction	Thorne and Williams (1988)
	Feral goats (*Capra pyrenaica hispanica*)	Mite	Reduction in population size	Leon-Vizcaino *et al.* (1999)
Reduced genetic variability	Cheetah (*Acinonyx jubatus*)		Increased susceptibility to diseases	OíBrien *et al.* (1985)
Independent transmission	Rabbit (*Oryctolagus cuniculus*)	Haemorrhagic disease	Potential extinction (model)	White *et al.* (2003)
Reservoir	White-tailed deer (*Odocoileus virginianus*), elk (*Alces alces*)	Nematode	Reduction in population size	Schmitz and Nudds (1994)
	Grey squirrel (*Sciurus carolinensis*), red squirrel (*Sciurus vulgaris*)	Virus	Reduction in population size	Rushton *et al.* (2000)

parasites show a transmission independent of host population size because of the dilution effect. Regulatory effects are maintained or increase with a reduction in the host population.

4. The reservoir effect and apparent competition: non-specific parasites are maintained on abundant populations but continue regulating all populations, even small ones.

6.12 Conclusion

Theoretical models and empirical studies have established that parasites can regulate host communities and host population dynamics. However, most classical studies model 'one host/one parasite' systems and probably do not correspond to the complex reality (Holt *et al.* 2003). More and more studies are now exploring host–community–parasite models.

Mechanisms explaining emerging pathogens are complex: to understand these and their consequences for human or animal populations we need a dialogue between epidemiological modelling, immunology, and evolutionary ecology.

Parasites and pathogens have an impact on endangered species because of apparent competition mechanisms and/or a decreased investment in immune defences, or a reduced genetic variability, especially for genes associated with the immune system. Parasites have to be considered in the objectives of conservation biology because they can compromise re-introduction programmes (Gog *et al.* 2002).

Important points

• Theoretical approaches, especially from epidemiological modelling, are necessary for an understanding of host–parasite interactions. Two key factors, parasite aggregation among host populations and virulence, should be taken into account when estimating parasite regulation and the destabilizing potential on host population dynamics.

• Simple epidemiological models allow us to explore the often complex interactions between hosts and parasites, and to produce testable predictions for ecology and management.

• Host–parasite interactions are co-evolutionary systems, with hosts investing in defence mechanisms (behaviour, immunity) and resistance, and parasites modifying some of their life-history traits affecting transmission and virulence. The adaptation (or non-adaptation) potential of host and pathogen will define future interactions in the domain of health (emerging diseases) or conservation biology (extinction).

Questions for discussion

• Aggregation is the result of the interaction between individual hosts and parasites. It depends on numerous factors affected by ecological and evolutionary pressures. However, aggregation is rarely considered as a dynamic variable in host–parasite models. How could it be incorporated into mathematical models?

• Hosts are not infected by a single parasite species (or a single strain) but by multiple enemies. How can we consider parasite (intra- or inter-specific) diversity in host–parasite interaction dynamics?

• There have been few experiments on the ecology of host–parasite interactions in natural environments. What experiments are needed to test methods for the ecological management of pathogen control?

Further reading

• Anderson, R.M., May, R.M. (1991). *Infectious diseases of humans: dynamics and control*. Oxford University Press, Oxford.

• Collinge, S.K., Ray, C. (2006). *Disease ecology: community structure and pathogen dynamics*. Oxford University Press, Oxford.

• Combes, C. (2001). *Parasitism: the ecology and evolution of intimate interactions*. University of Chicago Press, Chicago.

• Bush, A.O., Fernandez, J.C., Esch, G.W., Seed, J.R. (2001). *Parasitism: the diversity and ecology of animal parasites*. Cambridge University Press, Cambridge.

• Diekmann, O., Heesterbeek, H.J. (2000). *Mathematical epidemiology of infectious diseases: model building, analysis and interpretation*. Princeton University Press, Princeton, NJ.

• Ewald, P.W. (1994). *Evolution of infectious disease*. Oxford University Press, Oxford.

• Frank, S.A. (2002). *Immunology and evolution of infectious disease*. Princeton University Press, Princeton, NJ.

• Grenfell, B.T., Dobson, A.P. (eds) (1995). *Ecology of infectious diseases in natural populations*. Cambridge University Press, Cambridge.

• Hudson, P.J., Rizzoli, A., Grenfell, B.T., Heesterbeek, H., Dobson A.P. (eds) (2002). *The ecology of wildlife diseases*. Oxford University Press, Oxford.

• Poulin, R. (2006). *Evolutionary ecology of parasites*. University Press of Columbia, Columbia.

• Poulin, R., Morand, R. (2004). *Parasite biodiversity*. Smithsonian Institution Press, Washington, DC.

• Thomas, F., Guégan, J.-F., Renaud, F. (eds) (2004). *Parasitism and ecosystems*. Oxford University Press, Oxford.

CHAPTER 7

Parasitism and biological control

Éric Wajnberg and Nicolas Ris

7.1 Introduction

Parasites exert negative effects on their hosts, both at the individual and population level. This ecological feature can be used to develop **biological control** programmes, i.e. 'the use of living organisms [called biological control agents] to control the population density or impact on a specific pest organism, making it less abundant or less damaging than it would otherwise be' (Eilenberg *et al.* 2001). More than a thousand biological control agents are currently in use worldwide and most of them are **parasitoid** insects used to control phytophagous insect pests. This chapter aims to describe this particular form of parasitism, highlighting its huge diversity and the associated consequences for demographic interactions between hosts and parasitoids. More specifically, we will detail how the ecological and dynamic features of parasitoid species can be used to control noxious pests, and how recent advances may improve the efficacy of biological control programmes. We will conclude by emphasizing the complex scientific questions raised by the study of these particular species in terms of evolutionary ecology and the associated agronomic applications.

7.2 Ecological features of parasitoids

7.2.1 Overview

Parasitoids are organisms whose larvae develop to the detriment of a single host (Godfray 1994). Their mode of development lies between that of predators and true parasites since the host is generally killed and there is a tight physiological interaction between the two partners (Toft *et al.* 1991).

The adult parasitoids are free-living. According to recent estimates, parasitoids represent between 8% and 20% of all insect species. Most parasitoids are **Hymenoptera** (around 50,000 described species) or **Diptera** (around 16,000 species) (Feener and Brown 1997). Some species can also be found within Coleoptera, Lepidoptera, Trichoptera, and Strepsiptera (Quicke 1997). From an evolutionary point of view, 'parasitoidism' appears in a different way within the two main orders. More accurately, it seems that all hymenopteran parasitoids probably originated from a single mycophagous ancestor inhabiting dead wood. In the Diptera, however, parasitoids appear to have arisen independently numerous times from different saprophagous or predatory ancestors (Eggleton and Belshaw 1992). These different evolutionary origins may explain the important interspecific variations, but other factors (including ecological ones) must be taken into account to understand the processes of speciation and diversification.

7.2.2 Host range

Most parasitoids attack other insects, but some species attack other arthropod hosts or even hosts from other phyla (molluscs or even some chordates) (Feener and Brown 1997). Sometimes, the host is itself a parasitoid species leading to a tritrophic interaction between a host, a parasitoid, and a so-called **hyperparasitoid**. The parasitized host stage greatly varies according to the parasitoid biology but we can distinguish parasitoids of eggs, larvae, nymphs, or even adults. In some cases, **oviposition** (i.e. the deposition of an egg in (or on) the host) occurs at an early host stage (for instance the egg) but the development occurs in later stages (larvae or nymphs). Some parasitoid species are also able to infest more than one host stage. The host range greatly varies between species. For example, some tachinids are highly generalist, being able to develop successfully within several dozen species belonging to different families (Stireman *et al.* 2006) whereas numerous species are specialized and restricted to a limited number of host species.

There are numerous reasons for the particular host range (Godfray 1994; Stireman *et al.* 2006). For instance, dipteran parasitoids are generally more generalist than hymenopteran species, suggesting that some physiological constraints or pre-adaptation may favour or restrict the host range. The taxonomy of the potential hosts may also influence evolution of the host range since a parasitoid species can probably adapt more easily to new species that share similar physiological features and defence mechanisms with its original host. Similarly, parasitoids are more likely to infest hosts facing similar ecological constraints. However, the host range may evolve through time and space but, contrary to other organisms with a parasitic lifestyle (Fox and Morrow 1981; Jaenike 1990; Nosil 2002), only a few data are currently available on intraspecific variability in the number of potential hosts that can be attacked or, more generally, on ecological specialization phenomena.

7.2.3 Sex determination

Knowledge of the mechanisms of sex determination is an important point not only for understanding the ecology and evolution of parasitoid species but also for improving their use as biological control agents. Three different systems determine the sex of parasitoids: (1) **diplo-diploidy**, (2) **haplo-diploidy**, and (3) **thelytokous parthenogenesis** (Normak 2003). Diplo-diploidy is the main determinant of sex in dipteran parasitoids and is, for instance, similar to the way in which sex is determined in mammals. Each individual originates from a fertilized oocyte which contains both maternal and paternal genomes. The combination of the sex chromosomes determines the sex. In most of hymenopteran parasitoids, however,

haplo-diploidy, also called 'arrhenotoky', is the most frequent mechanism of sex determination. In this case, females originate from fertilized oocytes whereas males originate from unfertilized oocytes and are thus haploids. In this case, the parasitoid females can 'decide' whether or not to fertilize their eggs and thus produce females or males (cf. Section 7.3.1.5). Finally, numerous cases of **thelytoky** have also been described. Thelytokous females are able to produce diploid females without mating. In some cases, diploidy in the germinal cells may be caused by bacterial symbionts (see Box 7.3). Sexual and asexual species strains can sometimes coexist within the same parasitoid species (Amat *et al.* 2006).

7.2.4 Other life-history traits

Depending on the parasitoid species, immature parasitoids can develop either within or outside their hosts (endo- and ectoparasitoids, respectively). The ability or inability of several parasitoids to develop alone within a single host leads to a distinction between gregarious and solitary parasitoids. However, the distinction must be made with caution since the mode of development of some parasitoids may vary depending on host characteristics or status (species, size, quality). The duration of the interaction between parasitoids and their hosts also varies, and distinction can be made between **idiobiont** parasitoids, which quickly kill their hosts, and **koinobionts** that allow the host to survive for some time (Askew and Shaw 1986).

The availability of mature eggs in adult parasitoids also shows an important interspecific variability. Some so-called **pro-ovigenic** parasitoids have a complete stock of oocytes soon after adult emergence, whereas in **synovigenic** species new eggs continue to mature throughout adult life. Actually, the distinction is not always so clear-cut since a continuum can usually be observed between these two extremes (Jervis *et al.* 2001). Adult nutrition is also an important life-history trait in parasitoids. Numerous species do not feed or have access to sugars only (honeydew, nectar, etc.) (Wäckers 2003), while others 'host-feed', consuming host tissues or **haemolymph** (Jervis and Kidd 1986).

Such nutritional input appears to be necessary to prolong the adult lifespan or to increase egg production. Host-feeding behaviour, which has been described in 17 families of Hymenoptera as well as in some Diptera, may cause the death of the host and can thus deeply modify the host–parasitoid dynamics (Heimpel and Collier 1996).

7.2.5 Co-variations among life-history traits

One of the goals in the study of the evolutionary ecology of parasitoids is to understand their challenging diversity, and especially to try to identify more accurately some co-variations between the main life-history traits (Godfray 1994; Mayhew and Blackburn 1999; Jervis *et al.* 2001). Some studies also make a dichotomy between, on the one hand, species that are rather specialized, pro-ovigenic koinobiont endoparasitoids and, on the other hand, species that are generalist, synovigenic host-feeder idiobiont ectoparasitoids. However, the causes of such co-variation are still under discussion, and Godfray (1994), for instance, identified a set of selective pressures that could push one lineage to evolve towards one of the two strategies. Moreover, some phylogenetic constraints can limit certain transitions (for instance from **ectoparasitism** to **endoparasitism**) if some required pre-adaptations are lacking. Finally, it is worth noting that these general co-variational trends show numerous exceptions.

7.3 The different steps of parasitism

As shown in Fig. 7.1, several successive steps must be achieved for successful parasitism (Doutt 1959; Vinson 1975, 1976). These different steps can be divided into two main categories. The first one, so-called pre-ovipositional, depends on the behaviour of females (Vinson 1981). In this case, females use a variety of stimuli leading them to progressively reduce their search area until a host is discovered and attacked. The second category concerns the development of the immature parasitoids and depends on the physiological interactions between the two partners (Vinson and Iwantsch 1980a, b). These categories are described more fully in the following sections.

7.3.1 Pre-ovipositional steps

The mechanisms implied in these steps are based on the ecological and behavioural features of the two partners. At the individual level, they will determine the ability of female parasitoids to find and successfully attack their hosts. At the population level, they will influence host–parasitoid dynamics.

7.3.1.1 Habitat searching behaviour

A newly emerged parasitoid female is rarely close to its host. Thus, she must first go and find potential hosts to exploit. Numerous studies have demonstrated the important role of visual, acoustic, or olfactory stimuli in the detection of hosts or their habitats (Vinson 1976, 1981). As a consequence, some host species are more frequently attacked not because they are preferred but because their habitat is more actively searched for. For instance, *Leptopilina boulardi* females are attracted by ethanol produced by the fermenting fruits where its hosts (*Drosophila* larvae) develop (Boulétreau and David 1980). More recently, it has been demonstrated that some plants, when attacked by phytophagous insects, emits molecular signals called **synomones** that are used by parasitoids to accurately locate places harbouring potential hosts (Turlings *et al.* 1990). Such 'chemical communication' between different trophic levels raises numerous research questions both at the fundamental (identification of the proximate and evolutionary causes involved) and applied (the use of synomones for crop protection) level.

7.3.1.2 Host searching behaviour

When a host habitat is found, the female parasitoid must find the host through the use of sometimes acoustic or visual but more generally chemical or olfactory stimuli. For instance, sexual pheromones that hosts normally use for attracting or detecting mating partners can be involved. Female parasitoids can behave like spies, tracking host-emitted signals, called **kairomones**, that were originally emitted for another purpose (Noldus 1989). The mechanisms involved are complex and sometimes very highly specific. They are related to strong selective pressures and subtle

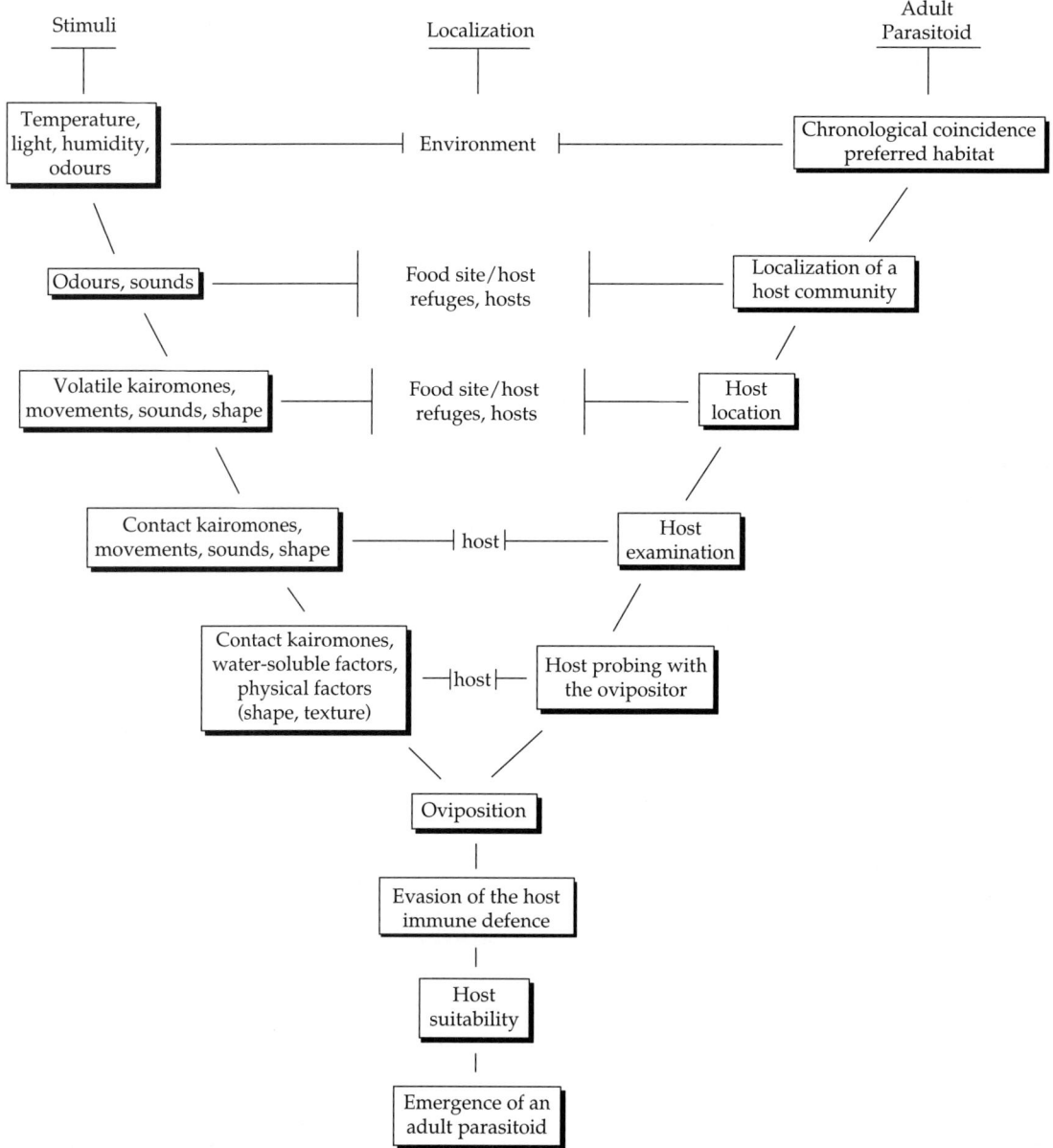

Figure 7.1 Flowchart showing the different steps involved in a host–parasitoid interaction, along with the different corresponding stimuli and their localization. After Vinson (1975).

adaptive mechanisms. For instance, the parasitoid *Cardiochiles nigriceps* is able to use as a cue the chemical 13-methyl dotriacontane that is present in the mandibular glands of its hosts, larvae of the lepidopteran *Heliothis virescens*. A single change in the methyl group of this molecule can entirely destroy its recognition by the females (Vinson *et al.* 1975).

7.3.1.3 Host acceptance

Once a host is found, the parasitoid female should ideally check whether it can ensure the good development of her offspring. Is it of the correct species? Has it reached its correct instar? Is it healthy or has it already been attacked? These questions are answered by examination of the external (shape, size, colour) or internal features of the host. Signals can be detected using the numerous receptors on the antennae (for external host features) or on the ovipositor (for internal host features). Here again, the level of specificity can be very high, leading the female to accept only a particular stage of a well-identified host. All of the mechanisms involved are, again, the result of strong selective pressures. Their study can lead to an explanation of some of the incredible richness in the diversity of parasitoids' behavioural strategies.

For numerous parasitoid species, the information collected during host examination also allows the female to detect whether the host has been parasitized before, either by the female herself or by a conspecific or heterospecific female. The decision to **superparasitize** a host (i.e. to lay an egg in a host that has already been parasitized) is usually risky for a solitary parasitoid, since only one parasitoid can develop in the host in this case. However, several theoretical models predict that the decision could be sometimes profitable under some ecological conditions (van Alphen and Visser 1990; Plantegenest *et al.* 2004). Such a prediction has been confirmed in various experimental works.

7.3.1.4 *Clutch size*

In the cases of gregarious parasitoids (see Section 7.2.4), the information obtained during examination of the host can be used by the female to accurately adjust the number of eggs she will lay. Bigger or more suitable hosts will enable the development of a greater number of progeny. It has been experimentally demonstrated that females of several parasitoid species are indeed able to estimate the size of their hosts. For instance, before ovipositing, females of the egg parasitoid *Trichogramma minutum* drum with their antennae on the surface of the egg while walking. During such an examination process, the bigger the host the smaller the angle between the first antennal segment and the head of the parasitoid. This angle has been demonstrated to be used by the females of this species as a proximate cue to adjust clutch size (Schmidt and Smith 1986).

7.3.1.5 *Sex allocation*

In haplo-diploid species, mated females are able to fertilize their eggs or not and thus choose the sex of their offspring (see Section 7.2.3). This may be influenced either by the quality (or suitability) of the host or by the number of other competing females on the host. In all cases, important selective pressures will lead females to optimize the sex ratio (i.e. the proportion of each sex in the offspring) they are producing in accordance with environmental constraints (Box 7.1).

Box 7.1 Optimal proportion of males and females in insect parasitoid progeny

Most parasitoid females use a special haplo-diploid sex determination system enabling them to control accurately the sex of each of the eggs they lay (Section 7.2.3). Since parasitoid females are under strong selective pressures, they most likely have to optimize the proportion of sons and daughters (i.e. the sex ratio) in their progeny in order to maximize transmission of their genetic makeup to the following generation. It has been known since 1930 that when mating between males and females occurs randomly the optimal sex ratio should be 50% males, 50% females (Fisher 1930). The reason for such an equilibrium is based on the fact that every mother obtains the same gain in fitness if she produces either

a son or a daughter. Moreover, such an equilibrium is known to be a so-called 'evolutionarily stable strategy' (ESS) (Charnov 1982). Indeed, if one of the two sexes is present in the population at a lower percentage, it would be advantaged, having a higher mating opportunity, and this would lead the global sex ratio in the population to return to 50% males, 50% females.

In insect parasitoids, however, mating does not usually occur at random. Indeed, most hosts are geographically concentrated in small patches of a few individuals, distant from each other, and colonized by a few parasitoid females only. Their progeny usually mate together before leaving to search for new hosts to attack. Since the

continues

Box 7.1 continued

number of parasitoids in a host patch usually remains low, mating occurs between progeny that can have some genetic relatedness. It can be shown that if n females are exploiting a host patch, the optimal proportion of sons they should produce should be $(n-1)/2n$. This is the so-called 'local mate competition' of Hamilton (1967) (see figure). When only one female colonizes a host patch ($n = 1$), she should lay, according to this model, only eggs producing daughters. More realistically, she should lay just enough sons to mate with all her daughters, which most likely means a small number of males since each of them can mate with several females. When the number of females exploiting a host patch increases, it becomes more and more worthwhile to produce sons that can then mate with females that are not produced by their mother. In this case, the situation progressively looks like random mating between males and females, and the Hamilton (1967) model indeed converges toward a 50% male,

50% female equilibrium, as in the original model of Fisher (1930).

A number of experimental works have shown that several parasitoid species accurately follow such predictions (King 1993).

References

Charnov, E.L. (1982). *The theory of sex allocation*. Princeton University Press, Princeton, NJ.

Fisher, R.A. (1930). *The genetical theory of natural selection*. Oxford University Press, Oxford.

Hamilton, W.D. (1967). Extraordinary sex ratio. *Science* **156**: 477–488.

King, B.H. (1993). Sex ratio manipulation by parasitoid wasp. In: *Evolution and diversity of sex ratio in insects and mites* (ed. D.A. Wrensch, M.A. Ebbert), pp. 418–441. Chapman and Hall, New York.

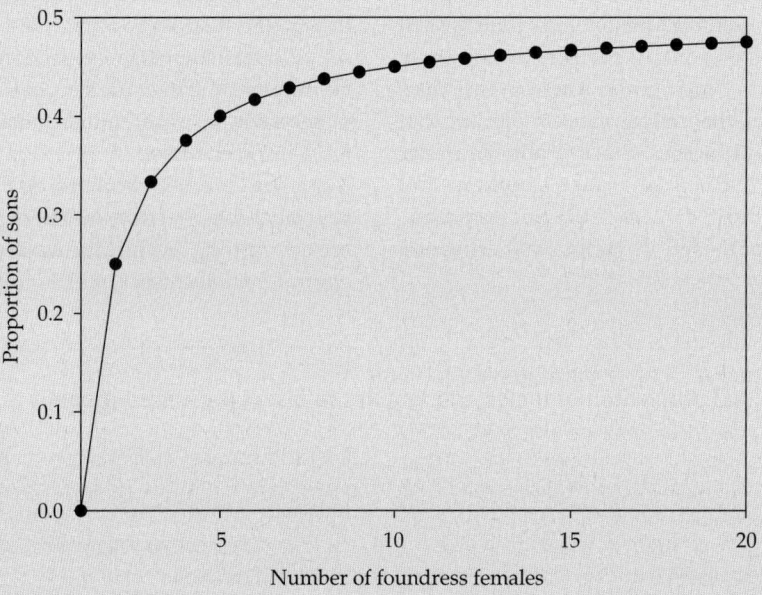

Box 7.1 Optimal proportion of sons parasitoid females should lay as a function of the number of females exploiting a host patch: model of 'local mate competition' from Hamilton (1967).

7.3.1.6 Patch time allocation

The adult lifespan of insect parasitoids is usually short, so females have to maximize their production of offspring per unit time. Moreover, most hosts are usually patchily distributed in the environment (Boxes 7.1 and 7.2). Females thus have to optimize their residence time on each patch they exploit before leaving it. Numerous theoretical works have dealt with this issue in an effort to predict the optimal residence time for a female according to environmental conditions (Box 7.2). Several experimental works indicate that females of different parasitoid species behave according to the corresponding theoretical predictions.

7.3.2 Post-ovipositional steps

After oviposition, the immature parasitoid faces a particular environment: its host. Indeed, the host not only constitutes protection and a source of nutritional resources but may also be hostile and defend itself against the intruder. In order to cope with these different features, several adaptive mechanisms have been selected through time including either the injection of specific products by the female during oviposition or particular adaptations of the immature parasitoids. The mode of action and availability of these different mechanisms vary greatly between species, but they have two main goals: (1) avoiding, destroying, or

diverting the immune defense of the host and (2) regulating the long-term development of the host in order to favour development of the parasitoid. Other complementary aspects are also included in so-called host suitability (Vinson and Iwantsch 1980a). For instance, the physiological state of the host greatly influences not only the survival of the immature parasitoid but also the adult phenotype (size, fecundity, adult lifespan, etc.). Moreover, the tight interaction between the two partners can also be modulated by environmental factors, either biotic like the presence of pathogens (Hochberg 1991) or abiotic like temperature (Ris *et al.* 2004).

7.3.2.1 Immune defences

One of the most frequent defence mechanisms of a host against endoparasitoids is the **encapsulation** of the immature parasitoids (Carton and Nappi 1997). The process implies the collaboration of different cell types called haemocytes that will successively identify the intruder, adhere to it, and recruit other cells to form a melanized capsule. The endoparasitoid is then killed, probably by asphyxiation or through the release of toxic compounds. Three different strategies have been shown to be used by parasitoids to evade encapsulation (Strand and Pech 1995). The first is to passively limit the interaction with the host, either through ectoparasitism or through the choice of unprotected instars (eggs for instance) or tissues (muscles, gut, glands, etc.).

Box 7.2 Optimal residence time on host patches

Most parasitoid species attack hosts that are distributed in the environment in patches that are distant from each other. In such a situation, parasitoid females are most likely selected to optimize the residence time on each of the host patches they are exploiting in order to maximize the number of progeny they can produce per unit time. When a parasitoid female exploits a host patch, the number of available, unparasitized hosts progressively decreases and there is a cost, in terms of both time and survival, in leaving the patch to try to find another one in the environment. Without competitors, the optimal time a female should remain on a host patch is provided by the so-called 'marginal value theorem' (Charnov 1976). In this

model, the female takes a time T to find the patch in the environment and exploits it during a time t, producing a cumulative number of progeny $f(t)$, a function having a decreasing slope since hosts are progressively depleted. If we call $R(t)$ the number of progeny produced per unit time, then $R(t) = f(t)/(T+t)$. The time t that maximizes this function is the one that leads to a null first derivate, leading to:

$$\frac{\partial R(t)}{\partial t} = 0 \Rightarrow \frac{\partial f(t)}{\partial t_i} = \frac{f(t)}{T+t}.$$

A graphical solution for solving this equation is presented in the figure.

continues

Box 7.2 continued

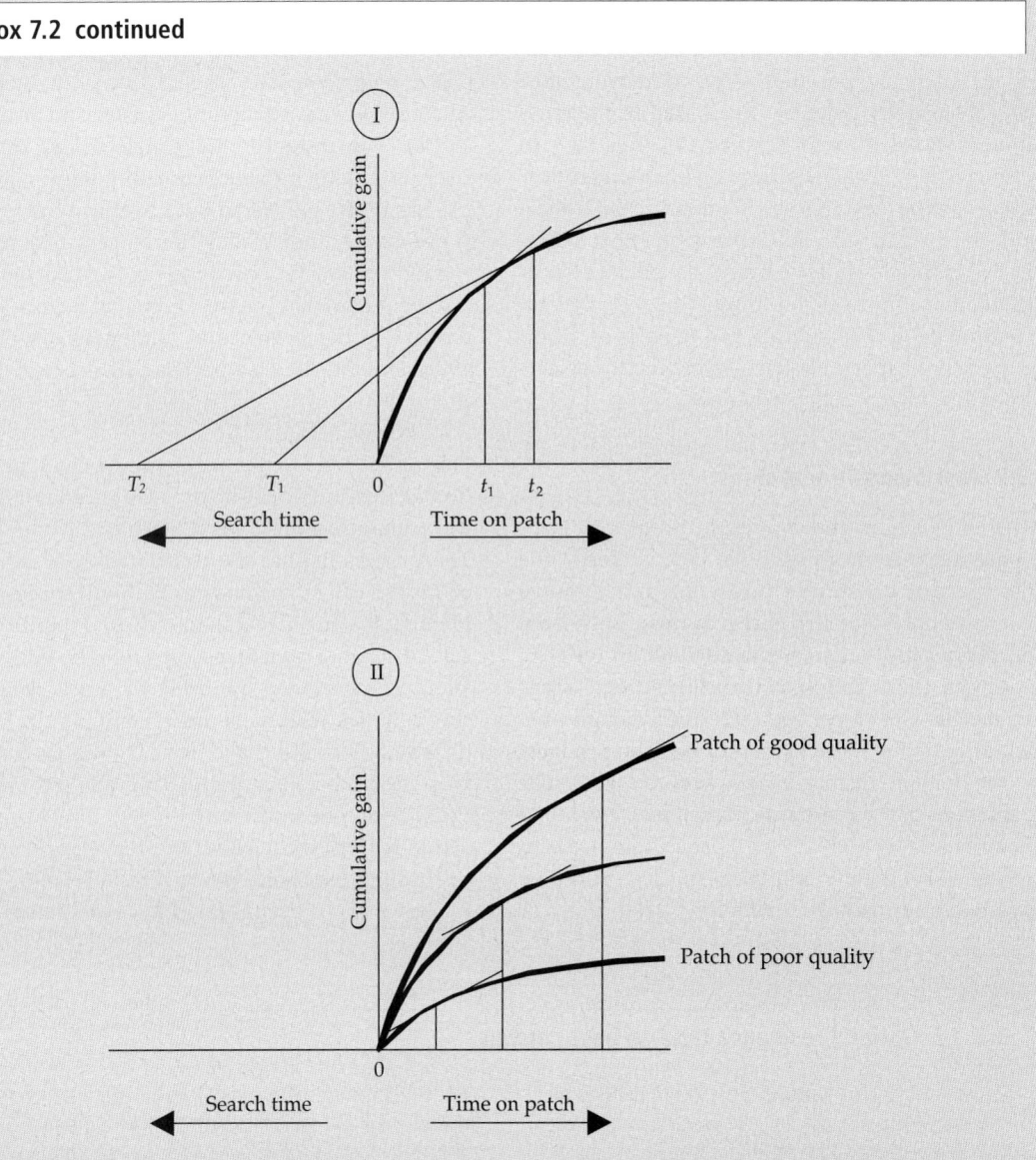

Box 7.2 Graphical representation of the 'marginal value theorem'. The time used by the parasitoid female to find a host patch to exploit is plotted as a negative *x*-value. Once the patch is discovered, the cumulative number of progeny produced (which corresponds to the cumulative number of hosts attacked) is represented by a thick curve. The optimal patch residence time corresponds to the *x*-coordinate of the contact point between the cumulative progeny production curve and a tangent to this curve originating from the time taken by the female to find the patch in the environment. This theoretical model predicts that the optimal residence time on a host patch should be longer when the female took longer to find the patch in the environment (case I). In a heterogeneous environment, this model also predicts that the female should leave the patch she is exploiting when her instantaneous fitness gain (slope of the curve) has decreased to a threshold value that corresponds to the average gain rate that can be achieved on all the patches available in the environment. Such a threshold value should not depend on the quality of the patch exploited, but it will be reached sooner on poorer quality patches. Hence, this model predicts that females should remain for a shorter time of poorer quality patches (case II).

continues

Box 7.2 continued

According to this theoretical model, all mechanisms that lead to a modification of the rate of host encounters per unit time should influence the optimal time females should remain on host patches. For example, if a female spends a lot of time finding a host patch in the environment, the number of hosts she will attack per unit time will remain low and the marginal value theorem predicts in this case that she should exploit such patch for a longer time before leaving it. Furthermore, if the quality of patches (expressed here as the number of hosts available) varies in the environment, the model predicts that females should remain for longer on better quality patches. A large number of experimental works have shown that parasitoid females usually follow these predictions rather accurately (Wajnberg *et al.* 2000, 2006).

References

Charnov, E.L. (1976). Optimal foraging: the marginal value theorem. *Theoretical Population Biology* **9**: 129–136.

Wajnberg, E., Fauvergue, X., Pons, O. (2000). Patch leaving decision rules and the marginal value theorem: an experimental analysis and a simulation model. *Behavioral Ecology* **11**: 577–586.

Wajnberg, E., (2006). Time-allocation strategies in insect parasitoids: from ultimate predictions to proximate behavioural mechanisms. *Behavioral Ecology and Sociobiology* **60**: 589–611.

Another is for the parasitoid to protect itself actively against the immune response. For example, numerous dipteran parasitoids circumvent the host's immune response by building a protective layer around their body while maintaining a respiratory channel to avoid asphyxia. Finally, many species have been shown to destroy the host's immune defences with the help of **virulence** factors. From an evolutionary point of view, all these weapons, defences, and counter-defences, can be viewed as the result of complex co-evolutionary processes and as an arms race between the two partners. This may lead to important intraspecific variation in virulence or resistance in the parasitoid or in its host, respectively (Carton and Nappi 1997; Kraaijeveld and Godfray 1999).

7.3.2.2 Host regulation

Because of the close physiological proximity between the two partners, synchronization of their development is necessary to ensure the survival of the parasitoid. Lawrence (1986) distinguishes two kinds of parasitoid: 'conformers' who use the hormones of the host to adjust their own development and 'regulators' who disturb the host's developmental schedule inducing either supplementary or precocious moults or by definitively blocking the host in an immature stage. These manipulations of the host's physiology imply several regulating factors of the two main hormones (**juvenile hormone** and **ecdysteroids**) that are involved in moulting and metamorphosis (Beckage and Gelman 2004).

7.3.2.3 Virulence and regulation factors

The study of virulence and regulation factors is currently a very active research area and several reviews have been published (Strand and Pech 1995; Beckage and Gelman 2004; Pennacchio and Strand 2006). Venom is probably the 'ancestral weapon' against host defences. Its primary role is to, temporarily or permanently, paralyse the host. Sometimes, it also contributes to host regulation by blocking the moults. In some species, venom can enhance the efficiency of other virulence factors like **polyDNAvirus** (PDV). PDVs are particular symbionts (called braco- or ichneumo-viruses) found in two families of hymenopteran parasitoids, Braconidae and Ichneumonidae, respectively. PDVs have a great similarity with viruses (Box 7.3) but their DNA is integrated as a provirus into the host genome and their replication only occurs in some particular tissues and at precise stages in the development of the female parasitoid. Viruses are injected into the host during oviposition, where they are expressed more or less rapidly and durably depending on the species attacked. PDVs are now known to play a major role in the suppression of the host's immune defence through the

Box 7.3 Parasitoid symbionts

In addition to their hosts, parasitoids interact in the wild with numerous other organisms: predators, pathogens, competitors, etc. Among these, **endosymbionts** have been neglected for a long time. However, their influence on the parasitoid phenotype and their impact on ecological and evolutionary processes have become more and more obvious (Boulétreau and Fleury 2005). These symbionts are mainly viruses or bacteria.

Viruses

Parasitoids, in particular hymenopteran wasps, harbour numerous DNA (e.g. Ascovirus) or RNA (Reovirus) viruses with variable influence (Renault *et al*. 2005; Stasiak *et al*. 2005). In numerous cases, the virus uses the parasitoid as a vector for transmission to its actual host. Then it quickly multiplies, leading to the death of the host and the parasitoid. In this case, the virus can be viewed as a parasite or as a competitor of the parasitoid since its development is detrimental to the parasitoid's offspring. In other cases, the relationship is rather commensal, the virus being either transmitted horizontally (via the parasitic behaviour of the parasitoid) or vertically (from the parasitoid female to its offspring) without having a significant impact on the parasitoid. Finally, the interaction between parasitoids and their viruses has sometimes seemingly evolved towards a real mutualism. In this latter case, the two partners have co-evolved so tightly that some researchers no longer consider the current virus to be a true one but rather an organelle like a mitochondrion (Federici and Bigot 2003). Indeed, they not only play a major role as virulence factors in the development of the wasp (Section 7.3.2.3) but also completely depend on the development of the wasp for their own transmission since their DNA is integrated into the wasp's genome.

More recently, the influence of a virus on superparasitic behaviour, i.e. the decision to lay an egg in an already parasitized host (Section 7.3.1.3), has been demonstrated in a wasp species highlighting the wide range of possible biological interactions between the two partners (Varaldi *et al*. 2003).

Bacteria

Parasitoids are also hosts of numerous bacteria. For instance, the alpha-proteobacteria of the genus

Wolbachia infect at least 16% of insect species and in particular numerous hymenopteran parasitoids (Werren and Windsor 2000). These rickettsia are endosymbionts and are located within the cytoplasm of their host cells. Their effects on their hosts are complex and variable. One of the most spectacular is undoubtedly the manipulation of parasitoid reproduction. In some haplo-diploid species (Section 7.2.3), some *Wolbachia* strains are able to induce thelytoky, the female parasitoids then being able to produce daughters without mating. Since the parasitoid males have no direct effect on the host populations, such thelytokous strains could be interesting for biological control in order both to limit mass-rearing costs and to increase the efficiency of releases (Stouthamer 2003). In other species, some *Wolbachia* variants induce **cytoplasmic incompatibilities** (CI). Female offspring are then reduced in some crosses involving partners with 'incompatible' variants. The consequences of CI are probably vast and range from reproductive isolation between populations with different variants to the disturbance of clutch size adjustment or sex allocation (Section 7.2.3). The outcome of endosymbiont–parasitoid interactions can be astonishing, as in the braconid *Asobara tabida* where *Wolbachia* has became necessary for oogenesis (Dedeine *et al*. 2001).

References

Boulétreau, M., Fleury, F. (2005). Parasitoid insects and their prokaryotic helpers: gifted parasites? *Bulletin de la Société Zoologique de France* **130**: 177–192.

Dedeine, F., Vavre, F., Fleury, F., Loppin, B., Hochberg, M.E., Boulétreau, M. (2001). Removing symbiotic *Wolbachia* bacteria specifically inhibits oogenesis in a parasitic wasp. *Proceedings of the National Academy of Sciences of the USA* **98**: 6247–6252.

Federici, B.A., Bigot, Y. (2003). Origin and evolution of polydnaviruses by symbiogenesis of insect DNA viruses in endoparasitic wasps. *Journal of Insect Physiology* **49**: 419–432.

Renault, S., Stasiak, K., Federici, B.A., Bigot, Y. (2005). Commensal and mutualistic relationships of reoviruses with their parasitoid wasp hosts. *Journal of Insect Physiology* **51**: 137–148.

Stasiak, K., Renault, S., Federici, B.A., Bigot, Y. (2005). Characteristics of pathogenic and mutualistic relationships of ascoviruses in field populations of parasitoid wasps. *Journal of Insect Physiology* **51**: 103–115.

continues

Box 7.3 continued

Stouthamer, R. (2003). The use of unisexual wasps in biological control. In: *Quality control and production of biological control agents: theory and testing procedures* (ed. J.C. van Lenteren), 93–113. CAB International, Wallingford.

Varaldi, J., Fouillet, P., Ravallec, M., Lopez-Ferber, M., Boulétreau, M., Fleury, F. (2003). Infectious behavior in a parasitoid. *Science* **302**: 1930.

Werren, J.H., Windsor, D. (2000). *Wolbachia* infection frequencies in insects: evidence of a global equilibrium? *Proceedings of the Royal Society B: Biological Sciences* **267**: 1277–1285.

modification or destruction of some targeted cells. In numerous cases, PDVs can also modify the endocrine system by either inducing precocious moults or metamorphosis or blocking the development of the host. In other species, 'virus-like particles' play a similar role to PDVs. However, they do not contain nucleic acids.

Some parasitoid species also exhibit teratocytes, particular cells originating from an extra-embryonic membrane. Soon after hatching of the egg they differentiate and are released freely in the host's haemolymph. Their role is still not totally clear but they could be used as a nutritional source for the parasitoid. Other functions, like host regulation, are also cited. Finally, some immature parasitoids can directly contribute, through their own secretions, to host regulation. Several experimental works have demonstrated that some parasitoids are also able to release hormones into the host in order to modify its development.

7.4 Demographic characteristics of host–parasitoid interactions

There are now a large number of works, some experimental but mainly theoretical, that have tried to describe, understand, and thus explain temporal fluctuations in the number of parasitoids and their hosts. There are several reasons why such a research effort has been maintained over several decades. The first is that the ecological features of parasitoids, as has briefly been described in the previous sections, are relatively easy to formalize mathematically. For example, in contrast to what can be observed in predators, only female parasitoids seek and attack hosts. Moreover, since a host

that has been attacked almost always dies, there is a direct relationship between the search efficiency of parasitoid females and the host mortality rate. Similarly, the reproductive success of female parasitoids is directly linked to the number of hosts attacked. Finally, hosts and parasitoids have similar generation times. Also, insect parasitoids can be used to control pests on different crops, and the goal in this case is to reduce the number of hosts (see Section 7.5). For this, an accurate understanding of the mechanisms involved in host–parasitoid dynamics is needed.

7.4.1 A basic demographic model

Several modelling approaches have been developed over the years (see Hassell 1978; Nisbet and Gurnet 1982; Begon and Mortimer 1986). Here, we will present the developments arising from the Nicholson and Bailey (1935) model only. All the corresponding models are based on the following basic equations:

$$\begin{cases} H_{t+1} = e^r (H_t - H_a) \\ P_{t+1} = H_a. \end{cases} \tag{7.1}$$

In this model, H_t and P_t are the size of the host and parasitoid populations, respectively, at generation t and H_a is the number of hosts attacked in this generation. Without parasitism, the host population shows exponential growth with rate r. At generation $t+1$, the size of the host population is the number of hosts escaping from parasitism in the preceding generation, taking into account its exponential growth, and each host attacked at generation t gives rise to a parasitoid at generation $t+1$.

Let us suppose now that female parasitoids search for their hosts randomly and that each host found is attacked. If R_t is the number of host–parasitoid encounters at generation t and A is the proportion of hosts encountered by each parasitoid at this generation (A is sometimes called the 'area of discovery' of the parasitoid), then $R_t = AH_tP_t$, and the number of host–parasitoid encounters per host is thus $R_t/H_t = AP_t$. The hypothesis that parasitoids are searching for their hosts randomly implies that the proportion p_0 of hosts that are not encountered by parasitoids corresponds to the first term of a Poisson distribution having the number of host–parasitoid encounters per host as an average value. Thus, $p_0 = \exp(-R_t/H_t)$ leading to $p_0 = \exp(-AP_t)$. The number of hosts attacked at generation t thus becomes $H_a = H_t[1 - \exp(-AP_t)]$ and, substituting this in the equation system describing the dynamics of host–parasitoid interaction, we obtain:

$$\begin{cases} H_{t+1} = H_t \exp(r - AP_t) \\ P_{t+1} = H_t[1 - \exp(AP_t)]. \end{cases} \tag{7.2}$$

This is the Nicholson and Bailey (1935) model whose properties are well known. More accurately,

for each value of r and A, there is an equilibrium demographic situation that is actually unstable since a slight disturbance from this equilibrium leads to divergent oscillations, leading both populations to disappear (see Fig. 7.2). In real situations, host–parasitoid associations are generally demographically stable since both partners can persist during the course of time. The aim of several theoretical works was thus to modify the original Nicholson and Bailey (1935) model in order to increase its stability.

7.4.2 Effect of competition between hosts

An obvious improvement of the Nicholson and Bailey (1935) model is to replace the exponential growth of the host population by a **density-dependent increase** resulting from intraspecific competition between hosts. For this, the constant growth rate r of the host population can be replaced by a growth rate that is inversely proportional to the size of the population: $r(1 - H_t/K)$. This growth rate decreases linearly from the value r, when $H_t = 0$, down to zero when $H_t = K$. K is the

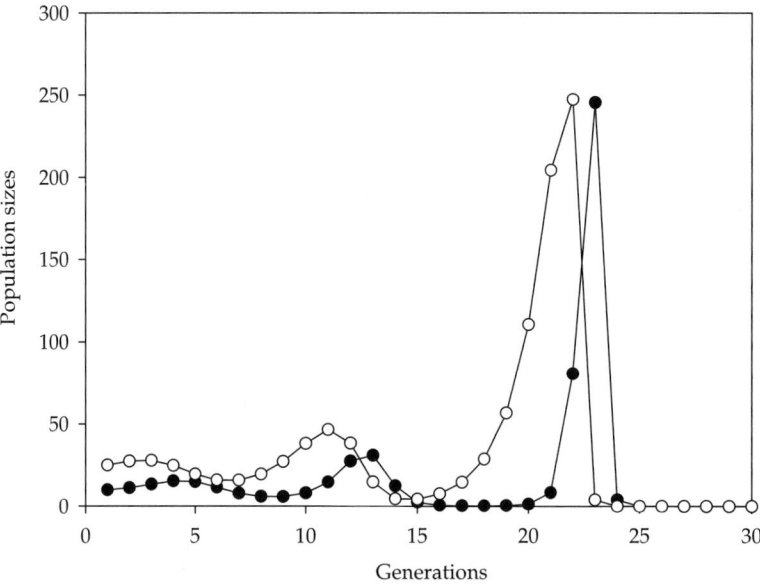

Figure 7.2 Example of demographic fluctuations between hosts (open circles) and their parasitoids (black circles), as computed from the Nicholson and Bailey (1935) model (Eqn 7.2), with $r = 0.693$ and $A = 0.6$.

'carrying capacity' of the environment corresponding to the maximum size the host population will reach without parasitism. The modification leads to the following equation system:

$$\begin{cases} H_{t+1} = H_t \exp\left[r\left(1 - \dfrac{H_t}{K}\right) - AP_t \right] \\ P_{t+1} = H_t[1 - \exp(-AP_t)]. \end{cases} \qquad (7.3)$$

This the Beddington *et al.* (1975) model that, in contrast to the original Nicholson and Bailey (1935) model, can produce stable demographic equilibria for specific values of the parameters, as can be seen in Fig. 7.3. More accurately, the conditions for stability depend on both the value of *r* and on the efficiency of the destructive effect parasitoids have on their hosts. The competition between hosts seems to be a non-negligible factor in the demographic stability of host–parasitoid interactions.

7.4.3 Effect of the number of hosts

The Nicholson and Bailey (1935) model assumes that the number of hosts attacked by each parasitoid increases linearly with host density, which is rather unlikely. Indeed, numerous experimental works have demonstrated that female parasitoids need a certain amount of time, that differs among species but that can be important, to attack each host that is encountered. When the number of hosts is important, such time constraint can limit the impact of parasitism on the host population, and the relationship between the number of hosts attacked by each parasitoid and host density (the so-called 'functional response') is concave down. Taking into account such a phenomenon leads us to modify the parameter *A* describing the proportion of hosts encountered by each parasitoid. If *a* is the instantaneous rate of host searching by each parasitoid and T_h is the time taken to attacked each host, then it can be demonstrated that adding a concave down functional response leads to $A = (aT)/(1 + aT_hH_t)$, where *T* is the total time available for each female. Such a new formulation of the parameter *A* leads to a modification of the initial Nicholson and Bailey (1935) model and the new

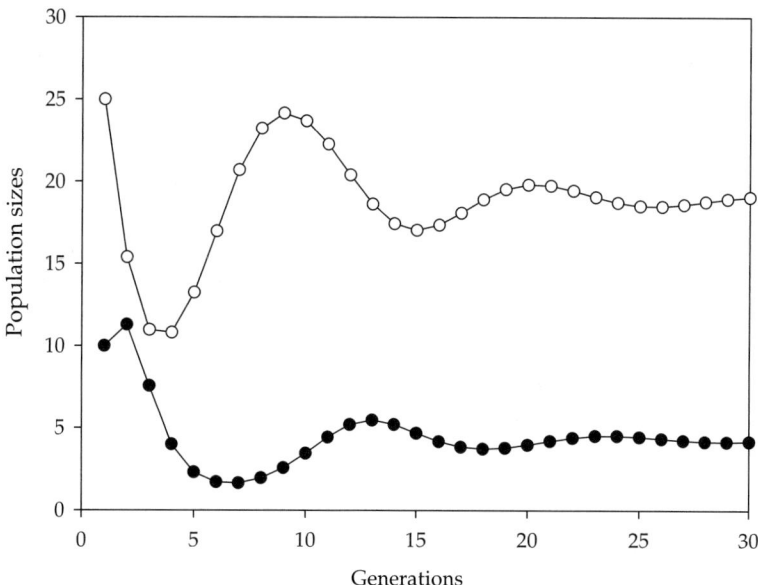

Figure 7.3 Example of demographic fluctuations between hosts (open circles) and their parasitoids (black circles), as computed from the Beddington *et al.* (1975) model (Eqn 7.3), with *r* = 0.693, *A* = 0.6 and *K* = 30. In this case, it is possible to demonstrate that both host and parasitoid population sizes converge to stable values, through oscillations that are getting smaller and smaller.

equation system:

$$\begin{cases} H_{t+1} = H_t \exp\left(r - \dfrac{aTP_t}{1 + aT_hH_t}\right) \\ P_{t+1} = H_t\left[1 - \exp\left(\dfrac{-aTP_t}{1 + aT_hH_t}\right)\right] \end{cases} \tag{7.4}$$

The dynamic properties of this model were examined by Hassell and May (1973). In all cases, such models are even less demographically stable than the original Nicholson and Bailey (1935) model, that can be regained when $T_h = 0$. This is due to the fact that the modification implies a destructive effect of parasitoids on their hosts that is weaker at higher host densities, in contradiction to what we would expect for a stabilizing mechanism.

7.4.4 Effect of host distribution

The Nicholson and Bailey (1935) model also assumes that host–parasitoid encounters occur randomly, despite the fact that numerous experimental works have demonstrated that female parasitoids are mostly aggregated where there is a high density of hosts. In this case, the use of a Poisson distribution to describe the number of host–parasitoid encounters is no longer correct. A so-called 'binomial distribution' seems to work better in describing observed situations (May 1978). In this case, the probability p_0 of escaping parasitism corresponds to $(1 + AP_t/k)^{-k}$, where k determines the level of aggregation of host–parasitoid encounters. The lower k is, the higher the parasitoid aggregation for higher host densities. Reciprocally, when k tends to infinity, we get back to a Poisson distribution describing randomly occurring host–parasitoid encounters. By simultaneously taking into account such an aggregation factor and competition between hosts (Section 7.4.2) we get the following equation system:

$$\begin{cases} H_{t+1} = H_t \exp\left[r\left(1 - \dfrac{H_t}{K}\right)\right]\left(1 + \dfrac{AP_t}{k}\right)^{-k} \\ P_{t+1} = H_t\left[1 - \left(1 + \dfrac{AP_t}{k}\right)^{-k}\right] \end{cases} \tag{7.5}$$

Analysis of the dynamic properties of this model shows that decreasing values of k (i.e. increasing the level of aggregation of host–parasitoid encounters)

increases the number of situations in which host–parasitoid demographic interactions are stable over the course of time. Aggregation of host–parasitoid encounters is now still considered to be the most powerful mechanism leading to stability of host–parasitoid demographic interactions.

More accurately, the analysis of this model globally shows that host–parasitoid demographic interactions should be stable as soon as $k < 1$ (May 1978). The mechanism of aggregation taken into account in this model is based on spatial heterogeneity in the distribution of hosts that are located in patches in the environment. May (1978) actually demonstrated that there is a tight relationship between the distribution of attacks on hosts and the spatial distribution of the hosts. The relationship leads to $k = (1/CV)^2$, where CV is the coefficient of variation of host density between patches. Hence, host–parasitoid demographic interactions should be stable as soon as $CV^2 > 1$, a condition described in the literature as the 'CV^2 rule' (Hassell and Pacala 1990). An important number of experimental and theoretical works have demonstrated that such a rule, which appears to be particularly simple, would be valid, although it is still being intensively debated (Bernstein 2000).

7.4.5 Effect of the number of parasitoids

The original Nicholson and Bailey (1935) model also assumed that the host searching efficacy of each parasitoid female remains constant, despite an important number of experimental works demonstrating that such efficacy apparently decreases when the number of parasitoid females increases. Such a decrease, which is the result of competition between females, is usually linear on a log–log scale. Thus, in an empirical way, $\log(A) = \log(Q) - m\log(P_t)$, leading to replacement of the parameter A by QP_t^{-m}, where Q is the intercept of the plot (searching efficiency of parasitoids when they are alone) and m is the slope describing the intensity of the phenomenon (Hassell and Varley 1969).

Taking account of this new modification leads to the following equation system:

$$\begin{cases} H_{t+1} = H_t \exp(r - QP_t^{1-m}) \\ P_{t+1} = H_t[1 - \exp(-QP_t^{1-m})]. \end{cases} \tag{7.6}$$

This model returns to the Nicholson and Bailey (1935) model when $m = 0$. Analysis of its stability conditions shows that competition between parasitoid females can represent an important stabilizing factor. It should be noted that this phenomenon is tightly linked to the mechanisms of aggregation of host attacks described in the previous section. This probably explains its important stabilizing effect (Hassell 1978; Begon and Mortimer 1986).

7.5 The use of parasitoids in plant protection

As shown in the previous sections, the successful development of parasitoids usually causes the death of their hosts. At the population level, this feature contributes to the regulation of host populations. Parasitoids can therefore be used as a means to reduce the impact of crop pests. This is the aim of biological control, broadly defined as 'the use of living organisms [called biological control agents] to control the population density or impact on a specific pest organism, making it less abundant than it would otherwise be' (OILB-SROP 1973). It is worth noting that parasitoid species are not the only biological control agents that can be used. Other natural antagonists can be used: predators, true parasites (like viruses), pathogens, or even competitors. Moreover, biological control is not restricted to the regulation of noxious arthropods but can also be used against weeds or even vertebrates. From an agronomic point of view, biological control constitutes one facet of crop protection programmes in addition to (or as a replacement for) other means like chemical and physical controls or the use of naturally resistant or transgenic plants. Four different strategies of biological control can be distinguished (Eilenberg *et al.* 2001), each of them being associated with specific scientific questions.

7.5.1 Classical biological control

This strategy aims to introduce an exotic biological control agent into the agrosystem, with the hope that it will become permanently established and provide durable control of the targeted pest. Historically, such a strategy seems to have been used for the first time at the end of the 19th century against the cottony cushion scale, *Icerya purchasi*. This insect had previously been inadvertently introduced from Australia and rapidly became a severe pest in Californian citrus groves. A screening of its natural enemies in Australia indicated the presence of a predatory beetle, *Rodolia cardinalis*. This beetle was consequently introduced to the USA, mass-reared in quarantine, and released into the citrus groves. Two years after its introduction, damage caused by the pest became almost negligible. After this first success, the introduction of exotic natural enemies has been repeatedly used worldwide with numerous other successes. As suggested by this case study, the pest is usually an exotic species which causes outbreaks in the absence of natural enemies or for other ecological reasons (Colautti *et al.* 2004). The introduction of one or several **sympatric** antagonists is thought to restore a demographic balance with a low level of the pest population in a new ecological context ('original classical biological control' *sensu* Eilenberg *et al.*, 2001). In other cases, an exotic biological control agent can be used against either an indigenous pest or an exotic but allopatric one. In such cases, the challenge is to create a new ecological interaction between two species which have never co-evolved ('new association classical biological control').

From an economic perspective, classical biological control is very interesting: for the durable control achieved, with no further need for human intervention, the associated costs of development and use are small. Nevertheless, it is noteworthy that the success of such strategies varies greatly. Greathead (1995) estimated that only 10% of the 4500 introductions of biological controls against noxious insects gave economically acceptable control. Consequently, one of the major goals for researchers and practitioners of biological control is to identify the causes of success or failure. This can be done by the use of specific databases (like BIOCAT) containing data from all realized biological control programmes (Greathead and Greathead 1992). Meta-analyses have been carried out to test the importance of numerous features of the biological control agent such as taxonomic origin (Lane *et al.* 1999), fecundity (Lane *et al.* 1999), host feeding behaviour (Jervis *et al.* 1996), functional response (Fernandez and Corley 2003), host specificity (Stiling and Cornelissen 2005), or the relevance of 'multiple species' versus 'single species' introductions (Denoth *et al.* 2002; Stiling and Cornelissen 2005).

Independently of the efficiency of the biological control programme against the targeted host, the impact of biological control agents on potential non-target species should also be studied in order to limit post-release ecological risks (Section 7.5.5.2).

7.5.2 Inoculation biological control

When pests attack non-perennial crops, the biological control agent released cannot usually become permanently established. In such cases, inoculation biological control aims at establishing a biological control agent for an acceptable but temporary period of time. This strategy is used worldwide, especially in greenhouses. Van Lenteren (2000) estimated that pests in 5% of the 300,000 ha of greenhouses are controlled using **integrated pest management**, which favours the use of inoculative releases of parasitoids. This proportion should reach 20% in the near future. The success of inoculation biological control can be explained by the wide range of available biological control agents (more than a thousand).

Contrary to classical biological control, inoculation biological control (and also inundation biological control) requires a continuous supply of large numbers of biological control agents. However, this leads to new difficulties that have to be solved. The production and storage conditions must protect the quality of the biological control agents as well

as their fecundity, longevity, dispersal ability, etc. Moreover, numerous authors have highlighted the possible genetic impact that mass-rearing conditions could have on biological control agents (van Lenteren 2003; Wajnberg 2004). As indicated in Table 7.1, the conditions during the mass-rearing step are usually very different from those encountered in real agrosystems. Knowing whether or not (and if so how) some identified selective pressures, drift events, or inbreeding can modify the genetic makeup of the initial strain of the biological control agent is also of utmost importance.

7.5.3 Inundation biological control

Here the aim is to control the pest population through a one-off impact of the released biological control agents. Unlike in the first two strategies, the primary goal is here a drastic and time-limited impact on the pest population. The impact of the offspring of the biological control agent can also be interesting, but is usually not a goal *per se*. This method greatly modifies the criteria used for the selection of efficient parasitoid species. For instance, idiobionts (Section 7.2.4) attacking their hosts at precocious stages or highly fecund species are very interesting for this purpose. These features explain the success obtained with the use of egg parasitoids in general, and more particularly with the inundative releases of *Trichogramma* spp. (Wajnberg and Hassan 1994). *Trichogramma* are minute Hymenoptera that lay their eggs within the eggs of Lepidoptera. Several species (e.g. *Trichogramma brassicae, Trichogramma cacoeciae,*

Table 7.1 Comparison between mass-rearing and post-release environmental conditions encountered by biological control agents. After van Lenteren (2003)

	Mass-rearing	Agrosystem
Abiotic factors (temperature, photoperiod, humidity, etc.)	Stable, homogeneous	Varying, heterogeneous
Host presence	High density. Uniform distribution and quality	Variable density and quality. Aggregative distribution
Interspecific interactions (competition, predation, etc.)	Limited	Important
Mating	Facilitated	More difficult
Dispersal	Limited	Necessary

Trichogramma dendrolimi, Trichogramma evanescens, etc.) are used for the biological control of pests of different crops (e.g. cereals, cotton, sugar cane, vegetables) over a total area of between 15 and 30 Mha (Li-Ying 1994; van Lenteren and Bueno 2003). For instance, more than 80,000 ha of corn are protected each year in France against the European corn borer, *Ostrinia nubilalis* thanks to the inundative release of *T. brassicae.* About 200,000 *Trichogramma* per hectare are required in this case (Frandon and Kabiri 1999). This clearly highlights the need for optimization of both the mass-rearing and the field release steps. This not only implies technological advances but also scientific work dealing with several fields of parasitoid biology. Ecophysiological studies are needed, for instance, to understand the role of abiotic (e.g. temperature, humidity, photoperiod) or biotic (host species) factors. It is worth noting that spectacular advances have been achieved in this field, leading for instance to the ability to produce some parasitoid species on artificial diets (Thompson 1999). The behavioural ecology of parasitoids should also be investigated in order to understand host searching behaviour or dispersal patterns. More generally, ecological work must lead to an accurate description of the demographic impact of the biological control agents on the target hosts as well as on related species in the community.

7.5.4 Conservation biological control

The aim of this strategy is to modify the agrosystem or agricultural practices in a way that will favour the action of a pest's native natural enemies. To date, this method is probably less developed than the others. Nevertheless, it offers some practical solutions and raises numerous scientific questions dealing with community ecology (Landis *et al.* 2000). Three non-exclusive tactics can be distinguished. The first one relies on the availability of shelters or microclimates that will favour the perennial presence of the biological control agent(s). Indeed, extreme temperatures are unfavourable for their activity or even dangerous to their survival. Neighbouring shaded areas (hedges for instance) can be used as temporary refuges. The use of sites for overwintering can also promote

the establishment of parasitoids during harsh conditions and favour more precocious pest control efficiency. The second tactic is to provide sources of food for the biological control agents. This can be done by growing some plant species producing pollen and nectar that can be used by the adult parasitoids. For instance, the influence of several flowers on the parasitoid *Diadegma insulare* has been investigated for biological control of the diamondback moth, *Plutella xylostella* (Idris and Grafius 1995). A last possibility is to provide or maintain a supply of hosts within or close to the crop. This will slow down the post-harvest escape of the biological control agent. The persistence of some generalist parasitoids can also be ensured through the use of plants that will be infested by other host species. These so-called 'plant banks' are currently used in greenhouses or near outdoor crops. Around 3 Mha are planted with spatially alternating wheat and cotton (Landis *et al.* 2000). The wheat is used as a source of natural biological controls in order to protect the cotton plantation more efficiently.

From an ecological point of view, all these practices favour both plant and animal biodiversity within the agrosystem. However, this is not a specific goal, and an increased biodiversity can even sometimes produce an unfavourable effect. For instance, Baggen and Gurr (1998) and Baggen *et al.* (1999) observed contradictory results between the positive influence of a flower on the parasitoid *Copidosoma koehleri* in laboratory conditions and the failure of its pest control efficacy during field trials. They finally observed that the selected plant was not only beneficial for the parasitoid but also for the pest. Only an accurate screening finally allowed the identification of a plant species with a differential impact on the two species. This shows how and why conservation biological control can depend on complex ecological interactions.

7.5.5 The limits of biological control

As previously seen, biological control can be an efficient way to protect crops, human health, and, in most cases, the environment. However, sometimes it is uneconomic. Moreover, unintended negative effects must also be avoided.

7.5.5.1 *Economic and social costs*

Both the economic and social costs of biological control depend on the strategy used. Unfortunately, most studies deal with classical biological control only (Tisdell 1990). The costs of inoculation or inundation biological control greatly vary according to different factors including: (1) the cost of the method itself (shipment, release, etc.); (2) the actual reduction in damage caused by the pest; (3) the correlated increase of yield and total production of the system; (4) the increase in harvest quality and corresponding price; (5) the relative gain compared with other methods; (6) the social or even health gains (Huffaker *et al.* 1976; Tisdell 1990). Despite this complexity, the ecological and economic benefits associated with biological control seem very high compared with chemical control (Table 7.2).

7.5.5.2 *Unintended effects*

During the last 120 years, more than 2000 arthropods have been introduced as biological control agents in around 200 countries or islands. Only 1.5% of these introductions have been followed by negative effects on the environment (van Lenteren *et al.* 2006). Nevertheless, the question of unintended ecological risks is currently being intensely debated from both a scientific and a political point of view. As a consequence, several countries have set up legislation governing the introduction of biological control agents (Wajnberg *et al.* 2001; Louda *et al.* 2003; van Lenteren *et al.* 2006). Potential risks associated with biological control programs can appear when the released biological control agents: (1) are attacking one or several non-target species

and are dangerously affecting their demography; (2) are competing with one or several species of the same trophic level; (3) are harbouring pathogens that are noxious for **endemic** species. Such unintended effects can also create economic costs if the released agents disturb other useful species (for instance, natural enemies of weeds). Health problems like allergies can also occur during the mass-rearing of some biological control agents. Although rare, such negative side-effects can tarnish the image of biological control and also limit its actual development. This is why the international scientific community has suggested different approaches in order to check, as far as possible, the innocuousness of the released agents. Accurate experimental designs and guidelines have been established for several key points.

The most sensitive point is of course the specificity of the biological control agent, which is not a simple point to resolve. Indeed, host range is not necessarily known as only partial data are usually available in the literature or museum collections. Laboratory experiments are thus required, during which several potential hosts are offered to the parasitoids. The presence or absence of parasitism is then checked in each case. In order to limit such a time-consuming task, Wapshere (1974) proposed a 'centrifugal phylogenetic testing method' or the successive testing of potential hosts from species closely related to the target pest to progressively more distantly related species until the host range has been adequately circumscribed. Such experimental trials can be done in choice or non-choice conditions which respectively correspond to cases

Table 7.2 Comparison of the relative performance of biological and chemical control of crop pests. Data for chemical control were established by the pesticide industry. After van Lenteren (1997)

	Chemical control	Biological control
Number of tested products or agents	More than a million	Around 2000
Success rate	1:30,000	1:10
Development costs	Around US$160 millions	Around US$2 million
Development time	10 years	10 years
Gain/cost ratio	2:1	20:1
Risk of acquisition of resistance	High	Very low
Specificity	Weak	Good
Noxious side-effects	Numerous and important	Almost nil

where several or only one species are offered during a trial. The collected data can be then adequately analysed using specific statistical tools (Prince *et al.* 2004; Babendreier *et al.* 2005). Field studies should also be realized in order to compare the dynamics of the target pest inside or outside the area where the biological control agent has been released. In some cases, non-target hosts are also released in order to estimate if or by how much they are attacked by the parasitoids (Mills 1997).

Since biological control agents could also outcompete species of the same trophic level, faunistic survey must be carried out in order to quantify their impact inside the release area in comparison with the pre-release state or surrounding areas. This work should also be linked to laboratory experiments in order to identify more accurately the mechanisms of interspecific competition. This is undoubtedly a difficult task, because the outcome of competition between two parasitoids varies greatly with the biological features of the two protagonists like host range, searching efficiency, host discrimination, competitive ability of the larvae, adult life-history traits, etc.

Finally, it is necessary to verify that the biological control agent will not disperse from the targeted crops to invade new environments. This can be done using experimental designs where traps are regularly placed around the release area. Powerful statistical methods have been developed to analyse results obtained from such studies (Mills *et al.* 2006).

7.5.6 Improving biological control through ecological- and population-based approaches

Optimization of biological control methods simultaneously implies improvement in the pest control efficacy, decrease in the associated costs, and minimization of potential unintended effects. Within this context, much progress on different aspects has been made during recent years: (1) mass-rearing conditions and quality control of the biological control agents; (2) their storage and methods of transporting; and (3) the release strategies. In addition to this empirical process, it seems more and more obvious that there is a need to accurately understand the underlying mechanisms. Biological control is thus nowadays recognized as a real ecology- and population biology-based method (Waage 1990). From this point of view, improved biological control using parasitoid species should be reachable using all the different concepts, information, and tools that have been introduced in this chapter. The main objective of this approach is to scientifically explore both the parameters linked to the efficiency of a parasitoid and their determinants. This is of course a central problem which is not easy to solve, since it varies with the features of the pest, the agrosystem, and more generally the abiotic and biotic environmental factors (Waage 1990).

As described in previous sections, the sequence of parasitism can be divided into successive steps from the search for a potential habitat with hosts to the emergence of the adult offspring. The pre-ovipositional steps are mainly determined by the ability of the females to detect cues from their biotic and abiotic environment. An adequate response of parasitoid females to this different information is necessary for finding a mate, for finding hosts, and finally for efficient control of the targeted pest. In this context, improvement of biological control must be based on ecological studies of the parasitoid, with the objective of analysing and understanding the different mechanisms and their sources of variability. Our brief description of the theoretical approaches to host–parasitoid demographic interactions has also shown how ecological and behavioural features of the parasitoids can affect the host population. Once the parameters of parasitoid efficiency have been defined, the next step is to choose the species to use as a biological control agent. It requires complementary and modern approaches or tools such as comparative analysis that can be used to identify the relative role of the **phylogeny** and the ecological determinants of the parasitoid life-history traits (Harvey and Pagel 1991; Martins 1996). The biological features of the parasitoids that affect their pest control efficiency should also be investigated at the intraspecific level, since strong genetic variability for these parameters can occur both within and between populations.

This shows how the aim of improving the efficacy of biological control programmes is deeply rooted in the disciplines of evolutionary biology and, particularly, behavioural ecology. The biological features of parasitoids are considered to be important components of their adaptive strategies under given environmental conditions. In the particular case of parasitoids, reproduction is directly linked to the destruction of the host, so that identification of factors and strategies that optimize their **fitness** should also improve their pest control efficacy in biological control programmes. Finally, the necessary comparisons of faunistic surveys before and after biological control agents have been released as well as the practices in conservation biological control are also directly connected with the current concern about biodiversity in natural or cultivated areas.

7.6 Conclusion

Parasitoid insects have several biological, ecological, and evolutionary features that make them an ideal model for addressing questions in population biology, from behavioural ecology to population dynamics. We have seen that the study of their reproductive strategies, which show important interspecific variation, raises several problems that can only be solved though the use of tools and methods from several diverse fields like molecular biology, biochemistry, and physiology, but also organism biology, population genetics, and even theoretical or community ecology.

Moreover, the particular mode of development of parasitoid insects enables us to use them in pest control strategies to protect crops. In recent decades this application has seen progressive uptake all over the world as the economics become favourable, leading to increasing avoidance of chemical pesticides that are noxious to both the environment and human health. The design of efficient biological control programmes against phytophagous pests on a given crop system implies the identification of potential biological control agents for production and release, along with an accurate understanding of their biology. This needs both pragmatic, empirical studies and a more formal approach based on evolutionary biology and ecology. Such combined fundamental and applied scientific work explains why a significant number of research laboratories worldwide are currently working on

insect parasitoids. These research laboratories are producing on a regular basis fascinating results demonstrating—contrary to what this chapter might suggest—that several important mechanisms involved in host–parasitoid associations still remain to be discovered. The aim of this chapter is to give the reader basic knowledge to enable an understanding of what parasitoid insects, their biology, and ecology are and how they can be used as biological control agents for controlling crop pests. It would be wonderful if readers were inspired to start a scientific research programme focusing on these fascinating organisms.

Important points

• Parasitoids are insects whose pre-imaginal development occurs by eating vital tissues from other living beings, mainly other insects. This usually implies that their development leads to the death of their hosts.
• Under the influence of strong selective pressures, different reproductive strategies and modes of interaction with their hosts have progressively appeared, leading to important interspecific variations.
• The use of mathematical models for analysing the dynamic interactions between parasitoids and their hosts has led to the identification of several key features (for both partners) that influence the demographic stability of their interaction and the ability of parasitoids to control the size of the host population.
• Due to the fact that parasitoids usually kill their hosts, they can be used for biological control programmes to control phytophagous pests on crops. It is now fully recognized that improving the efficacy of such a pest control strategy requires the use of concepts and tools developed in ecology and evolutionary and population biology.

Questions for discussion

• What are the differences between parasitoids and 'real' parasites, both in their biology and in the corresponding ecological consequences?
• According to the theoretical models in this chapter describing the demographic interactions

between parasitoids and their hosts, what are the behavioural mechanisms influencing the stability of such interactions and those that should not have any effect?

• According to the ecology of insect parasitoids, what could be done to improve their efficacy to control insect pests when they are released for biological control programmes.

Further reading

• Begon, M., Mortimer, M. (1986). *Population ecology: a unified study of animals and plants*, 2nd edn. Blackwell Scientific Publications, Oxford.

• Mackauer, M., Ehler, L.E., Roland, J. (1990). *Critical issues in biological control*. Intercept Ltd, Andover.

• Pennacchio, F., Strand, M.R. (2006). Evolution of developmental strategies in parasitic Hymenoptera. *Annual Review of Entomology* **51**: 233–258.

• Quicke, D.L.J. (1997). *Parasitic wasps*. Chapman and Hall, London.

• van Lenteren, J.C. (2003). Quality control and production of biological control agents: theory and testing procedures. CABI Publishing, Wallingford.

• Wajnberg, E., Hassan, S.A. (1994). *Biological control with egg parasitoids*. CABI Publishing, Wallingford.

• Wajnberg, E., Scott J.K., Quimby, P.C. (2001). *Evaluating indirect ecological effects of biological control*. CABI International, Wallingford.

• Wajnberg, E., Bernstein C., van Alphen, J.J.M. (2008). *Behavioral ecology of insect parasitoids—from theoretical approach to field application*. Blackwell Publishing, Oxford.

CHAPTER 8

Health ecology: a new tool, the macroscope

Guillaume Constantin de Magny, François Renaud, Patrick Durand, and Jean-François Guégan

8.1 Introduction

Epidemiologists and parasitologists studying transmissible diseases tend to take an interest in direct causes, the visible modes of spread, and disease prevention. Usually, these disciplines show little interest in the physical, ecological, or evolutionary processes which occur on larger time and space scales. Epidemiologists and parasitologists have for a long time also considered the individual host as the only frame of reference for understanding the causes and consequences of infections. However, recent studies on the impact of global environmental changes and the dynamics of pathogens and their geographical distributions perfectly illustrate the relevance of large-scale studies for a better understanding of the persistence and dynamics of micro-organisms in host populations of humans, animals, or plants. As in classical mechanics, changing our frame of reference allows us to highlight important but previously unsuspected phenomena for a better understanding of the modes of transmission of pathogens within and between host populations.

In this chapter, using several recent examples, we illustrate the need to use a new instrument in **epidemiology** and parasitology, the 'macroscope' (de Rosnay 1975), in parallel with the conventional use of the microscope, to analyse on a more global scale the forces and parameters implicated in the emergence and spread of infectious or parasitic agents. As the scale of study (spatial or temporal) grows, one begins to see the involvement of parameters and mechanisms more usually thought to belong to academic disciplines other than epidemiology

and parasitology. Solution of the world's health problems requires an opening up the conventional frames of reference in which host–pathogen interactions are usually analysed; the importance of non-causal or distal factors or processes having a direct, or more often indirect, influence on the nature and quality of the associations cannot be ignored.

8.2 Interactions between host populations and natural systems marked by exponential human growth

The current exponential growth in the human population, and the consequent huge and increasing impact that humans have on natural ecosystems, is one of the major problems facing human society today. The need for ever-increasing amounts of food has led to the rapid intensification and industrialization of agricultural production systems. This has had negative consequences for the viability of natural ecosystems (Aron and Patz 2001; Aguirre *et al.* 2002). The development of industrial-scale ruminant or poultry breeding in periurban areas has significant consequences for the quality of life and health of urban populations. Modern breeding plants pollute the water table and act as 'time bombs' for the spread of pathogens, which may be dangerous to humans. The present fears about propagation of the H5N1 bird flu virus in human populations will be even more justified if we offer such viruses favourable conditions for their spread; indeed, a large concentration of poultry (often genetically homogeneous) close to an urban community offers ideal conditions for H5N1. We are, in effect, reorganizing natural systems to make them less favourable to our quality of life and at the same time more dangerous as conditions because more suitable for the spread of numerous diseases. Growth of the human population also leads to the search for new areas of land to colonize and on which to build new communities, structures, and infrastructures; again, most of these settlements are established at the expense of natural ecosystems. The example of the Amazon forest in Brazil is illustrative; deforestation of the primary forest for the purpose of farming and stock breeding has led to the emergence of several **infectious diseases**, including malaria and leishmaniasis, because the environmental destruction has provided suitable habitats for the **vectors** and/or **reservoirs** for these diseases.

Socio-economic measures and political decisions often have unforeseen negative environmental consequences—affecting soil and plants and increasing pollution and the proliferation of pathogens. The structure and dynamics of the networks used by micro-organisms to move from a reservoir to a vector, then from a vector to another reservoir, and even to tangential hosts including humans, are integral parts of ecosystems, each having its own mechanics. We are therefore compelled to consider in a more integrated way the research on communicable diseases: a single fact about the molecular features of a virus, for instance, is not sufficient to explain an epidemic disease. The word 'macroscope' comes from the famous book by Joël de Rosnay (1975), and we have chosen to use it in the title of this chapter. It has become necessary nowadays to develop this concept in public, veterinary, and plant health. We recommend here the use of a 'macroscope' together with the more usual 'microscope' in epidemiology and parasitology. The microscope and macroscope are both useful and valuable, but are used to answer some very different questions and investigate mechanisms occurring at different levels of organization.

8.2.1 Complex relationships between ecosystems, hosts, and pathogens

Recent years have seen the emergence or re-emergence of an increasing number of transmissible diseases. But why? Is there a single common explanation for all these epidemic diseases or, alternatively, is each unique regarding its mechanism of transmission/spread? Are there any common factors behind the Chikungunya epidemic on Réunion Island in 2006, and the intense more or less recurrent bursts of **Ebola fever** observed in Central Africa? The first epidemiologists were accused of being naïve because they acknowledged the major importance of a pathogenic organism and its distinctive features in the spread of a new epidemic even if the organism was already present before the disease outbreak! Emergences and recent re-emergences (some examples are given in this chapter) show that many factors, and even some interactions between these factors, can favour the outbreak of a pathogen in a host population which previously seemed to be disease free. A wider vision of what pathogenic micro-organisms are, of the interactions they have with their hosts, reservoirs, or vectors, and of the influence that the environment can have on the interactions between such partners is necessary. This integrative, interdisciplinary approach is necessary if one wants to try to solve the problems of population health. To paraphrase Morange (2005): '[a] fuller vision of what an epidemic is…takes into account the existence of other organisms which are their natural reservoirs, and of changes in the ecological environment or the behaviour of human populations which are bound to create new niches for the bacteria or the virus, and new ways of spreading'.

Numerous infectious diseases, particularly those that spread through a vector and/or a reservoir, depend on particular environmental conditions to develop. For many this can be the proliferation of a species of rodent, encouraged by meteorological and ecologically favourable conditions, as in the case of the **hantavirus** spread by rodents of the Muridae family in South America (Suarez *et al.* 2003). Other diseases are spread by tick vectors: large numbers of tick larvae hatch during the rainy season, followed by a peak in the adult population, leading to increased incidence of tick-borne disease (Gardon *et al.* 2001). **Lyme disease** is one such tick-borne disease present in Europe, but there are many others. The recent example of cases of **legionnaires' disease** in France is interesting because this involves neither a reservoir *per se* nor a vector. The *Legionella* bacterium, responsible for occasional outbreaks of legionnaires' disease, is naturally present in water and soil, and can spread via plumbing and hot water systems when the conditions are favourable, particularly at temperatures varying between 25°C and 45°C. Industrial and domestic hot water systems and cooling systems using water make artificial ecosystems suitable for the development of these bacteria. Death occurs in 15% of cases of legionnaires' disease.

We shall now illustrate the complex relationships between the environment, micro-organisms, and hosts with some examples.

8.2.2 Different explanations

Even if we have learned much about transmissible diseases, their natural potential hosts, their reservoirs and vectors, and their interactions with the environment, there is no denying that many questions remain unanswered.

First let us take the example of cholera, a disease which until recently we thought we fully understood. The epidemiology of cholera can be looked at from two very different viewpoints. One explanation gives particular importance to the transmission of the two pathogenic strains (O1 and O139) of *Vibrio cholerae*, the bacterium responsible for the disease, and the consequences of these strains for **morbidity** and mortality in human populations. These aspects are the ones generally studied and analysed by molecular biologists who identify the **virulence** genes on isolated bacteria, and by epidemiologists who describe the evolution of outbreaks and intervene to stop their effects. A second type of explanation, a more recent addition to our understanding of this human disease, dates from the beginning of the 1980s. This considers that the two strains that are pathogenic in humans cannot be distinguished in a larger environmental context, in which a large number of *V. cholerae* strains coexist in the aquatic environment. Here, we need to understand the interactions between the environment and the bacterium, humans being a possible host for some of the bacterial forms. What is our current knowledge of the ecology of *V. cholerae*?

Cholera, or 'blue death', is an infectious diarrhoeic bacterial disease with an epidemic character. The pathogenic agent of cholera, as already mentioned, is *V. cholerae*, a Gram-negative motile bacillus belonging to the family Vibrionaceae. Not all strains of *V. cholerae* are responsible for cholera. In fact, the *V. cholerae* strains can be classified according to the O **antigen**, and, at present, nearly 200 serogroups are known; only the strains belonging to serogroups O1 and O139 have been linked to major cholera outbreaks. *V. cholerae* strains belonging to the other serogroups can cause sporadic diarrhoea, abscesses, or **septicaemia**. Cholera is the result of the absorption, by ingestion, of the cholera bacterium in water or food, but can also be transmitted from person-to-person by contact with pathological products (faeces, vomit, sweat). The experimentally determined **infectious dose** is of the order of 10^8 to 10^{11} bacteria, but it can be less, of the order of just 10^4 to 10^6 bacteria. After the passage through the gastric barrier, the bacteria become fixed to the proximal part of the small intestine, cross the mucous layer, and exude the choleric toxin. It is this toxin which modifies water and electrolyte exchange preventing the entry of sodium into the body cells; this results in the passage in the lumen of the digestive tract of a huge quantity of water, leading to severe dehydration of the patient. After an incubation period varying from a few hours to a few days, cholera manifests itself with violent diarrhoea and vomiting, but without fever. The water leakage leads to acute cramps propagating throughout the whole body, making the eyes sink in their sockets and contracting the orbicular muscle of the mouth, giving a cyanotic aspect to the patient's face ('blue death'). The expression 'to get a blue fear' (meaning to be scared out of one's wits) also originates from these symptoms.

It is the production of the cholera toxin from the *V. cholerae* O1 and O139 strains which is responsible for the expression of the disease. The **pathogenicity** of these strains is the result of the joint action of a set of factors favourable to the colonization of the intestine and the production of toxins by the bacterium. An environmental filamentous **bacteriophage** called CTXΦ, a virus attacking the cholera bacterium, has the ability to lyse the bacterial cell wall and integrate its genome into the bacterial genome. An inserted bacteriophage gene *ctxAB* produces the cholera toxin. Another important factor implicated in the virulence of the pathogenic strains of *V. cholerae* is slightly co-regulated with the toxin. Its role was first identified in the colonization by *V. cholerae* of the host intestinal wall, but seemed to be more directly linked to the development of pathogenicity, being used as a receptor by the CTXΦ bacteriophage.

Recently, Faruque and his collaborators (Fig. 8.1) offered a new synthesis to account for the re-emergence of cholera cases and the pathogenicity

Human compartment

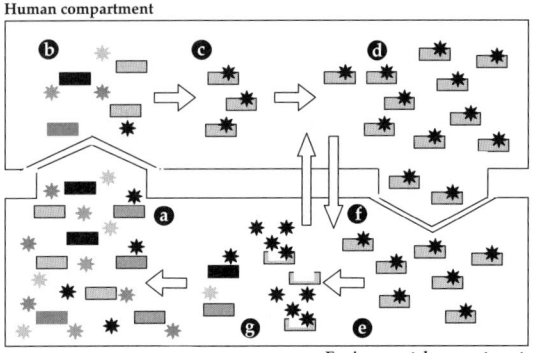

Environmental compartment

Figure 8.1 Diagram showing the way in which the epidemiology of cholera is currently interpreted. The aquatic environment maintains the phenotypic and genetic diversity of the strains of the bacterium *Vibrio cholerae* that are not very pathogenic for humans. Aquatic ecosystems also host a multitude of bacteriophages (viruses that naturally attack bacteria). Human populations form a selective filter, responsible for the better adaptation of the two pathogenic strains of the *V. cholerae* bacterium (O1 and O139) to the intestinal tract, by acquiring virulence genes from the bacteriophages. In some ways, humans are responsible for the disease that can be fatal to them. (a) Different environmental strains of *V. cholerae* (rectangles in different shades of grey and black) and of bacteriophages (stars) coexist in nature. (b) Bacteria and bacteriophages in the human small intestine; the bacteriophages can insert virulence genes into the bacterial genome. (c), (d) The now pathogenic bacteria are better adapted to the intestinal tract and spread; they are discharged with the diarrhoea to the external environment. (e) The aquatic environment is preferably loaded with pathogenic strains coming from the infected humans that 'selected' them. (f) An epidemic cycle is set up in which human populations, when drinking the water, take in the pathogenic strains which are the most abundant in the environment. (g) The spread of the pathogenic bacteria is followed by a population burst of lysogenic bacteriophages, a process that reinstates the initial conditions of the cycle. Adapted and interpreted from Faruque *et al.* (2005a,b).

of this disease in human populations (Faruque *et al.* 2005a, b). Locally, for example in the Bay of Bengal, the coastal and estuary waters, and even fresh water, naturally shelter a great variety of environmental strains of *V. cholerae*. So there is good reason to think, even if we have no direct evidence yet, that other tropical areas house a greater variety of these strains than waters situated further north. Several environmentally occurring non-pathogenic strains of *V. cholerae* can be ingested during contact or by swallowing water, and CTXΦ bacteriophages, naturally present in the water, can also be ingested. Besides, bacterial populations and bacteriophage dynamics can be observed and they are identical to what has been observed in prey–predator systems, i.e. an increase in bacteria in the aquatic environment is followed by a peak abundance of the bacteriophage (Faruque *et al.* 2005a,b). The small intestine therefore becomes a 'partners' crossroads' where contacts and exchanges between bacteria and their bacteriophages are greatly facilitated. Two strains of *V. cholerae*, O1 and O139, have indeed been more successful in infecting human populations, with tragic consequences. But why is this so? And how can we account for the visible antagonism between this increased ability of the O1 and O139 strains to develop, but also to bring about morbidity and mortality in human populations?

Let return briefly to what happens in the small intestine when *V. cholerae* strains and bacteriophages have been ingested. Some contacts and associations occur between some bacterial cells and the 'predator' bacteriophages, by chance or following some principles we are not yet aware of. In any event, the insertion of bacteriophage genes into the bacterial genome (a process called transduction) leads to an advantage for those bacteria carrying the pathogenicity gene *ctxAB*. From then on, these bacteria start to produce the gel-like toxin which has a devastating effect on the intestinal walls as well as on the natural intestinal flora, but forms a protective cocoon for the *V. cholerae* bacteria enabling the bacterial population to increase inside it. The significant diarrhoea provoked by the toxin then ejects the bacteria into the external environment which gradually becomes loaded with bacteria selected in the intestines, i.e. the pathogenic O1 and O139 strains exclusively. A 'reactor' is created, and humans eat or absorb some elements which are contaminated by the pathogenic strains which are ever more present in the environment as the cholera epidemic spreads. In other words, the human population, and the miniature ecosystems formed in thousands of small intestines, generate the pathogenic bacteria that are responsible for their own misfortune. A sad fate for the hosts, but just a survival strategy for the micro-organisms which have to adapt to survive, or die. But in that case, why doesn't the epidemic continue to make progress as more and more pathogenic bacteria are poured into the external environment? Faruque and his colleagues once more have a rational explanation for this.

The pathogenic strains of *V. cholerae*, better adapted to the environment of the intestinal tract, are much less viable in the external environment: there, they meet other environmental strains of the same species, but also other species of potentially competitive micro-organisms, and even predators. The relatively rapid input of the O1 and O139 strains to the external environment during an epidemic also leads to the proliferation of bacteriophages in the aquatic environment which attack the bacteria by lysis. Progressively, or more suddenly, the O1 or O139 strains in the aquatic environment disappear, and a bacterial microfauna consisting of environmental strains returns. The Vibrionaceae are Gram-negative bacteria, and this bacterial group includes bacteriophages associated with the different species. There is good reason to believe that many diseases caused by these Gram-negative bacteria could also follow such a scenario (Fig. 8.1).

There are of course many other explanations for the outbreak and spread of cholera within human populations; interestingly, the explanation given by Faruque and colleagues matches the more traditional view of cholera epidemiology. The different explanations are not antagonistic but complementary, the traditional explanation finding its place within a novel and more integrated approach. The continual coming and going between *V. cholerae* bacteria and bacteriophages would, according to Faruque and colleagues, be the explanation accounting for the very marked seasonal variation of the epidemic episodes of cholera in the endemic zones.

We are not that certain, but we do admit that such an ecologically based explanation is quite attractive. It could also be that particular oceanographic and climatic conditions are responsible for the exponential growth of the bacteria and their bacteriophages, as suggested in several recent works by Colwell and colleagues in the USA (Colwell 1996). Other factors also affect the impact of cholera, such as sanitary facilities and the level of poverty, the public health services, population demography, changes in the use of soil, and urbanization and travel. These causes are those that are usually cited to explain the outbreaks of cholera epidemics and their dispersal within populations. The model of Faruque and colleagues provides interesting possibilities: an appreciation of the existence of two compartments, on one side the environment that shelters a large diversity of *V. cholerae* strains and on the other human populations that unintentionally select the virulent strains; this model readily allows us to understand where action is needed to control or prevent the disease—in other words, within the environmental compartment that is upstream in the causal chain. We are obviously not questioning the critical contribution that medicine and public health care make in fighting outbreaks of communicable diseases, but we think that consideration of this kind of integrative approach in epidemiology can help us find alternative solutions to existing problems.

This observation leads us directly to the second example we wish to address in this section, which concerns the importance of economic exchange in the transmission of contagious diseases. Many examples feature the importance of sea and air travel as a driver behind the initiation and dispersal of epidemics (Ferguson *et al.* 2005). We shall mention a few of them here. One quite conclusive example concerns the seemingly trivial importation into the USA of several specimens for the pet trade that almost caused a major health crisis.

During the spring of 2003, several cases of monkey pox—caused by a simian orthopoxvirus provoking a disease clinically undifferentiated from smallpox—were reported in small pets in Texas (Di Giulio and Eckburg 2004). A few sporadic cases of monkey pox had already been reported in humans living in tropical forest areas in Central and West Africa, where the major reservoirs of this disease are squirrels of the genus *Funisciurus* and *Heliosciurus*, Gambian rats of the genus *Cricetomys*, and several species of monkeys. Monkey pox had never previously been reported in the Northern Hemisphere. The introduction of this virus into the USA took place when some small African mammals were imported. Within the six different classes of rodents introduced, at least one Gambian rat, two squirrels, and three dormice (genus *Graphiurus*) were identified by the American medical authorities as having been infected with the monkey pox virus. Native prairie dogs cohabiting with Gambian rats in pet shops were then infected, and several people who bought these pets were infected in turn. Fortunately, none of the humans identified became seriously ill and the outbreak of a new **zoonosis** was averted. The occurrence of an **epizooty** originating from a few prairie dogs infected in a pet shop could have had dramatic consequences for the wild animals of the southern USA if they had come into contact with the virus. Undoubtedly, the transfer of animals, and thus probably of the pathogens hosted by them, is very disquieting given our current understanding of the susceptibility of other animal species and humans to infectious or parasitic agents of animal origin. This is also the case for the trade in food products and plants.

In order to deal with these 'new biological invasions', it will be necessary for there to be greater cooperation between national and international health services, doctors, veterinarians, and scientists. Imagine the consequences of the introduction—intentional or unintentional—to a conurbation of an animal that is a reservoir for a virus such as the one causing Ebola fever. Indeed, in Central Africa, where cases of Ebola fever are often reported, the spread of the virus is nonetheless limited by the small number of humans living in the forest villages.

The importance of worldwide exchanges of goods and humans today makes the rapid dispersal of infectious agents from their original area almost inevitable. Infectious diseases are very likely to be one of the priorities for the 21st century, and we must plan in order to better foresee the potential hazards and deal with the expected consequences.

8.2.3 Ecosystem dynamics and infectious diseases—or the snowball effect

Each new human or animal epidemic (surprisingly we seldom hear about the true devastation caused by infectious diseases of crops) is followed by many public and media arguments regarding its origin. The example of the epidemic outbreak of avian flu in Asia at the end of 2003 and beginning of 2004 is enlightening regarding the fact that wild birds, especially aquatic birds, were said to be responsible for the disquieting spread of the disease in Asia, while domestic ducks were blamed for having played a major role in the origin and the persistence of the flu virus (Li *et al.* 2004). The Thai political authorities even ordered the selective slaughter of the Asian openbill—a species of migratory bird of the same family as the stork—even though the decision was later revoked. Facing panic, the Food and Agriculture Organization of the United Nations (FAO) valiantly advised the Asian public authorities against the slaughter of wild birds, since there was no direct evidence of their being responsible for spreading of the disease between different epidemic centres and given that these wild birds represent a major component in the dynamics of aquatic ecosystems (see http://www.fao.org/newsroom/en/news/2004/48287/index.html). Recent observations in birds, but also in humans, of avian flu in Europe, Asia Minor, and Africa have also illustrated our total ignorance regarding the biological cycle of the viruses responsible. It is now thought that certain avian flu viruses—without knowing exactly which ones—have an exceptional capacity to survive for several months in the cold water of ponds and marshes; this upsets our knowledge about this zoonosis to such a degree that the role of the migratory birds or the domestic bird trade seems questionable. To illustrate this, imagine that a political decision had been taken in Thailand to exterminate the Asian openbill under the pretext that it was responsible for the spread of the avian flu virus. What would have been the epidemiological and ecological effects of such a suppression of predatory birds from the ecosystem?

One of the best documented examples today on the relationship between ecosystem dynamics, animal communities that coexist in the ecosystem, and microbial agents is that of Lyme disease (Ostfeld and Keesing 2000a,b). Lyme disease is a disorder conveyed by a tick, *Ixodes scapularis*, in North America, and caused by micro-organisms called *Borrelia burgdorferi* (a spirochaete bacterium). Most often, it gives rise to a ring-shaped skin rash and **erythema**, accompanied by non-specific symptoms such as fever, discomfort, headache, or muscle and joint pain. Lyme disease is considered as an emerging disease in Western countries. In the USA, it has been shown that the risk of transmission of the pathogen is closely linked to the local diversity of small vertebrate species, certain ones of which can act as reservoirs for its growth and circulation. It seems that a small rodent, *Peromyscus leucopus*, is the reservoir species most likely to host and re-transmit the bacterium when a tick bites humans and other species. The environmental conditions, in particular the reforestation taking place in several north-eastern regions of America, have greatly modified the biological diversity of the resident species. Richard Ostfeld and collaborators have studied the risk of transmission of Lyme disease to humans according to the animal species composition of the forest fragments of different areas. In those forest segments with sufficiently large areas they found a large biological diversity, including predators and competitors of the rodent *P. leucopus* which regulate the population dynamics of this small mammal. On the other hand, in small forest systems, the lack of predators and competitors favour larger populations of *P. leucopus*. The increase in the numbers of *P. leucopus*—an excellent reservoir for *B. burgdorferi*—in this environment causes an increase in the general circulation of the pathogen within the ecosystem, and thus an increased probability of transmission to other animals and humans. American scientists have called this phenomenon the 'dilution effect'—a true functional role played by biological diversity in the transmission of a zoonosis to human populations.

According to this very original study, the main question that immediately arises is how does an infectious agent spread within a community of reservoir species? Is this situation also true for vector-borne diseases that show different transmission capacities? The work on Lyme disease represents a fascinating first study, that is not only focused on the relationship between a pathogen and a reservoir or vector species, but also tries to understand the circulation of a micro-organism within the ecosystem (Fig. 8.2). At the moment we generally lack

Community with high biological diversity

Community with low biological diversity

Farming situation

Figure 8.2 Depending on the diversity of species within a community the same infectious agent will present very different transmission modes (for example bird species that are reservoirs for a virus such as the one responsible for West Nile fever). (a) In the presence of diverse reservoir species with very different virus-transmitting capabilities, the pathogen has many possible modes of transmission with a resulting dilution of its effects. The dilution phenomenon increases with increasing proportions of weak reservoir species. (b) In communities with fewer reservoir species, where one or several strong reservoir species have taken advantage of the environmental conditions to reproduce, the transmission of the infectious agent is facilitated and its effects are more apparent. A similar situation exists when bird migrations occur with a flow of some strong reservoir species into a local community where the virus did not previously exist. (c) This illustrates an extreme situation where the individuals of a single bird species, often very capable of transmitting a virus, or alternatively have weak resistance to its invasion, are concentrated in large numbers, as in farms.

this kind of wholesale understanding of the ecological and evolutionary mechanisms implied in the emergence of pathogenic agents. Future work should focus on other examples of infectious diseases for which the dynamics of biological diversity and ecosystem change play a major role.

Undoubtedly, it is now obvious that human interaction can easily upset the equilibrium of ecological systems to a greater extent than in the past (Lebel 2003). As shown by many examples, only a few of which are given here, the emergence of infectious diseases or of certain epidemics with a zoonotic or vectorial origin is often the result of upsetting the equilibrium between a pathogen, the vector or reservoir populations, and the human populations that are often unfortunately the target but also actors in the situation. Epidemic phenomena are themselves complex and cannot be reduced to a simple explanation involving the presence and virulence of the pathogen, as was stressed early on by Duclaux (1902). An understanding of the dynamic equilibria occurring within ecosystems, where micro-organisms are a fundamental part of these giant puzzles, and to describe their changes to anticipate potential impacts in terms of health risks should today be part of our scientific culture from school upwards, and should receive more attention in undergraduate medical and scientific college courses.

8.2.4 A new school of thought: ecomedicine or health ecology

A new field of 'conservation medicine' has developed in recent years, almost exclusively in Anglo-Saxon countries (see http://www.conservationmedicine.com/). Its main objectives are to better understand the relationships and interactions between human health, animal health, and ecosystem dynamics (see Fig. 8.3), and also, of course, to promote these ideas to policy-makers, economists, and other leaders at national and international levels (Aguirre *et al.* 2002). The examples presented in this chapter are representative of this expanding line of research because the health of human populations cannot currently be dissociated from environmental conditions on the one hand ('ecosystem health') and animal health on the other. The example of avian flu is a perfect illustration of this. The reader will find much information on conservation medicine in the literature of Aguirre and colleagues (e.g. Aguirre *et al.* 2002) and of Lebel (2003; a free copy of Lebel's book is available in French from the site of the Canadian Development and Cooperation Agency http://www.idrc.ca/openebooks/012-8/). A programme initiated by the present authors in France contains sufficient material to keep several young and brilliant scientists in work. The stakes for

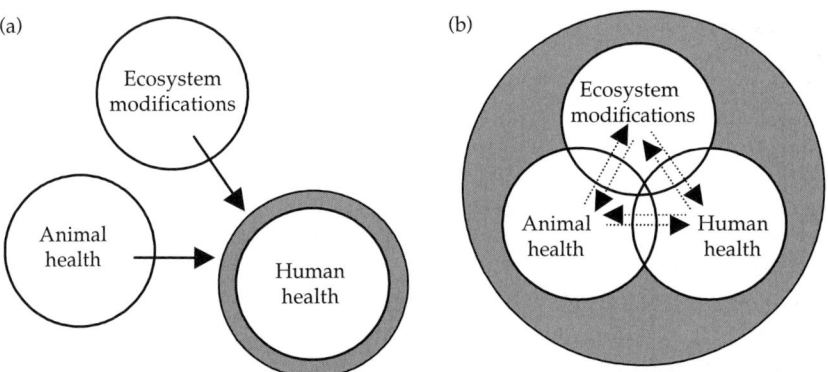

Figure 8.3 (a) The classical epidemiological view of the relationships between human health, animal health, and ecosystem modifications. Human health problems are interpreted according to a view where only the effects of changes in the ecosystem on human health are considered (see the arrows). (b) Potential multiple relationships between human health, animal health, and ecosystem health, where all the components interact with one another; a serious epidemic affecting whole populations of large ruminants in the East African savannah, for example, can have major consequences on the stability of this particular ecosystem as a whole.

population health are too high for this synthesis to be dismissed. Surprisingly enough, the field of epidemiology is probably one that is most in need of multiple explanations (Morange 2005), and yet where real joint research between disciplines is still very hard to implement.

8.3 The ecology of infectious diseases

For the reasons discussed in this chapter, we have a limited knowledge of the geographical distribution of the major pathogens in human populations (Jones 1990). This is also the case for animal and plant pathogens. What are the factors and mechanisms that are responsible for the present distribution of pathogens in human populations around the world? What role does climate and its variability play in the distribution of these microorganisms? What are the quantitative relationships between biological diversity and pathogens? What are the main factors responsible for the emergence, re-emergence, and dispersal of a pathogen, excluding the socio-economic reasons generally put forward? Are the globalized economy and modernization still paramount in explaining the spread of diseases and, if so, of what diseases? We have not yet found convincing answers for many questions of this type. The article by Tony McMichael (2004) gives an overview of the major causes of the emergence/re-emergence of current human diseases, and below we give a few examples of situations illustrating the relationships—often inextricable, but always complex—between ecosystem mechanics and population health.

8.3.1 What happened first: biology or socio-economics?

Two questions seem to dominate the traditional epidemiological approach to the study of communicable diseases—Where do they develop? Why do they develop there? Parasitology has the same questions. Determination of the 'where' has needed a considerable amount of observation, identification, and description. Understanding the 'why' has focused on the importance of social, economic, and technological organization to explain the occurrence, development, and distribution of infectious and parasitic diseases. There have been few works in the literature with an ecological or biogeographical approach.

Recent statistical studies of the large-scale spatial distribution of human infectious and parasitic diseases have provided interesting predictions concerning the different mechanisms that are responsible for the observations. Through a modelling approach based on a large set of data (332 human pathogen species with a worldwide distribution), Guernier *et al.* (2004) have shown, after correction for disturbance variables, that the species diversity of human pathogens is highly correlated with latitude: human communities that live in the intertropical regions generally host a larger diversity of pathogens than populations living in temperate regions (Fig. 8.4). In other words, the species diversity of human pathogens is consistent with the rule that applies to other free-living species (Hawkins *et al.* 2003; Chown *et al.* 2004; Hillebrand 2004). Is this geographical distribution the rule for all groups of pathogens? The study showed that pathogens that are transmission via a vector or a reservoir have a distribution based on latitude. It is the presence and the distribution of the vector, or of the reservoir, that determines that of the pathogen. Thus, many vectorial or reservoir diseases are present in intertropical regions because the diversity of the vector or reservoir hosts is very high in these regions. A single group of pathogens does not follow this general trend—pathogens with direct transmission, i.e. those for which infection takes place from one human to another and for which no animal reservoir exists. Many viruses and bacteria fall into this group, for instance the viruses responsible for measles and chickenpox and the bacterium responsible for whooping cough. As a matter of fact, these were originally animal diseases that were transmitted to the first organized human populations in the Neolithic age. These diseases, with progressive selection for forms adapted to humans, were then distributed by humans during colonizations, conquests, and displacement and economic exchange. Nowadays these pathogens are present almost everywhere on Earth apart from the two poles. They take advantage of large-scale transportation by air, rail, or sea to disseminate worldwide and their host resource is global, i.e. the human population.

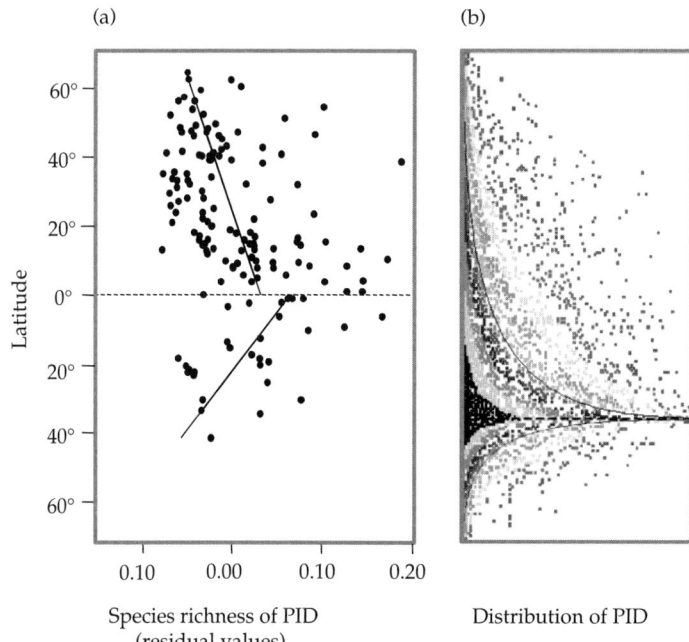

Figure 8.4 (a) Trend of the species diversity of pathogens for human parasitic and infectious diseases (PID) according to latitude. We can see a negative correlation between pathogen species diversity, after correction for disturbance factors to estimate diversity, and latitude for both the Northern and Southern hemispheres. (b) Presence/absence distribution matrix for 332 PID pathogens in both hemispheres. The spatial distribution of the aetiological agents, based on real data, was computed using random permutation statistics (Monte Carlo tests). The presence/absence matrix provides information on the distribution of the presence of a species in a given location (indicated by a dot) and deduces its absence elsewhere. Part (b) shows that the PID diversity decreases from the equator to the north or to the south. Furthermore, the PID pathogens present in a temperate region also tend to be present in human populations of the intertropical region. The opposite situation, where a pathogen is present in temperate regions and absent in intertropical regions, is very seldom observed. The spatial distribution of human pathogens follows an interlocked spatial model called a nested pattern (Guernier *et al.*, 2004).

Apart from that group of pathogens transmitted only from human to human, this study shows that the worldwide spatial distribution of human pathogens strongly depends on climatic conditions. Two factors are paramount: the temperature and especially the variation in precipitation during the year, which statistically explains the observed spatial distributions (Guernier *et al.* 2004). Undoubtedly, the particular ecological conditions acting on vector or reservoir species are those that limit the spatial distribution of these pathogens. Other pathogenic micro-organisms that, strictly speaking, lack reservoir or vector species also depend for their survival on the bioclimatic conditions that prevail in the ecosystem. In the next section we will discuss the example of the bacterium *V. cholerae*. We can then speculate about how these disease agents will

evolve, either directly or indirectly, when faced with the expected changes in global climate.

8.3.2 Global climate change and the spread of infectious diseases

For efficient disease control it is essential to have a grasp of the ecology of infectious diseases and particularly of the role of environmental parameters in the evolution of interactions between hosts, vectors or reservoirs, and parasitic or microbial agents. By the contingent displacement of vectors and reservoirs, global climate change will undoubtedly contribute to major changes in the distribution of infectious diseases. A direct consequence will be the exposure of new human populations to exogenous pathogens against which they will have no protection. Several recent examples show a trend toward an increase in the **incidence** and a change in the spatial distribution of infectious diseases of plants and animals in response to local climate change (Harvell *et al.* 2002). But what about diseases affecting humans? Are there any convincing demonstrations of the action of climate and its variability on the behaviour of microbial and parasitic agents? Let us look at the example of cholera epidemics in Bangladesh, which to date is one of the best documented studies in the literature (Rodó *et al.* 2002).

Through an analysis of the historical data on the incidence of cholera in Dhaka, the capital of Bangladesh, Rodó *et al.* (2002) found a close relationship between the El Niño phenomenon [more particularly the set of physical parameters measured by oceanographers and climatologists in the Indian Ocean (the Southern Oscillation Index, or SOI)] and the temporal dynamics of cholera cases in Dhaka. Figure 8.5 shows that the dynamics of cholera cases tend to change with the SOI, showing a strong match between the cholera peaks and the SOI minima for the period 1980–2002 (on the right of the figure). The nearly quadrennial cycle (between 4 and 5 years) of cholera epidemics in Dhaka since 1980 is interpreted as the result of the major climatic role played by the El Niño/Southern Oscillation (ENSO) phenomenon for the last two or three decades. But what links the cases of cholera in Dhaka to the ENSO phenomenon in the Southern Pacific Ocean as measured by the SOI? Physicists explain that the Earth's water masses are linked by teleconnections and that there are large-scale

energy transfers from one ocean to the others. The disruptions linked to the ENSO phenomenon in the South Pacific then have consequences for the Indian Ocean. As was said above, environmental bacteria of the genus *Vibrio* are very sensitive to the physico-chemical conditions of the water they live in. For instance, temperature modifications at the surface of the Indian Ocean tend to favour these bacteria in coastal environments where human populations are concentrated. The work of Rodó *et al.* (2002) is one of the first contributions based on a mathematical/statistical approach to demonstrate a link between climate change (under way since 1976 according to climatologists) and the temporal dynamics of cases of an infectious disease (see also Speelmon *et al.* 2000).

In the same vein, let us take another example—flu epidemics in France. Using two independent sets of data for the period between 1971 and 2002, Viboud *et al.* (2004), showed a statistical relationship between mortality and morbidity due to flu epidemics in France, and the ENSO oscillations discussed above. During the study period, flu mortality in France was significantly higher for the 10 winters marked by cold ENSO periods compared with the 16 more temperate winters. Another study on the same subject confirms the overall influence of climate on the number of flu (due to the influenza virus) and viral pneumonias (other types of virus) hospitalizations in Sacramento, California (Ebi *et al.* 2001).

Although no hypothesis about the mechanism linking climate to the dynamics and size of flu epidemics has been put forward, the authors suggest

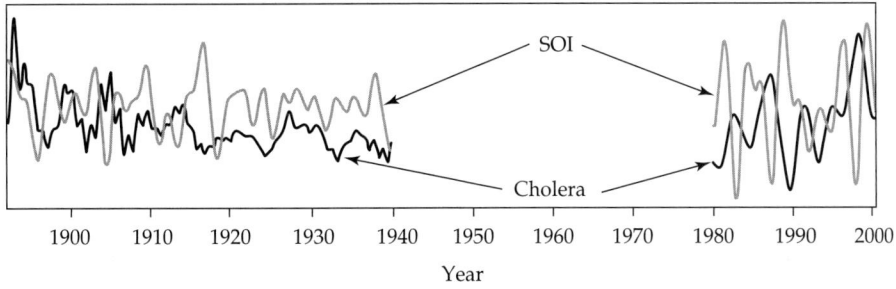

Figure 8.5 Relationship between the Southern Oscillation Index (SOI) and cholera cases in Dhaka (Bangladesh). Since the beginning of the 1980s, cholera epidemic peaks have occurred when the SOI index is at its lowest levels, and vice versa (see Rodó *et al.*, 2002). ©*PNAS* (2002).

that climate change could favour the emergence of new **genotypes** of flu viruses that are better able to survive and spread in the new conditions. Another hypothesis, that does not exclude the first, suggests that lower temperatures and higher humidity rates, as observed during the very cold ENSO periods in Europe, could affect the physiology of individuals, making them more vulnerable to certain diseases due to immune deficiencies. There is currently no study demonstrating the better capacity of certain genotypes of flu virus to spread in particular environmental conditions, reflecting our present lack of knowledge about the ecology and evolutionary biology of the large majority of pathogens. This observation should generate new paths for research and draw young talent into this expanding discipline.

Climate and its variability have well-known consequences for living organisms, and ecologists would not omit bioclimatic parameters from their studies. Unfortunately, and somewhat surprisingly, very little importance has been attached to climate in explanations of the survival and viability of micro-organisms. The present unprecedented rate of global warming underlines the urgent need to develop adequate research in order to understand the adaptations taking place in microbes. Only such research will allow us to understand this process and predict the responses in the face of climate and anthropogenic changes. Further information on this topic can be found in a report from the American Society for Microbiology (King *et al.* 2001).

8.3.3 Ecosystem changes and health

Many examples of the emergence of infectious or parasitic diseases are consequences of human-induced ecosystem changes and habitat disruption. The epidemics of infectious or parasitic diseases, such as Lyme disease, schistosomiasis, and infections caused by Hantaviruses in Latin America, represent striking examples of the way in which ecosystems can affect the emergence of new diseases. This emphasizes the importance of developing pictures of ecosystem dynamics, the evolution of biological diversity, and their respective effects, or synergy, on health (see Section 8.2.3). We

will illustrate this chapter with two new examples: one concerns the recrudescence of schistosomiasis in an East African lake, and the other the increase in malaria cases following the unprecedented deforestation of a major part of the Amazon plain in South America. Let us start with Africa.

Most infectious diseases have complex life cycles that require a reservoir host or a vector. One such disease, schistosomiasis, or bilharziasis, is one of the most serious tropical diseases. The helminths causing this disease are transmitted by gastropods (intermediate hosts) that are re-emerging in different African and Southeast Asian countries despite an improvement in the sanitary and socioeconomic conditions of these regions. The worms in the larval stage (called cercaria) leave the snails and enter humans in contact with fresh water. The adult worms live in the human blood system and feed on blood cells. Their eggs are deposited in several human organs and tissues (liver, bladder, gut, etc.). The blood vessel damage they cause, and the corresponding physiological complications, give rise to the symptoms of the disease (Combes 2001). A key element in the re-emergence of schistosomiasis is the increasing number of habitats that are favourable to the different snail species that are compatible with the transmission of the parasite. Snails develop in rice fields, dams, and areas set aside for fish farms. For example, widespread large water reservoirs in Africa, e.g. the Aswan Dam that gave rise to Lake Nasser, have greatly increased the parasite's transmission, and consequently human morbidity and mortality in the region.

Changes in trophic webs due to the introduction of new species have also facilitated the increase in snail populations that are responsible for the transmission of schistosomiasis. For example, let us discuss the epidemic of schistosomiasis on the banks of Lake Victoria in East Africa. Lake Victoria has hosted a great diversity of cichlid fish species for several thousand years, and this exceptional diversity has led to spectacular specializations in several fish species. In particular, several cichlids have adapted to feed exclusively on molluscs, thus regulating gastropod populations and keeping their densities low enough to slow down the transmission of the parasite to humans. However, the Nile perch, *Lates niloticus*, was introduced to stimulate the local economy—with such 'success' that we can now find Nile perch on sale in nearly all fish markets or supermarkets in France some 10 years after it was introduced into Lake Victoria. The consequences of the introduction of the Nile perch are many and have been the subject of a documentary film called *Darwin's Nightmare*. Two such consequences have been a drastic reduction in the biological diversity of native cichlid species and the consequential explosion of the mollusc populations on the lake's banks. The increase in the human population seeking to take advantage of the new local economy linked to fishing thus generated new centres for the transmission of schistosomiasis (Ogutu-Ohwayo 1990). If the introduction of the Nile perch in Lake Victoria did indeed lead to direct short-term economic benefits for the local population, the loss of biological diversity that it caused has led to considerable health problems for humans in the long run.

Another example that illustrates the way hosts and agents of infectious diseases interact with their ecosystems is that of malaria in the Amazonian equatorial forest. The deforestation associated with agricultural development and the construction of the trans-Amazonian highway has favoured the development of the mosquito *Anopheles darlingi* that is the main vector for two pathogens, *Plasmodium falciparum* and *Plasmodium vivax*, responsible for malaria in South America (Tadei *et al.* 1998; Conn *et al.* 2002). *A. darlingi* is known to occupy a specific ecological niche in the canopy of the humid forest. Deforestation has opened new habitats favouring the irruption, development, and establishment of populations of this mosquito species (Vittor *et al.* 2006). The arrival of human communities to the newly developed agricultural areas was accompanied by the introduction of the *Plasmodium* parasite by infected persons. The presence of larges numbers of *A. darlingi* then efficiently contributed to the creation of new centres for the transmission of malaria. In Peru, the areas deforested for agriculture have an almost 400-fold higher risk of malaria transmission than the forest zones (Vittor *et al.* 2006). The ecological disruption of the equatorial forest ecosystem due to deforestation for agriculture, the proliferation of the vector, and the settlement of human communities form a complete

picture for understanding the spread of centres of malaria within the Amazonian plain. The role and importance of environmental and socio-economic factors are easy to identify in this example, but they are also closely related. Another example of the role of deforestation on the emergence of diseases is given by the recrudescence of cases of **hookworm**, an important human pathogen in Haiti (Lilley et al. 1997).

Over the past 50 years, industrial and agricultural changes, economic and social changes, and the rapid growth of the human population and international travel have all contributed to modification of the occurrence profiles and spatial distribution of infectious and parasitic diseases. However, for a large number of known pathogens, and probably also for thousands of other unknown ones, the role of the ecosystem is fundamental in explaining the emergence or recrudescence of diseases, as shown here. It is crucial that national and international organizations start to implement an ecosystem approach to health problems (Rapport and Lee 2003).

8.3.4 Land use, agricultural development, intensive farming, and health

The huge increase in population that began in the 20th century necessitates a permanent increase in the production of food products. The soil and the ecosystems are already being vastly over-exploited in order to satisfy these needs: to feed the 6.6 billion people that currently live on Earth (Aron and Patz 2001; Patz *et al.* 2005). (For further information seven very interesting seminars on 'Medical Ecology: Environmental Disturbance and Disease' by Dickson Despommier of Columbia University may be accessed at http://ci.columbia.edu/ci/eseminars/1111_detail.html.)

What might be the ecological consequences for human health of these land use practices, intensive farming, and agronomic systems? Are our practices also re-creating new types of ecosystems which will facilitate the development and circulation of new pathogens? Let us look at several examples of infectious diseases whose recent emergence is linked in part to land use or intensive practices in agronomy.

The Japanese encephalitis virus has represented a serious public health problem in several Southeast Asian countries since its emergence at the beginning of the 1970s. This virus is transmitted by mosquitoes of the species *Culex vishnui* which preferentially breeds in rice fields. The intensive cultivation of rice in the Tamil Nadu region of India, as in many regions of developing countries, has strongly determined the recrudescence of these insects in rice fields (Sunish and Reuben 2001). We know today that the use of fertilizers such as nitrates has a positive effect on the abundance of mosquito larvae in rice fields, namely on the propagation of aquatic micro-organisms on which the mosquito larvae feed. Likewise, for the same agrosystem but in north-eastern Argentina, the abundance of molluscs of the genus *Biomphalaria* (possible vectors for the transmission of schistosomiasis) is linked to concentrations of nitrates and nitrites in rice fields (Rumi and Hamann 1990). Similar scenarios have been observed for Korean haemorrhagic fever, caused by a virus of the Hantaan group, and for the Argentinean haemorrhagic fever caused by the Junin virus (Morse 2004). The virus causing Korean haemorrhagic fever infects the vole, *Apodemus agrarius*, in several Southeast Asia countries and in particular in the People's Republic of China. This pathogen can infect humans. The expansion of rice fields has created favourable conditions for an explosion in the vole population, which thus increases the risk of transmission of the disease to humans, mostly farmers. Moreover, it is supposed that the conversion of prairies into corn fields in several regions of Argentina has facilitated the proliferation of a reservoir of host rodents for the Junin virus (Morse 2004).

Ecological change subsequent to agricultural development are is of the factors most frequently identified with the emergence of a new disease. Avian flu caused by the influenza virus is another example of the way in which intensive animal production can play an important role in the origin and circulation of a disease. Indeed, the intensive production of birds and pigs represents an artificial ecosystem favourable to the circulation and transmission of this type of virus, because the organisms are present in high densities and they are often physically compromised. One can imagine how a viral epidemic could spread through the intensive concentrations of such farm animal in periurban areas of megacities such as Bangkok or Mexico City, for example. Let us also recall the recent outbreak of severe acute respiratory syndrome (SARS) in the south of China, caused by a coronavirus from a small civet-like mammal sold in markets and eaten by the local population (Guan *et al.* 2003). The intensive production of cattle and the methods used to feed them have facilitated the transmission of prion agents causing bovine spongiform encephalopathy (BSE) in ruminants, linked to a new variant of Creutzfeldt–Jakob disease in humans (Ghani *et al.* 2000; Valleron *et al.* 2001). Intensive farming and related industrial production methods favour the increase of contamination and accidental amplifications, as was observed in the case of BSE. Indeed, animal by-products—mainly ovine and bovine—were used to feed the cattle, thus creating a new 'artificial food chain' and an occasion for cannibalism in otherwise strictly herbivorous animals.

The intensification of agriculture and animal rearing represent major sources of new infectious agents. Given that many pathogens have displayed a natural ability to overcome the species barrier, we must also reflect on the consequences, in terms of public health, of allowing the spatial coexistence of thousands or even millions of animals for human consumption in the vicinity of large cities where millions of inhabitants are sometimes concentrated. This is what the next section deals with.

8.3.5 Human population growth and the evolution of infectious diseases

In their respective conclusions, McMichael (2004) and Morse (2004) underline the fact that the human demographic explosion and its consequences are responsible for the many emergences and/or re-emergences of infectious or parasitic diseases. What is in store for the future then? What does theoretical and empirical epidemiology have to say?

There were 2.5 billion people on Earth in 1955, 6.5 billion in 2006, and a predicted 9 to 10 billion 50 years from now. This increase in the 'host population' can only favour the persistence and evolution of a greater diversity of pathogens (Guégan and Broutin 2008). Indeed, even if we do not have a

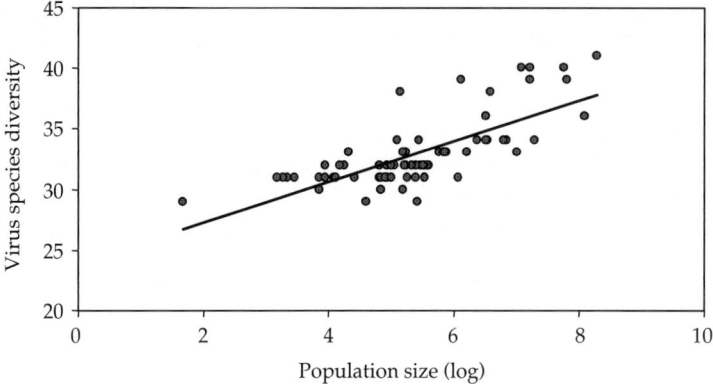

Figure 8.6 Linear relationship between the size of a human population (number of individuals) and the diversity of viral species in 71 human communities living on oceanic islands. The linear regression is $y = 1.67x + 23.97$, $r^2 = 0.551$, $P < 0.0001$. The 'population size' variable is log-transformed. From Guégan and Broutin (2008).

comprehensive grasp of the biological diversity of pathogens hosted by human populations, it is quite easy to understand that larger human populations with more individuals will display increased hosting possibilities for a larger number of pathogens than smaller and less numerous populations (Fig. 8.6).

There are still many uncertainties about the mechanisms that regulate pathogen diversity, particularly in human communities. The examples above give quite a convincing demonstration that simple laws of population and community ecology are perfectly adaptable to communities of pathogenic micro-organisms. Indeed, recent works tend to show that free-living micro-organisms such as planktonic algae or microscopic fungi (Finlay 2002; Green *et al*. 2004) as well as bacteria and viruses infecting human populations (Guernier *et al*. 2004; Guégan and Broutin 2008) are not randomly distributed, but display organized and predictable spatial distributions. This type of results offers new possibilities for a synthetic understanding of the way in which pathogens organize and become structured in host communities.

The population size, i.e. the number of people within a community, is an essential factor for the persistence and spread of a contagious pathogen, such as a virus or bacterium. Many studies have thus shown the existence of a critical threshold for the number of individuals in human populations

(critical community size, CCS) below which the pathogen cannot persist (Grenfell and Harwood 1997; Grenfell *et al*. 2001; Broutin *et al*. 2004). In the case of the virus responsible for measles, the critical threshold is estimated to be 250,000–300,000 inhabitants, and in the case of the bacterium responsible for whooping cough, the threshold value is 400,000–450,000 individuals. Inspired by the **metapopulation** theory, Rohani *et al*. (1999) showed that the measles virus and whooping cough bacterium spread in the United Kingdom from certain urban centres such as London, Liverpool, and Birmingham toward peripheral rural areas. Several urban areas function as sources or reservoirs that allow these infectious diseases to persist and give rise to new epidemics (Fig. 8.7).

As shown by that work on the spread of the measles virus and whooping cough bacterium, it is critical to consider the regional scale (in this case England and Wales) in order to understand events at the local scale, i.e. a city or a town. Once again, the use of the macroscope allows a broader vision of the way infectious agents spread within a web of cities and towns. These ideas, developed from the concept of metapopulation ecology, are crucial in epidemiology and public health because they allow, once the special dynamics of a pathogen are understood, new means of control and vaccination. Indeed, wouldn't it be better to vaccinate populations in several 'source' areas when the case

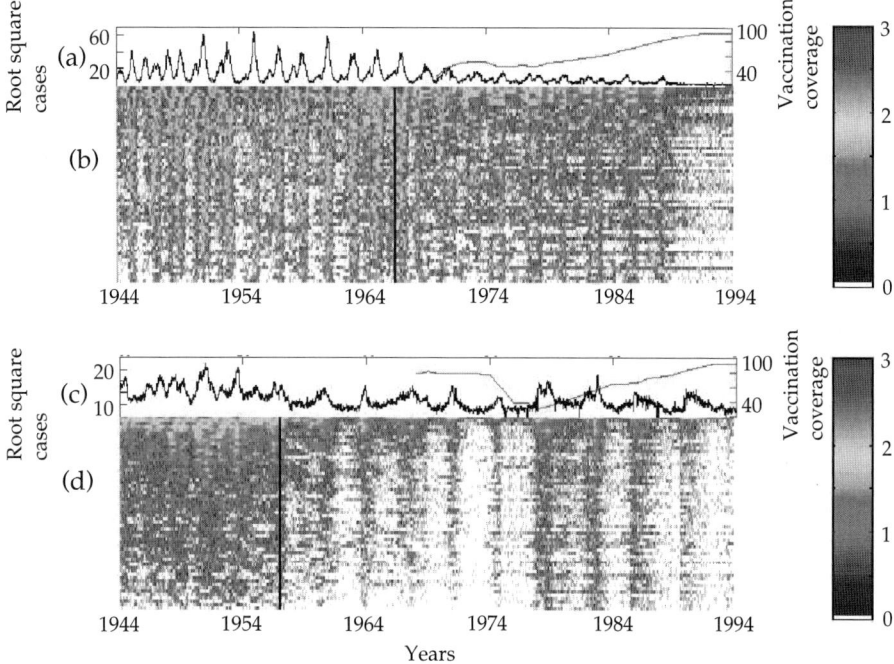

Figure 8.7 Parts (a) and (c) show the dynamics of the number of individuals infected by the pathogens that cause measles (a) and whooping cough (c) in England and Wales. The grey curve shows the evolution of the vaccination coverage (per cent) for these two diseases. Parts (b) and (d) show the spatial organisation of the number of cases from the largest city (London (on top) to the smallest town Teignmouth (on the bottom). The greyscale show the incidence rate of the disease. The spatial distribution of the cases in (b) and (d) show that both these early childhood diseases persist in time in the largest towns of England and Wales (red). This spatial organization is particularly clear in the cases of measles (b). Modified from Rohani *et al.* (1999) with the permission of the first author.

dynamics are at the inter-epidemic stage, i.e. when the number of cases is reduced? This is an important issue, and the question remains as to whether or not these ideas will be tested or adopted in the near future. We have analysed the dynamic behaviour of a pathogen at the scale of a country; the following section analyses what happens on a worldwide scale.

8.3.6 International travel and trade

Of all the species on the planet, humans probably have the most extraordinary abilities to adapt and survive. Modern technologies have facilitated access to and colonization of geographical areas that had remained isolated for a long time. Transcontinental flights and sea transport, and economic and business exchanges have contributed to this. This almost total colonization of the

planet by human populations has also been profitable to 'aliens', i.e. free riders, such as parasites and microbes (Spielman *et al.* 2004). The development of air transport now favours the transmission of pathogens between opposite sides of the globe (Fig. 8.8). This is an unprecedented situation in human history. Pathogens such as many viruses and bacteria that are transmitted in this way can readily spread and infect human populations that were previously free from infection. The recent avian flu epidemic in Southeast Asia should teach us many lessons; by generating worldwide panic, it showed that global action was needed to manage an issue with a global outcome (Li *et al.* 2004). Instead of choosing one or two particular diseases with which to finish this section and chapter, we will present new data provided by Smith *et al.* (2007) regarding the degree of homogenization of nearly 320 human pathogens. The issue here is to

Regular   Small world   Random

(a)   (b)   (c)

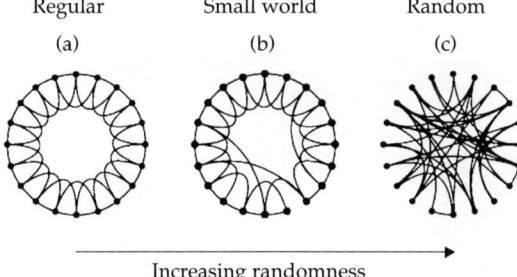

Increasing randomness

Figure 8.8 Diagram of the levels of homogenization of human populations. (a) In ancient times human communities were only in contact with their nearest neighbours. (b) When transcontinental exchanges were still limited several decades ago, local human populations were in contact with other neighbouring ones, and occasionally with other communities farther away (notion of a *small world*). (c) Each local community is in contact with the others, illustrating a globalized world. Infectious diseases could take advantage of such a strongly interconnected world to spread and develop in host populations. Modified from Watts and Strogatz (2004).

determine which human pathogens display global dynamics as opposed to more local mechanisms (Fig. 8.8).

Based upon a set of data from 317 infectious diseases affecting human populations, Smith *et al.* (2007) first sought to classify the different **aetiological agents** according to the type of reservoir hosts: (1) pathogens that are strictly specific to humans; (2) pathogens that use human and non-human reservoirs (called multi-reservoir pathogens); and (3) pathogens whose reservoir is an animal, i.e. a zoonosis, and that are occasionally transmitted to humans. Based on a similarity index, identical to those commonly used in community ecology, and measuring the degree of homogenization between geographical areas, the authors showed that contagious agents of the first category displayed the highest degree of similarity between regions, followed by the infectious multi-reservoir agents, and finally zoonoses (Fig. 8.9).

The homogenization of directly transmitted pathogens suggests that their host resource is global. The local and global diversities of this type of pathogen and their specific types are almost identical; a locality like Montpellier displays both the same number of pathogen species and the same species composition as another locality far away, such as Los Angeles, for example. Furthermore, the richness and diversity of these pathogens in Montpellier are not very different than what is observed at the global scale. The immigration/colonization/extinction processes are the ones that are in a large part responsible for the local diversities observed. Colonizations originate from a continental or even global colonization pool. On the contrary, aetiological agents having more complex cycles do not achieve this degree of homogenization. This is more obvious for zoonoses (Fig. 8.9). Two adjacent geographical areas can shelter of very different pathogen communities. In this chapter, we have already discussed the reasons for this phenomenon. It is the endemic nature of the host reservoir species, and very often the related pathogens that explain this situation. These pathogens display very marked local or regional spatial distributions, whereas directly transmitted pathogens have a global spread. The processes that explain their spatial distribution and those of their hosts are more related to habitat heterogeneity and whether it is favourable or not to their development. Since for directly transmitted pathogens the host resource is global and 'finite', we think that the reservoir pathogens are those that today represent the biggest danger to human populations if they are given the chance to spread (like the monkey pox example given earlier). The increasing rate of introduction of exotic species, as already pointed out above, greatly facilitates the arrival of these new 'aliens'. What should we do? How can we prevent these new invasions? Shouldn't we exert stricter control on the introduction of species by targeting animal groups that are known to be important disease reservoirs, such as rodents, birds, or ungulates for example?

International traffic and trade, like the tyre trade responsible for the spreading of disease vectors (Renaud *et al.* 2005) or the pet trade, are today especially alarming. They have played—and will probably still play in the future—an important role in the spread of pathogens in regions that were previously free from them. This issue in itself requires quick decisions on trade regulations and worldwide transportation.

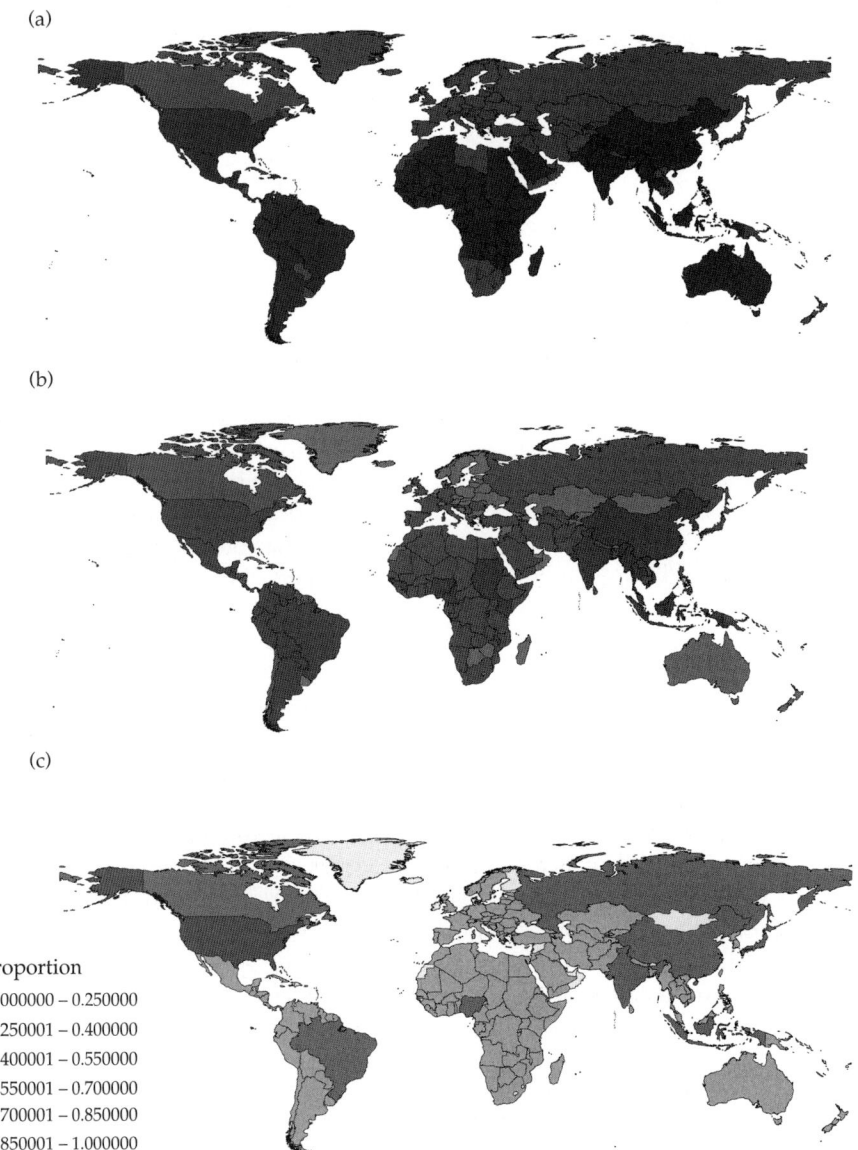

Figure 8.9 Degree of homogenization of human pathogens at the global scale, according to their life cycle: (a) for pathogens with direct transmission; (b) for pathogens that have both human and non-human reservoirs (multi-reservoirs); and (c) for pathogens with an animal reservoir. The classification according to the degree of homogenization is based on the Jaccard similarity index which varies between a low value (lightest) and a high value (darkest). From Smith *et al.* (2007).

8.4 Conclusion and proposals for new research perspectives

We have analysed and summarized in this chapter several issues concerning the relationships that exist between environmental changes and health problems, and the challenges arising from this new global context in terms of public health issues. This contribution is by no means an exhaustive review of the scientific literature on the subject. In fact we have deliberately selected just a few examples that give a good illustration of the scientific and health situation. Our selection was intended to give the reader a broader vision of the health issues, using a new tool we call the 'macroscope'. The same kind of reflection and approach should of course be followed in plant health and agronomy (due to lack of space our selection of the literature has omitted to include many situations of interest in these fields).

Many epidemiologists think that we have learnt enough about infectious diseases; this chapter seeks to convince even the most sceptical that we must improve our understanding of infectious diseases and change the way we think and work. The challenge for modern epidemiology is to change scales and adopt a global ecological perspective on health issues. Lawton (2000) has identified four fields in which he thinks that community ecology has not yet found a happy medium. These fields can easily be transposed to modern epidemiology: (1) too big an investment in field experiments or short-term laboratory research, often with a reductionist character; (ii) too much importance given to research in processes involved at a local scale rather than processes taking place at larger spatial or temporal scales; (3) lack of synthesis between molecular ecology, populations genetics, and population and community ecology to better understand infectious and parasitic diseases; and (4) a lack of statistical and mathematical models.

What can we do to change this research trend? We can summarize the possible actions and priorities as follows:

1. Develop a multidisciplinary approach. Develop research favouring interdisciplinarity and an exchange of views. Reconcile explanatory molecular–mechanistic, neo-Darwinian schemes and non-causal physical mechanisms with one another. Avoid setting disciplines against one another, but rather use their complementarity to express different points of view on the same phenomenon.
2. Carry out containment research in pilot zones. We must concentrate our research efforts on a given number of geographical areas, or observatories, and on studies of biological systems. Research on infectious diseases cannot do without such an approach, which will ultimately allow a comparative intersite analysis.

3. Develop long-term epidemiological study sites, particularly in intertropical regions where the impact of the environmental changes on health issues is apparent, and where the risk of infection is greatest.
4. A consequence of point (3) is the need to standardize the protocols in the different epidemiological disciplines. The inability to standardize protocols during recent decades has virtually excluded any attempts at meta-analysis and comparative analyses in epidemiology. Quite surprisingly too, the reason mentioned to justify this was the lack of standardization.
5. More coordinated research programmes at the national and international level. National and international scientific institutions must promote coordinated multidisciplinary research. For example, in the USA, the experience with the joint project of the National Science Foundation and the National Institutes of Health represents an important effort toward better collaboration between medical scientists and ecologists/evolutionists.
6. Promote population and community epidemiology. In epidemiology as in other disciplines, there is an actual trend to consider molecular biology as the only true science, namely because it takes the molecular-deterministic approach. The same trend can apply to field epidemiology with regard to theoretical epidemiology. We must encourage a diversity of approaches that will provide different points of view and information that is essential to the understanding and modelling of how a local outbreak may happen. We must also take interest at the population level, which is the fundamental basis of the approach in public health, and the community level, more in phase with the natural circulation of pathogens within ecosystems.

At the end of this chapter, we are convinced that by broadening the approaches of our studies, i.e. using the 'macroscope' as much as we do the microscope, we will be able to answer some of the questions about the issues concerning the emergence or re-emergence of pathogens. Our ability to react and to adapt our research largely depends on the eventual answers to these questions. Additional or complementary information on this topic is available in the United Nations Environment Programme Geo Year Book (2004/5) *Emerging challenges—new findings. emerging and re-emerging infectious diseases: links to environmental change.*

Important facts

• Each species of bacteria or virus in the environment has an important phenotypic and genetic diversity, but only some rare forms are pathogenic for humans (see Faruque *et al.* 2005a, b).
• Potentially infectious organisms are often present in the environment before the onset of an epidemic (see Morange 2005).
• The health of humans, animals, and ecosystems are closely related (see Lebel 2003).

Questions for discussion

• How do we explain the current outbreak of epidemics in populations?
• Does the determination of the virulence genes of an infectious organism help us understand its ability to spread?
• How can we make a compromise between the traditional vision of cholera epidemiology and a more integrative approach considering the aquatic environment as a natural reservoir for bacteria?
• Why is it important to switch spatial and temporal scales in epidemiology and infectious disease ecology?

• Why is it important to adopt an approach based on community ecology when studying the emergence of new pathogens?

Further reading

• Further information on the effects of environmental global changes on the health and well-being of humans can be on the World Health Organization website at http://www.who.int/globalchange/en/
• Information on the policy of the Canadian International Development Research Centre regarding health ecology can be found at http://www.idrc.ca/ecohealth.
• Chivian, E. (ed.) (2002). *Biodiversity: its importance to human health.* Harvard Medical School, Center for Health and the Global Environment, Cambridge, MA.
• An introduction to health ecology: Guégan J.F., Renaud F. (2004). Vers une écologie de la santé. In: *Biodiversité et changements globaux* (ed. R. Barbault and B. Chevassus-au-Louis), pp. 100–135. ADPF/MAE Editions, Paris (in French).
• Collinge S.H., Ray C. (2006). *Disease ecology.* Oxford University Press, Oxford.
• Patz J., Confalonieri, U. (coordinators) *et al.* (2005). Human health: ecosystem regulation of infectious diseases. In: *Ecosystems and human well-being: current state and trends: findings of the Condition and Trends Working Group,* Millennium Ecosystem Assessment Series, pp. 393–415. Island Press, Washington, DC.

CHAPTER 9

Parasitism, biodiversity, and conservation biology

Camille Lebarbenchon, Robert Poulin, and Frédéric Thomas

9.1 Introduction

Biological diversity—or biodiversity—can be defined as the variety and variability of all living organisms. It includes the genetic variability within species and their populations, the variability of species and of their life-forms, the diversity of related species complexes and of their interactions, and that of the ecological processes that are influenced by the organisms or of which organisms are the actors. Besides **systematics** and **phylogeny** that describe and classify the diversity of living organisms, population and community ecology are two of the main subjects for ecologists who study the dynamics of biodiversity.

Many factors influence the structure and diversity of species assemblages (Barbault 1995; Blondel 1995). Among these, the physical properties of the environment undeniably play a major role: generally speaking, the richness and diversity of species are positively correlated with the structural complexity of the **biotopes** (Barbault 1981). However, although the physical structure of the environment has a crucial influence on the organization of populations, it also is the 'setting' in which other important phenomena take place, such as disturbances (fires, storms, droughts, etc.), extinction/colonization processes, and interactions between species (competition, predation, parasitism, and mutualism). Compared with the interactions involving free-living organisms (i.e. competition, predation) those concerning parasites have long been neglected by ecologists. In this chapter, we present the main mechanisms through which parasites affect species assemblages, ecosystem

stability, and intraspecific variability. We also discuss the implications of this knowledge for conservation biology.

9.2 Parasitism and apparent competition

9.2.1 Overall mechanism

Interspecific competition occurs when several organisms make use of common resources that are present in limited quantities. In this case, the rate of growth of one or several species is reduced because one common limited resource is used by one or several other species. When common resources are not limited, the competing organisms can still harm one another. In this case, the type of competition is called interference, and one species actively excludes the other. Ecologists have known for a long time that the interactions between two populations or species can be influenced by a third species. This phenomenon, which can be called 'arbitration', is well known with predators. When one or several species compete for the same resource, it often occurs that these species coexist through the action of predators. Indeed, predators often attack the most abundant prey, therefore limiting the demography of the most competitive species. This phenomenon prevents the extinction of the least competitive species. The result is thus apparent competition between the prey species concerned. In the first part of this section we will see that parasites, just like predators, can play the role of arbitrator among competing species.

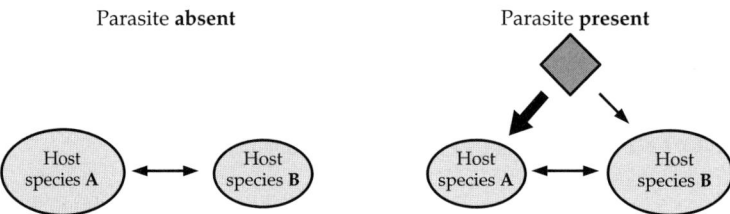

Figure 9.1 Parasitism and apparent competition. When the parasite is absent from the ecosystem, species A is more competitive than species B. However, given species A's greater susceptibility to infection than species B, the presence of the parasite changes the outcome of the competition. The size of the ellipse represents the abundance of each species in the community and the thickness of the arrows represents the intensity of the interactions between the different host species.

Inequality in the face of parasitism is present at all levels of biological organization: between individuals of the same population, between sexes, populations, species, etc. These differences can be expressed through the infection parameters (**prevalence, abundance, intensity**) and/or through the related pathological disorders. When two species, A and B, compete for the same ecosystem resources and one parasite mainly affects the **fitness** of species A, this species is disadvantaged with regard to species B (Fig. 9.1). When species B resists infection, differential regulation by the parasite gives B an advantage with regard to species A in the exploitation of resources. When species B tolerates the parasite with less damage than species A, the competition can then be considered as interference competition: species B affects species A either directly by transmitting the disease (contagious parasite) or indirectly by favouring the parasitic cycle via which the infecting forms will sooner or later come into contact with individuals of species A. Depending on the original dominance relationship (without the parasite) between the competing species and on the magnitude of the variance in susceptibility to the parasite (inequality when facing the parasitosis), the following situations can occur: exclusion of the dominant or dominated species or coexistence of both taxa.

The direct or underlying effects of this type of interaction can significantly affect the community structure, the viability of the species within a community, or even the capacity to be colonized by non-native species (Combes 1995). The effects on biodiversity can also be positive or negative. For instance, a parasite infecting different host species in a frequency-dependent fashion, or a parasite that

will preferably affect a host that is competitively superior to other species, will allow the diversity of host species to be maintained. In these situations, the parasite concerned should be considered as a 'keystone' species that maintains the specific diversity of the ecosystem by its presence and its effects. One could also consider the opposite case, i.e. the parasite can be unfavourable to the conservation of biodiversity if it preferably attacks the least competitive species, and thus, in the long run, entails the local extinction of the host species (Combes 1996). Phylogenetically related species, which are close from an evolutionary point of view, are generally more likely to be infected by the same parasite species. Therefore, the influence of the parasite on the outcome of the competition between two hosts is stronger when they are phylogenetically close (Freeland 1983; Holt and Pickering 1985).

9.2.2 Examples

There are many examples of parasitic arbitration, but only few cases will be given here. Park (1948) was one of the first to experimentally show that parasites can modify the outcome of competition between two host species. He worked on the interactions between two flour beetle species (*Tribolium castaneum* and *Tribolium confusum*). When both these species are held together in captivity, most often *T. confusum* becomes extinct, suggesting that *T. castaneum* has a competitive advantage. However, when a parasite is added to the same environment (the sporozoan *Adelina tribolii*), this trend is reversed and *T. confusum* seems to be the best competitor. *T. castaneum* is in fact more vulnerable to the parasite than *T. confusum*, and this on its own leads to the reversal of the outcome of competition between the two beetles. In the same vein, the work of Park was resumed when Yan *et al.* (1998) showed that this scenario can be changed if *A. tribolii* is replaced by another parasite, the cestode *Hymenolepis diminuta*. Indeed, *T. confusum* is the most sensitive host to the parasite in this case. When the parasite and both species of beetles are kept together, *H. diminuta* allows *T. castaneum* to be even more competitive by quickly pushing *T. confusum* to extinction. The outcome of the competition between these two beetles is thus totally influenced by the parasite species that is present.

The mechanisms that operate in the apparent competition are not limited to differential mortality phenomena; they can also involve the fecundity of the hosts. The nematode *Howardula aoronymphium* causes a 100% reduction in the fecundity of the vinegar fly, *Drosophila putrida*, and reduces that of *Drosophila falleni* by half (Jaenike 1992). As before, the outcome of the competition is reversed in the mixed cultures where the parasite is present (Jaenike 1995).

In many host–parasite associations, the pathology related to the infection leads to an overall weakening of infected individuals. Given that predators partially prey on sick individuals, a differential susceptibility between the two host species can thus lead to a preferred predation on the more susceptible species. The work of Anderson (1972) has shown that the nematode *Protostrongylus tenuis* causes serious neurological disorders in many species of ungulates, except the Virginia deer. This difference in susceptibility has played an important role in the spatial distribution and abundance of the different species. The selective disadvantage of the susceptible species has been attributed on the one hand to the direct consequences of the pathology and on the other hand to higher predation of these populations.

In certain situations, the predation of the host is a crucial step for the completion of the parasite's life cycle. There are indeed many parasites that have a parasitic cycle that is complex or heteroxenic (with several successive hosts), with at least one trophic **transmission** step, i.e. the parasite is transmitted from one host to another though a predation event. Since the early 1970s, it has been known that many of these parasites are able to manipulate their

hosts' behaviour in order to increase the risk of predation by targeted predators who will become the final hosts. Behavioural changes are part of the many mechanisms referred to as 'favourization' (see Combes 1991) selected during the evolution of parasites and that help to increase the probability of transmission in parasitic cycles (see also Chapter 4). These parasite manipulations can cause deep inequalities between related species with respect to predation risk. The mediation role of parasites can be detected in the field using indirect methods (detailed in Chapter 6) based on the distribution of parasitic loads in the host population (Anderson and Gordon 1982; Rousset *et al.* 1996).

Parasitoids (Chapter 7) can also be involved in the apparent competition process. Boulétreau *et al.* (1991) have shown that the coexistence of *Drosophila melanogaster* and *Drosophila simulans* in captivity is only possible and stable when the parasitoid *Leptopilina boulardi* (Cynipidae) is present. The preference of the parasitoid for *D. melanogaster* could explain this phenomenon.

In natural ecosystems the effects of parasitic arbitration on biodiversity are more important when the species most affected by the parasite is strongly competitive. The work of Ayling (1981) in the **intertidal zone** illustrates this phenomenon. The intense development of some sponges locally leads to the total extinction of other invertebrate species that inhabit this biotope. However, in some localities, there are areas where the sponges died because of infection by a specific fungus. A suite of invertebrate species become established within these gaps.

Even if we only list a few examples here, there are today many studies that show or suggest that parasites play significant arbitration roles within ecosystems. For many ecologists, the conclusion is obvious: parasitic arbitration is a significant interaction within ecosystems, as important as direct competition or predation.

9.3 Parasitism, ecosystem stability, and cascade effects

When a parasite is responsible for the local extinction of one or several species (acting upon the survival and/or the fecundity of its hosts), not only will it also affect the community of competitors but also that of prey and predators. For instance, if the species infected by the parasite is a predator, there will sometimes be an explosion of the prey populations that are usually regulated by this predator. Such population explosions represent important ecosystem destabilization hazards because of cascade effects, often resulting in negative effects on biodiversity. A spectacular example of this phenomenon is given by the sarcoptic mange (skin infection caused by the mite *Sarcoptes scabiei*) **epizooty** that decimated fox populations in Scandinavia. The decline of these predator populations strongly disturbed the populations of rodents these foxes used to prey upon. Another example involves the introduction of the myxomatosis virus in Great Britain that was responsible for the rapid decline of rabbit populations, which in turn caused important changes in the vegetation and in the invertebrate and vertebrate assemblages of affected areas. The onset of elm blight in Great Britain was just as spectacular. This disease affecting elms is caused by a fungus (*Ophiostoma ulmi*) transmitted by bark beetles. Because the trees died one after the other, habitat availability strongly declined for many bird species. On the other hand, the abundance of dead trees allowed the emergence of large numbers of beetle larvae that were preyed upon by other bird species.

Besides the introduction of parasites, their extinction (where they are normally present in the ecosystem) can also cause major disruptions within the ecosystem. One of the most famous examples of this phenomenon is the rinderpest outbreak in East Africa (Plowright 1982; Dobson 1995a,b). This highly infectious viral disease (caused by a virus of the genus *Morbillivirus*) can kill whole populations of bovines, buffalos, and other wild species (wildebeest, warthogs, etc.). Substantial efforts were made to eradicate this virus. In regions that depend on cattle for meat, milk products, and draft power, rinderpest caused widespread famine and inflicted serious economic and political damage even though the virus did not affect humans directly. In the 1950s, a relatively efficient vaccine was developed. Even if only cattle were targeted for vaccination, the outcome was significant since the disease was eradicated from wild species (which means that cattle were the main **reservoir**).

In the Serengeti Park, the gnu population grew from 200,000 to 1.5 million, and buffalo numbers increased by four- or five-fold and their area of distribution expanded tremendously. This increase in the ungulate population was accompanied by a steady growth in carnivore populations, especially lions and hyenas. At the same time, the increase in predation pressure harmed antelope populations. The changes in the herbivorous communities were accompanied by changes in the vegetation. Populations of wild dogs became extinct, obviously through competition with hyenas. In light of the complete destabilization caused by the eradication of the virus it is clear that it was a key species within the ecosystem. Its effects were major even if its total biomass was negligible.

9.4 Parasites and ecosystem engineers

9.4.1 Ecosystem engineers

The term 'ecosystem engineer' was introduced by Jones *et al.* (1994) to define organisms—plants and animals—that directly or indirectly modulate resource availability for other species, and cause physical changes in the biotic or abiotic environment (Jones *et al.* 1994, 1997). Some species can change the environment by their own physical structure: they are called autogenous engineers. For example, during their growth, trees naturally create cavities that provide habitats for other organisms such as insects or birds. Other species can modify living or non-living matter from state A to state B (allogenous engineers), state B being a resource for other organisms. For example, many woodpecker species transform trees from state A (with no holes) to state B (with holes), and state B represents a habitat for many birds, bats, insects, etc. These two types of engineers modify, maintain, or create habitats within ecosystems (Jones *et al.* 1994). In contrast with keystone species, engineering species do not have an obviously positive effect on biodiversity. In other words, the passage from state A to state B is not obviously favourable for the maintenance of biodiversity (see Jones *et al.* 1994 for more details). Whatever the case may be, the (auto- or allogenous) 'engineer' label always applies to the phenotypic traits

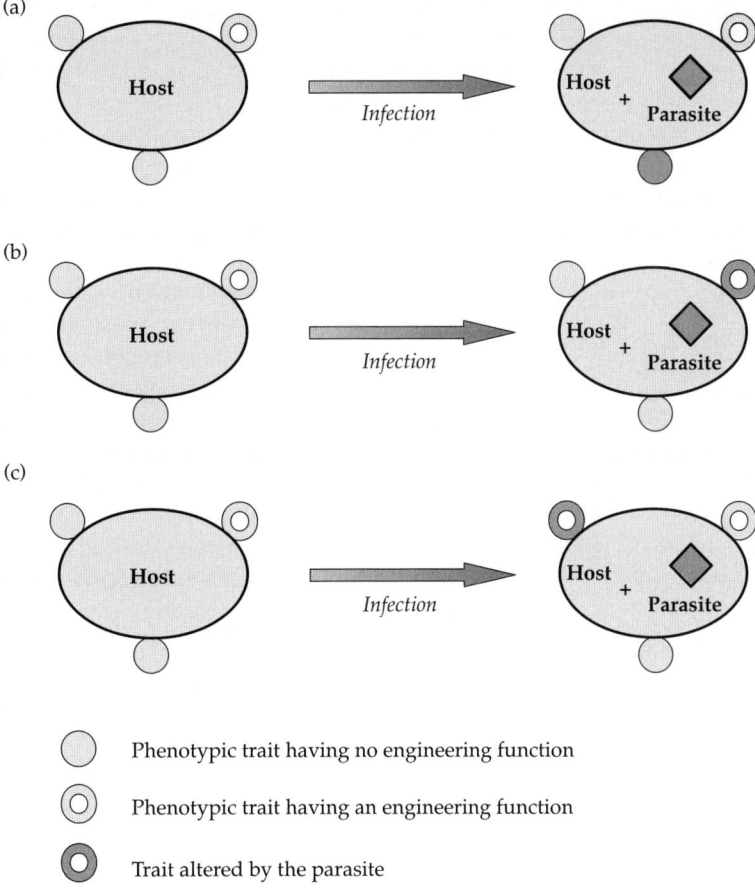

Figure 9.2 Potential interactions between the host's phenotypic traits, those modified by the presence of the parasite, and those involved in the host's engineering role (Thomas *et al.* 1999a). (a) The infection of an individual with an engineering role has no effect on the community if the modified trait is not directly involved in or closely related to this role. (b) The parasite modifies a host trait directly involved in the engineering role. (c) By modifying a trait that has no engineering role, the parasite creates a novel habitat and thus attributes to the trait an engineering role that did not exist before.

of organisms (size, behaviour, morphology, etc.). Inasmuch as parasites can modify the phenotypic traits of their hosts, one can easily understand how they can simply interfere with the engineering processes. Through these interactions, parasites can directly affect the engineering functions of their hosts, or act as engineers themselves (Thomas *et al.* 1999a). Figure 9.2 highlights, from a theoretical point of view, the different potential interactions between the phenotypic traits of a host, those that the parasites change, and those that are eventually involved within engineering processes.

9.4.2 Infection of an ecosystem engineer

The infection of an organism that has an engineering function within the ecosystem does not always have consequences for the rest of the community. The trait modified by the parasite is not always linked to an engineering function, as shown in Fig. 9.2(a). In other cases, the parasite can change one or several traits of the host that are directly involved in an engineering function (Fig. 9.2b). In these cases, the parasites will indirectly affect the community. For example, many species have the role of an autogenous engineer because they provide, via their development, space that can be used by other species (trees, shells of marine molluscs, etc.). Any parasite that changes the properties of this habitat, for instance by affecting the growth of the infected host (stunting or gigantism) will also have indirect impacts on the **epibiont** community. Within the same context, many infections normally entail a decrease in host activity. By reducing the activity of its host, a parasite can affect its engineering function (for example, beavers or woodpeckers; Jones *et al.* 1997). By changing the appetite of their hosts, the gastrointestinal parasites of herbivores can have an indirect impact on plant communities. By changing the behavioural phenotype of its hosts from one state to another, a parasite can create a new resource or a new habitat for other species. The association between the cockle *Austrovenus stutchburyi*, the trematode parasite *Curtuteria australis*, and several invertebrates using the shell of the cockle as a substrate illustrates this type of interaction (Thomas *et al.* 1998; Mouritsen and Poulin 2005).

9.4.3 Parasites as engineers

By changing a particular trait of the host, a parasite can create a novel habitat, that is, it may generate within the host an engineering function that was not there before (Fig. 9.2c). This occurs for example in the association between the crab *Carcinus maenas* and the crustacean parasite *Sacculina carcini*, in which the parasite interrupts its host's moult. Unlike infection-free crabs that moult regularly, the infected crab's carapace now becomes a stable substrate allowing the establishment of many organisms.

In nature, it seldom happens that a free-living organism is only infected by a single species of parasite. Within this context, any parasite that changes the properties (physiological, behavioural, etc.) of its host also modifies the ecological properties of the other parasites' habitat. For example, one of the major constraints acting upon parasites when they try to establish themselves within a host is the latter's defence reaction (the immune system of vertebrates). Parasites that cause immunosuppression control the availability of resources for a large number of other parasite species: the tragic

example of human immunodeficiency virus (HIV) reminds that the implications of the potential establishment of pathogen communities are serious.

As previously mentioned, and as discussed in Chapter 4, many parasites can manipulate their host's behaviour to facilitate their transmission to the predator that serves as their final host. For the group of parasites that share the same host, this phenomenon is far from trivial. The infection of a host by a 'manipulative' parasite can be considered as an engineering act that makes the hosts pass from being in a state of 'low predation risk' (state A) to a 'high predation risk' (state B). Depending on the shared or conflicting interests of parasites, different parasite strategies have evolved via **natural selection**. When a non-manipulative parasite completes its cycle within the same final host as the manipulative parasite, it is more advantageous to infect hosts with phenotype B (already manipulated) than those of phenotype A (with normal behaviour). Indeed, by targeting hosts with phenotype B, the non-manipulative parasite can take advantage of the transmission mechanism implemented by the manipulative parasite, without having to pay its energetic cost. This 'pirate' transmission strategy has been called 'hitch-hiking' (Thomas *et al.* 1997). It can be found for example within trematode communities that infect gammarid amphipods in the brackish waters of southern France (Fig. 9.3). The trematode *Microphallus papillorobustus* is a manipulative parasite since it brings the infected gammarid (second intermediate host, the first one being a mollusc of the genus *Hydrobia*) towards the surface, where the risk of predation by aquatic birds is higher. The trematode *Maritrema subdolum* displays the same parasitic cycle, using the same hosts in the same sequence, but is unable to manipulate the gammarid's behaviour. The work of Thomas *et al.* (1997) shows that the invading larvae of *Maritrema subdolum* nevertheless have a behavioural trait that allows them to preferably target the 'crazy' gammarids: by swimming high in the water column, they are much more likely to encounter photophilic (manipulated) than photophobic (healthy) gammarids. Within these same gammarids, one can find another species—the nematode *Gammarinema gammari*. This species lives within gammarids for its whole life, since it

(a)

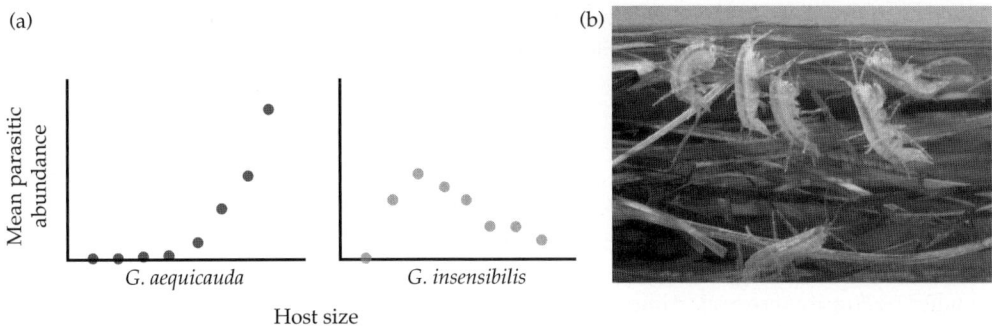

(b)

Mean parasitic abundance

G. aequicauda

G. insensibilis

Host size

Figure 9.3 (a) Pattern of infection of *Microphallus papillorobustus* in relation to host size in two closely related host species (*Gammarus aequicauda* and *Gammarus insensibilis*) (modified from Thomas *et al.* 1995). Body size is a good correlate of age in these two host species. In *G. aequicauda*, the parasite has no significant effect on survival compared with *G. insensibilis* as illustrated by the continuous acquisition of parasites through time. As a result, the parasite abundance increases steadily with age in *G. aequicauda*. Conversely, in *G. insensibilis*, a maximum parasite abundance is reached in hosts of intermediate age, a pattern predicted by theoretical models (Anderson and Gordon 1982; Rousset *et al.* 1996) in the case of parasite-induced host mortality. (b) The crustacean amphipods *Gammarus insensibilis* manipulated by the trematode *Microphallus papillorobustus* (photo P. Goetgheluck); infected gammarids display positive phototaxis, negative geotaxis, and an aberrant evasive behaviour. Indeed, instead of hiding under stones, they swim towards the surface and are preferentially eaten by aquatic birds (definitive hosts) (Helluy 1983).

parasitizes the gammarid only and is not transmitted to birds. Therefore, there is a conflict of interest between this nematode and the manipulative trematode. The study of Thomas *et al.* (2002) suggests that the nematode avoids, and even 'tampers' with the 'crazy' gammarids: the manipulated gammarids hosting nematodes display less pronounced behavioural changes than those that are nematode free. These findings clearly show that parasites that are able to modify the behaviour of their hosts affect the structure of the parasite communities that feed on the same host. The responses observed are clearly of an evolutionary nature.

Under natural conditions, interactions between species are numerous and the phenomena can overlap. A study by Mouritsen and Poulin (2002) gives the example of a parasite that, by eliminating a species with engineering roles from the ecosystem, causes dramatic changes in the invertebrate community of the intertidal zone. After unusually heavy infection by a trematode parasite, the amphipod host *Corophium volutator* completely disappeared from Denmark's mudflats for several months. This crustacean plays a major allogenous engineering role in the ecosystem because the burrows it makes in the sediment tend to stabilize the substrate. Its local extinction led to the quick

erosion of the sediment. Major cascade effects followed, such as changes in the sedimentary composition of the mudflats with direct consequences for invertebrate diversity and, on a larger scale, on the species feeding on the crustacean.

9.5 Life-history traits and coexistence

Besides the properties of the ecosystem itself (productivity, complexity, stability, etc.), many traits related to the organisms themselves (body size, fecundity, reproductive period, etc.) are key variables for the potential coexistence of species. Depending on their life-history traits, some species are more likely to coexist than others. This can be simply explained by the fact that differences in life-history traits favour specialization. For instance, morphological differences (such as body size) in phylogenetically close species often entail the use of different resources which in turn favours the coexistence of these species. Similarly, a staggering of the reproduction periods of **sympatric** species (temporal segregation) can decrease the possibility and/or intensity of competition.

As shown in Chapter 2, parasites can influence the life-history traits of their hosts, such as survival, reproductive effort (Richner and Tripet 1999), dispersal (Sorci *et al.* 1994; Heeb *et al.* 1999), or growth (Agnew *et al.* 1999), in many ways. In some cases, they can even affect the life-history traits of their host's progeny (Sorci *et al.* 1994). By their effects on the life-history traits of their hosts, parasites can thus influence whole species assemblages. When the traits concerned are fecundity or survival, we are back to the concept of parasitic arbitration mentioned at the beginning of the chapter. On the other hand, the processes are different when other traits are concerned. Although there is no empirical evidence to explain these phenomena, some scenarios are theoretically possible. Insofar as parasites can influence several parameters of the reproductive biology of their hosts (e.g. the reproductive age and/or period), they can theoretically affect the degree of overlap between reproductive periods and subsequent competition within host communities (Fig. 9.4).

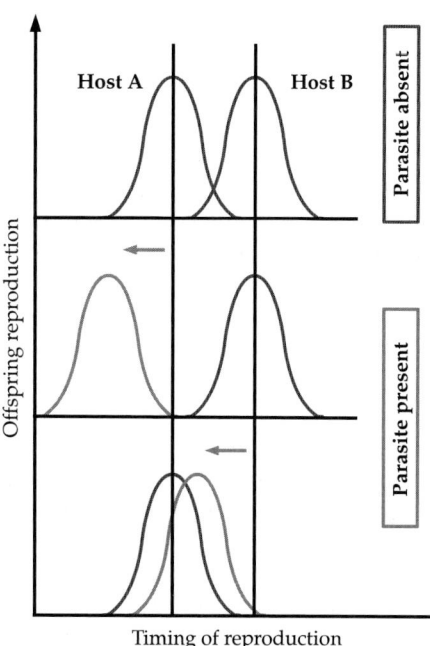

Figure 9.4 Effects of a parasite on the reproductive period of two hosts. The two species A and B can coexist due to different reproductive periods that do not overlap in time. A parasite that moves the reproductive period of species A forward will favour the coexistence of both species. On the contrary, if a parasite affects species B, also by moving forward its reproductive period, the coexistence of both species is jeopardized.

9.6 Invasive species

Human activity is directly or indirectly responsible for the frequent exchange of living organisms between one continent and another. This is slowly leading to global homogenization of the biosphere. Each introduced species, if it propagates successfully, inserts itself within a well-established network of biotic interactions and can potentially unbalance the ecosystem after a series of cascade processes. For example, if the invasive species is a predator, it can decimate local prey species that have not co-evolved with it and thus do not have specific adaptations to protect themselves against the new enemy. Parasitism can also play a very important role in the impact of invasive species on the new ecosystem: the invasive species can either transport parasites that can infect the local species, or can serve as a reservoir for one or several pathogen species already present in the area it is introduced to (Daszak *et al.* 2000; Cleaveland *et al.* 2002; Tompkins and Poulin 2006).

Generally speaking, invasive species show less parasitic infection in their new environment than in their native area (Torchin *et al.* 2003). The introduction of a species in a new area, by accident or not, often involves only a limited number of individuals. As we saw at the beginning of this chapter, individuals of the same population are not equal in the face of parasitism. It is thus not surprising to see that the founding individuals of an invasive population can lack or carry only very few parasite species. In this case, the invasive species are then initially 'free' from their parasites and other traditional enemies. This phenomenon explains the great success of several invasive species in their new environment. However, when free-living invasive species are accompanied by their parasites, these can be transmitted to new host species. The consequences are often quite tragic. For example, it is now becoming clear that the gradual extinction of the red squirrel, *Sciurus vulgaris*, in Great Britain can be largely attributed to a viral disease (parapoxvirus) introduced by the grey squirrel, *Sciurus carolinensis*, at the beginning of the 20th century (Tompkins *et al.* 2002, 2003). The pathogen has no visible effects on grey squirrels, whereas it strongly decreases the fitness of red squirrels. This is a typical case of parasitic arbitration: the differential regulation by the virus gives the grey squirrel an advantage in its competition for resources with the red squirrel. A true 'Trojan horse', the grey squirrel has quickly invaded all of Great Britain and progressively eliminated its competitor using the viral weapon it transported.

Another typical example is given by the work of van Riper *et al.* (1986, 2002) on the extinction of **endemic** bird species in Hawaii. Following James Cook's discovery of the islands in 1778, the geographical isolation of the Hawaiian Islands was no longer an obstacle to colonization by bird species from Europe and elsewhere. These invasive birds have introduced avian malaria (caused by the protozoan *Plasmodium relictum*) and fowlpox (*Poxvirus avium*). Due to its isolation, the Hawaiian bird fauna, rich in unique species, had never been exposed to these two parasites and lacked defences against these diseases. The turning point was the accidental introduction in 1827 of a potential vector for these two pathogens—the mosquito *Culex quinquefasciatus*. From then on, transmission between invasive and native species was possible, and the result was terrible: more than half the native and unique Hawaiian species became extinct. The endemic avian species that survived extinction now live in the mountains where the mosquito vector cannot survive at altitudes higher than 600 m. The same scenario is actually taking place with the West Nile virus imported into North America in 1999, carried by invasive bird species that come from the Middle East. Only 3 years after its introduction, the virus has spread over most of Canada and the United States, infecting more than 70 species of birds (Enserink 2002).

In all these examples, what determines whether a parasite introduced with its host in a new area can perturb the ecosystem or not is its ability to infect new host species, even if they are not phylogenetically close to the original host. The copepod parasite *Lernaea cyprinacea* and the cestode *Bothriocephalus acheilognathi*, both introduced all over the planet with the carp *Cyprinus carpio*—their original host—now infect many other freshwater fish species. More than 100 host species, belonging to 16 different orders, have been recorded for *Lernaea cyprinacea* (Poulin and Morand 2004).

Therefore, generalist parasites such as these, that are able to 'jump' from one host species to another following their arrival in a new environment, are the biggest concern for biodiversity conservation.

Even when it comes without parasites, an invasive species can become a host for one or several pathogens that are already present in the environment in which it is introduced. Here again, these tend to be generalist parasites that are capable of infecting new arrivals. For example, the toad *Bufo marinus*, introduced to Australia in 1935, now hosts 20 or so metazoan parasite species that are not present in its native distribution area in America (Barton 1997). All are generalist parasites that *B. marinus* 'shares' with amphibians native to Australia. The same has happened with several species of salmonid fishes that have been subjected to many transcontinental movements: particularly, the brown trout, *Salmo trutta*, native to Europe, and the rainbow trout, *Oncorhynchus mykiss*, native to the west coast of North America, that are now present on all continents except Africa and Antarctica. Having been introduced as fertilized eggs, the original populations of these fish lacked parasites in the environment into which they were released. Today, all these invasive populations host metazoan parasite faunas that are taxonomically as rich as those in their native area, although the species are themselves different (Kennedy and Bush 1994; Poulin and Mouillot 2003).

The addition of an invasive species to the range of hosts that a parasite uses does not, however, systematically entail important changes in the mechanics of the ecosystem. For this, it would be necessary for the new host to change the dynamics of the parasite population, for instance by generating an epidemic explosion. In such a case, the invasive species must serve as a reservoir for the pathogen or be the 'best' host where the pathogen can proliferate at an abnormally high rate. This phenomenon is clearly illustrated by the work of Rauque *et al.* (2003) on the acanthocephalan *Acanthocephalus tumescens*. This worm generally infects the guts of several species of freshwater fish in Argentina. Since the introduction of salmonids to that country, it seems that not only can these new hosts be infected by the parasite, but they can also allow the worms to reach larger sizes and achieve higher fecundity than those found in the original hosts. Under the current conditions, the total number of parasite eggs discharged with the fish faeces is greatly increased by the use of salmonids as alternative hosts, and the per capita infection rates are increased in the native species. The invasive species thus allow the production of additional parasites, at the expense of local species: another case of apparent competition!

Settle and Wilson (1990) reported an original example of amplification of a local parasite through an invasive species. In normal conditions, the two homopteran species *Erythroneura variabilis* and *Erythroneura elegantula* are **allopatric**. Both leafhopper species are infected by the parasitoid hymenopteran *Anagrus epos*. However, *E. variabilis* is less affected because it lays its eggs deeper into the leaf parenchyma than *E. elegantula*. When *E. variabilis* was introduced to the habitat of *E. elegantula*, the difference in susceptibility gave an advantage to the incoming *E. variabilis* and its population increased at once. Even if *E. variabilis* is less affected by *A. epos*, the population increase of *E. variabilis* positively affected that of the parasitoid. This phenomenon, combined with a greater vulnerability of the resident homopteran, secondarily contributed to the extinction of the latter's population.

We only mention here a few examples among the many studies that suggest, or even prove, the interaction between biological invasions and parasitism. The message is clear: geographical displacements of free-living organisms often cause major disruptions in the targeted ecosystems, and parasites are important mediators for these changes.

9.7 Parasitism and intraspecific diversity

Biodiversity not only includes the diversity of species but also the diversity of genes within a species. The study of the mechanisms that generate, maintain, or, alternatively, decrease intraspecific diversity are very important in many fields, including agriculture, pharmacology, and conservation. Beyond radical events, such as outbreaks that lead to local extinctions or at least to strong **bottleneck** effects, parasites can influence this intraspecific diversity in several ways. For

instance, Buckling and Rainey (2002) experimentally investigated the evolutionary diversification of bacterial populations with and without **bacteriophages** in their environment. Their results were quite spectacular: in the absence of bacteriophages, a strong sympatric (intra-replicate) diversification of bacteria occurred, whereas allopatric (inter-replicate) differentiation was limited. The opposite result occurred in the presence of bacteriophages. What can explain these findings? In the absence of bacteriophages, the strong growth of the bacterial population gives rise to intense intraspecific competition that favours the local diversification of bacteria in spatially structured microhabitats. Since the same phenomenon occurs in all the replicates, the diversification observed is comparable from one replicate to the other. In the presence of bacteriophages, the bacterial populations are demographically controlled and remain under the threshold at which competition has noticeable effects. Sympatric diversification is therefore weak. On the other hand, diversification increases with time because of the selection for resistance against bacteriophages within each replicate.

Gene dispersal (via individuals that carry these genes) is one of the major factors that determines the degree of genetic structuring of wild populations. It has been proved that many parasites cause a drop in the activity levels of infected hosts. This phenomenon alone can modify the dispersal capacity of individuals (Thomas *et al.* 1999b) and influence the levels of genetic structure between populations. In other cases, natural selection has selected in the hosts a tendency toward greater dispersal when the risk of parasitism is locally high (Loye and Caroll 1991; Sorci *et al.* 1994). In this case, one would expect the opposite result, i.e. greater mixing between populations and thus some genetic homogenization. Finally, in other cases, the mean dispersal level of the hosts could be a trait directly manipulated by parasites as part of their own dispersal strategy (Lion *et al.* 2006). The consequences of these changes in dispersal in the hosts on the genetic structure of populations still need to be empirically quantified.

The genetic **polymorphism** of populations can also be directly affected by parasitic pressures. The human leucocyte antigen (HLA) system (also called the major histocompatibility complex or MHC) is a gene complex that plays a crucial role in the discrimination between 'self' and 'non-self' and the presentation of non-self components to the cells of the immune system. Surprisingly, the genes of the HLA system are among the most polymorphic genes known in humans. For the gene *HLA-B* alone, there are nearly 400 known alleles in the global population. Why are these genes so polymorphic? It is now certain that these genes undergo strong balanced selective pressures, i.e. a type of selection that favours the preservation of polymorphism within populations. Several ideas have been put forward to explain this selection, from avoiding cross-breeding between related individuals (the HLA system works as a relatedness marker for individuals) to the selection imposed by the pathogen diversity found in the populations. This latter type of selection is expected if each allele is maintained within the population because of its ability to provide protection against one or several particular pathogens. Several studies seem to favour this idea of balanced selection imposed by pathogens, particularly that of Prugnolle *et al.* (2005). These authors have indeed analysed the genetic diversity of HLA system genes in 61 human populations living in very different parasitic environments. After correcting for the portion of the genetic diversity linked to human population demographics (and not to selection), they noticed that the residual diversity of the HLA genes, only explained by selection, was positively correlated with the pathogen diversity to which the human populations were exposed. Populations living in areas where the parasite diversity is higher are thus submitted to a stronger balanced selection pressure from the pathogens and actually display the highest polymorphism in the HLA gene system. This study clearly shows the way in which pathogens (and their diversity) can shape the diversity of certain human genes.

9.8 Parasitism and conservation

As we have just seen, parasites play a fundamental role in the mechanics of ecosystems. As they can strongly affect biodiversity, it is obvious that they are of importance for conservation biology (Poulin 1999; Torchin *et al.* 2002; Lafferty and Kuris 2005).

Despite advances in our knowledge of the role of parasites within ecosystems, their integration in natural environmental management and conservation programmes is still limited.

Artificial supplementation of food for a wild population can help overcome a lack of food in the ecosystem and favour an increase in its rate of reproduction. However, this measure can lead to an increase in the transmission of viruses, bacteria, and other parasites related to the food and/or transmitted by contagion. Wright and Gompper (2005) showed that, in racoons, food input favours the concentration of individuals around the resource, a direct consequence being an increase in the rate of infection by parasites (nematodes) directly transmitted from one individual to another.

Animal translocation is a frequently used measure in conservation biology. It allows in particular the protection of individuals of an endangered species by placing them in a favourable habitat. Today, it is clear that this practice can fail and paradoxically amplify the rate of extinction of the protected species if parasites are not taken into account. Apart from results observed directly in the field, recent experiments have been made on populations under controlled conditions in order to estimate the cost of parasitism for populations that have undergone translocation (Sasal *et al.* 2000; Collyer and Stockwell 2004). All these studies reached the same conclusion: it is important to measure the risk related to parasite infections before designing species translocation programmes (Cunningham 1996).

The removal of a key species (competitor, predator, or introduced species) is a strategy used to conserve a target population. Predator control programmes are generally implemented to protect cattle or increase prey populations. Packer *et al.* (2003) studied the consequences of the removal of a predator from an ecosystem on the populations of prey naturally regulated by infectious diseases. Through theoretical studies, they showed that this practice could frequently lead to an increase in the parasite infection rate, a reduction in the number of healthy individuals, and a decrease in the size of the prey population.

The consideration of parasitism in conservation biology is a theme that is still at an early stage

(Cleaveland *et al.* 2002). Many issues and problems have been clearly identified but still remain unexplained. For example, what are the effects on or of parasites of the measures taken to protect natural habitats? Let us consider the example of wetlands during the winter. The number of birds (particularly ducks) in these areas is strongly affected by the environmental protection status and by the frequency of disturbance. In areas where hunting is permitted bird density is highly variable. In protected areas, the same marsh can support tens of thousands ducks. These differences in bird concentration must have important effects on the dynamics of the pathogen community, whether they be ectoparasites or viruses (the transmission success of which often depends on the density of hosts) or helminths (like trematodes and cestodes) whose eggs are discharged from the final host (bird) in faeces. The quantification of these phenomena is important for several reasons. In terms of conservation biology, it would indeed be paradoxical if nature reserves turned out to be areas of high parasitic risk for the species they are supposed to protect! On the other hand, the massive input of infective helminth stages from the birds, added to a high predation pressure exerted by the latter, should lead to important differences in the structure of the invertebrate population. Finally, knowing that, by definition, parasites divert part of their hosts' energy reserves, what are the effects of parasitism on the nutritional value of invertebrates as prey for the predators? On the scale of a marsh, does this phenomenon correspond to an important loss of nutritional value for the whole area? At the moment we have no answers to these few simple and well-founded questions.

Important points

• Parasites influence species assemblages by interfering with processes as varied as competition, invasions, life-history traits, and engineering processes. These effects may be amplified by cascade phenomena.

• Parasites can also affect intraspecific diversity through their effects on intraspecific competition, dispersal, or by acting directly as a selective pressure.

• Parasites must be integrated into conservation biology programmes.

Questions for discussion

• Given their effects on ecosystems, should some parasites be on the list of species to protect? What arguments could be proposed?

• When a parasite is at the same time harmful to an individual but beneficial to the species (for example, by giving it an interspecific competition advantage), there is a selection conflict. How can this conflict evolve?

• Some ecosystems are richer than others in parasite species. Formulate hypotheses for the potential causes and consequences of these differences.

Further reading

• Lafferty, K.D., Gerber, L.R. (2002). Good medicine for conservation biology: the intersection of epidemiology and conservation theory. *Conservation Biology* **16**: 593–604. Relating key epidemiological principles to conservation biology.

• Lafferty, K.D. (1997). Environmental parasitology: what can parasites tell us about human impacts on the environment? *Parasitology Today* **13**: 251–255. How parasites can provide information on the impact of human activities (pollution) on the environment.

• Thomas, F., Renaud, F., Guégan, J.F. (2005). *Parasitism and ecosystems.* Oxford University Press, Oxford. A comprehensive book that identifies the interactions of parasites and ecosystems.

Conclusion

Jacques Blondel

The metaphor of evolution as a tinkering machine, making new living beings from old ones, has been well known since it was proposed by François Jacob; as Stephen Jay Gould put it concerning parasites, evolution is sometimes a matter of magic or spells. This book is a wonderful anthology of the 1001 examples of odd solutions devised by evolution to allow parasites to flourish in circumstances which are made more and more difficult as their hosts oppose all sorts of strategies. This is of course a legitimate behaviour, but it inspires parasites to act shrewdly and to devise a lot of trickery to get their way in the end.

If it is true, as is commonly claimed, that around half of all living animal species are parasites and that not a single one of them is parasite free, and since parasites are at odds with their hosts because they directly or indirectly take part of their energy requirements from them, then parasitism becomes an inescapable component of research in many fields of biology, especially evolutionary ecology. The chapters of this book offer a beautiful illustration of this, and I will comment on three points as a matter of conclusion.

My first point is that parasites and pathogens are always harmful in some way to their hosts, even if they appear to be 'silent' or almost inoffensive. This unquestionable reality is illustrated in this book by a myriad of examples. But it is true, also, that without parasites the living world would be completely different. Would it be more beautiful? Not necessarily; but in any case it would be much more monotonous because parasites always introduce some kind of heterogeneity, diversity, and trickery, as shown by the myriad of defence mechanisms their hosts have invented for outsmarting their tricks,

manipulations, and facilitation mechanisms that these engineers, as they are sometimes described, are capable of evolving. To illustrate the infinite variety of devices parasites use for getting their way we use metaphors describing them as pirates, saboteurs, manipulators, hitch-hikers, and cheaters. Trying to imagine a world without parasites would be to build a utopia where everything would be completely different because their presence and importance has shaped all living beings and their life histories. Many behavioural traits would be useless or different. Sexual selection—assuming that sex would still exist—would have other functions, leading to different mating systems and feeding habits. The opportunity to economise on immune defences would provide organisms with new opportunities for managing energy, modifying **trade-offs** among traits that allow them to keep themselves alive without being tempted by Darwin's devil. The genetic diversity of hosts would either increase or decrease, drastically changing the responses of populations to selection regimes. Even the morphology of many species would be deeply modified as well as many interspecific interactions within trophic chains in ecosystems. A world without parasites would certainly include quite different species assemblages, sometimes enriched but more often impoverished because, like predators, parasites and parasitoids act as mediators as a result of their differential effects on growth rates and therefore on the coexistence of competitive species which otherwise would be excluded through competitive exclusion. Interactions such as competition and predation would also be completely different, thus modifying the structure and dynamics of communities.

How would populations of pest organisms be controlled without a large diversity of parasitoid insects? To what extent would some populations experience demographic outbreaks if the predators which control them were exterminated by a zoonosis? Even speciation mechanisms could be modified, because evolutionary differentiation is often mediated by parasites in natural hybridization between closely related species. Botanists demonstrated a long time ago such hybridization processes in plants submitted to variable pressures of herbivory. By enhancing or preventing the chances of closely related taxa to hybridize, parasites play a role in the differentiation processes, including reproductive isolation depending on whether gene flow between parental taxa is increased or decreased. Thus, parasites may have a strong influence on the diversity of communities, just as they may have strong effects on the genetic diversity of populations. In short, we would be completely confused in a world without parasites because such a world would be severed from a part of what makes it diverse, insecure, but also charming...!

The second point I would like to address is the enormous responsibility, which will undoubtedly increase in the coming decades, faced by researchers working on parasites and pathogens. Indeed, many chapters of this book consider the consequences that human-induced changes of the environment will have for interactions between parasites, pathogens, and human populations. In our world with so many global and often unpredictable changes, the combination of concentrations of humans and domestic animals, globalization of trade and travel, and uncontrolled introduction of invasive organisms are real threats to human health. This combination of factors entails serious risks because **vectors** of infectious diseases against which naïve populations of humans are not protected, combined with industrial-scale breeding of fish, mammals, and birds, could result in the emergence and spread of devastating pandemics. Such a cocktail of factors is a time bomb, and the associated risks for human welfare must be a major priority for 21st century public health. Indeed, it is not possible to avoid the problem of emergence of new pathologies or the re-emergence of ones which

were considered to be eradicated but which could reappear as a result of huge human-induced environmental changes.

Many diseases, for example avian influenza H5N1, malaria, leishmaniosis, Lyme disease, West Nile fever, SARS, AIDS, Ebola fever, blue tongue disease, and foot and mouth disease to name only a few, obviously raise questions of evolutionary biology and **epidemiology**. These questions cannot be addressed by classical medicine alone, either because the community of practitioners in medicine does not have the expertise to address questions concerning the mechanisms of ecosystem and population dynamics it is not familiar with, or because it does not feel concerned with the physical, ecological, and evolutionary processes that operate at scales of space and time other than those at which it usually works. It is symptomatic that not a single doctor of medicine contributed to this book. Many problems related to human health are clearly associated with or dependent on issues of global change and should therefore be approached by specialists in environment-related diseases who have a good expertise in population dynamics and evolutionary ecology. The consequences of several types of environmental degradation, which result in a dramatic increase in vectors and/or reservoirs of infectious agents that may escape from their areas of endemism and prompt devastating zoonoses, are a clear example of this. There is also the question of the functional role of biodiversity on the transmission of diseases to human populations. Besides the indisputable importance of classical medicine in understanding epidemiological issues, the environmental component must be carefully considered because it is at the root of the chain of causal events that are involved in developing an 'ecology of health'—mostly a matter for evolutionary ecologists. Such a new approach, which is strongly advocated in this book, must integrate specialists in medicine and ecology. Ecologists are particularly well qualified to provide alternative methods for fighting infectious diseases which, unfortunately, are likely to become more and more widespread in the forthcoming decades. Evolutionary ecologists are certainly well placed to answer fundamental questions on the origin of diseases, the mechanisms of their development, their spatial and temporal spread, variation of the **virulence** of pathogens, and finally the impact of environmental and climate change on their development. All these questions should be the subject of long-term research projects which must be developed soon if we are to understand, predict, and control the dynamics of pathogen populations. The challenges of these problems affecting human health are crucial in the context of global warming, which will undoubtedly result in changes in the distribution of pathogens and their reservoirs, exposing new human populations to new sets of diseases against which they are not immune. An ecosystemic approach combining medicine, ecology, evolution, climatology, and metapopulation dynamics must be devised without delay for tackling these new challenges raised by the globalization of health problems. The recent idea of using macroscopic components to understanding how global change affects the dynamics of pathogens in host populations is an illustration of advances in macroecology and the development of an epidemiology of populations and communities. Anticipating the effects of global change on public health using modern tools of ecology and evolutionary biogeography as well as new programmes of ecological epidemiology and new systems of health surveillance is certainly one of the most crucial and difficult challenges for the scientific community.

My third and final point is conceptual and methodological. An enormous epistemological jump has been made during the past two decades or so in the scientific approach to parasitism by incorporating modern concepts and methods of population biology and evolutionary ecology in research projects. Just as ornithologists and botanists are no longer just bird watchers or plant lovers but scientists who utilize birds and plants as study models to address the fundamental problems of basic biology, parasites and pathogens have become a raw material for addressing fascinating problems of evolutionary biology, community ecology, and issues related to human health. Therefore it is not surprising that whatever their biological model, scientists use a similar set of methods, tools, tricks, and gimmicks to mimic as far as possible the functional mechanisms which make their study organisms tick. Both in the field and in the laboratory they take advantage of the vast number of modern tools and

techniques which are becoming more and more sophisticated and miniaturized. These new technical packages include biochemical approaches which allow researchers to penetrate the depths of immune mechanisms or to address new methods of global molecular analysis for understanding molecular dialogues and conflicts. They also include new approaches for understanding complex processes such as, for example, how enzymes work to transform the viral RNA genome into DNA which is integrated into the genome of host cells. Another example is the recent emergence of proteomics which has been made possible thanks to the development of mass spectrometry, adding to the fantastic armamentarium available to modern scientists. All these approaches make it possible to go back and forth along scales of space and time as well as up and down various levels of biological complexity, opening a window onto the wonderful methods parasites use to succeed in exploring an infinity of possibilities. This new ecological thinking, combined with endless methodological tinkering, offers exciting opportunities for deciphering, analysing, and dissecting what parasites do to the morphology, physiology, development, and behaviour of their hosts, as well as the sometimes unbelievable means they use to reach them. Populations biology, whether of free-living or parasitic organisms, has been progressively structured as a highly integrative multidisciplinary framework incorporating in its paradigms various disciplines using the microscope and 'macroscope', molecular approaches and global climates, behaviour and physiology, immune approaches, and population dynamics. The recent emergence of immuno-ecology is an excellent example of these new developments. As mentioned by several authors in this book, this new way of thinking goes far beyond the traditional proximate view of the immune function and puts the rationale of immunocompetence in the framework of evolutionary ecology. The ultimate goal is to understand in terms of **fitness** how **natural selection** shapes the investment of organisms in immune functions.

As a final word, in these times of uncertainty about the future of biodiversity, another challenge is to wonder whether parasites are useful for its maintenance and functioning. By definitely answering 'yes' to this question, I would like to advocate that parasites are not mere stowaway passengers in the game of life but are necessary players in this game, even if they may be terribly harmful. Yet, we still have to demonstrate that parasites and parasitoids are obligate mediators in many processes such as the coexistence of species, engineering functions, biological competition etc. We must never forget, however, that besides functions that are useful to ecosystems and biodiversity, many parasites are really or potentially a permanent threat to their hosts. This is why any programme for monitoring and managing biodiversity must carefully consider the parasitic component of communities, especially in projects involving reintroduction or reinforcement of populations.

This multi-author book which uses a wide range of biological models, from prokaryotes to humans, should contribute to making the unbelievable diversity of parasites as well as their interactions with free organisms more accessible. It should certainly attract young scientists and students to engage themselves in wonderful adventures in basic research and research applied to the welfare of human societies.

APPENDIX

Methods

Thierry Lefèvre, Nicolas Ris, and Guillaume Mitta

A.1 Introduction

A major goal in modern parasitological research is to understand the influence of parasites on the physiology, development, and behaviour of their hosts, and vice versa. The biological processes that characterize host–parasite interactions rely on a multitude of cascading biochemical reactions that involve chemically different molecules such as DNA, RNA, and proteins. Consequently, the researcher who wishes to analyse the biological processes involved in host–parasite interaction must be able to detect, quantify, and identify these biomolecules with precision. Several techniques that allow for this type of analysis have already been developed in molecular biology, the field of research that focuses on this type of interaction at a molecular level. It is not our objective here to provide an exhaustive review of the techniques currently in use in molecular biology. Instead, we will shed light on the principles underlying the techniques that have been mentioned in this book. These techniques can be used in a targeted (or candidate) approach or can address a biological system as a whole. Naturally, the choice of the appropriate method also depends on the chemical nature of the molecule of interest (e.g. nucleic acid, protein, polysaccharide). We have therefore decided to divide the next section of this Appendix into two parts: we will deal with the candidate approach and the holistic approach separately. Each time, the techniques will be grouped according to the targeted molecules. In the candidate approach, it is possible to identify molecules that are associated with a particular biological process. The holistic approach permits the characterization of all molecules that belong to a biological system, such as a set of mRNAs (transcriptome) or proteins (proteome). The final section of this Appendix will address the question of genetic variability, and the influence of different evolutionary forces. We will elucidate the difference between genetic markers, that are considered to be neutral, and phenotypic characters that are subject to **natural selection**.

A.2 How to study a biological process

A.2.1 Candidate approach

Even without detailed knowledge about a biological mechanism, it is often possible to identify markers that are tightly associated with a particular biological function. In general, we can take advantage of the vast amount of knowledge that has been accumulated by studying the molecular genetics of model organisms such as the mouse (*Mus musculus*), zebrafish (*Brachydanio rerio*), fruitfly (*Drosophila melanogaster*), the nematode *Caenorhabditis elegans*, or the model plant *Arabidopsis thaliana*. Based on what is known in these model organisms, a hypothesis can be proposed for certain biological processes in less studied species. If we suspect the involvement of a particular molecule (a candidate molecule), this molecule can be selectively labelled or revealed in a specific way. This approach is particularly promising for evolutionarily conserved pathways (e.g. those involving neurotransmitters or hormones). According to the chemical nature of these molecules, two basic methods are used:

- immunological methods (for proteins, polysaccharides, hormones);
- methods for the detection and quantification of nucleic acids (DNA and RNA).

A.2.1.1 Immunological methods

Immunological methods are based on the specific interaction of an **antibody** (Ab) with a molecule of interest, the **antigen** (Ag). These methods allow for the detection and/ or quantification of the Ag in biological samples. Abs are generally obtained from animals (e.g. mouse, rabbit, goat) that have received multiple injections of the Ag of interest. Often, two different Abs are used consecutively: the first Ab (primary Ab) binds to the Ag, and a second Ab (secondary Ab) binds to the primary Ab. This procedure leads to an amplification of the signal since several secondary Abs can bind to a single primary Ab. The detection of the

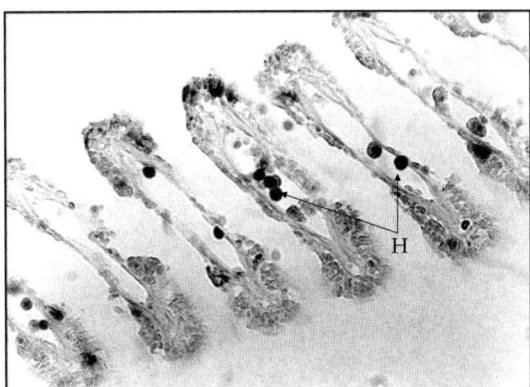

Figure A.1 Schematic representation of an immunohistological reaction. Labelling of an antibacterial peptide (mytiline) in the haemocytes (H) of *Mytilus galloprovincialis*. Indirect immunohistological labelling: the primary antibody binds to mytiline and the secondary antibody binds to the primary antibody. An enzyme is covalently linked to the secondary antibody that catalyses the conversion of a chromogen substrate into a dark pigment that precipitates at the site of the colour reaction.

Ab *in situ*, i.e. inside a tissue (immunohistology) or inside a cell (immunocytology), relies on the use of tracer molecules (fluorophores, enzymes, etc.) that are covalently linked to the Ab. These tracers are observed using light, electron, or fluorescence microscopy. An example of an immunohistological reaction is shown in Fig. A.1.

It is also possible to detect and quantify Ags in biological fluids (e.g. plasma, secretions) or in cell extracts. In this case, the Ab is fixed on a solid support. Two main types of detection principle are used, involving either enzymes or radioactive isotopes. The first is used in a technique called ELISA (enzyme-linked immunosorbent assay; Fig. A.2), the latter in the so-called RIA (radioimmunoassay; Fig. A.3). In the ELISA technique, the tracer is an enzyme that is either covalently linked to a primary Ab (direct ELISA; Fig. A.2a), or to a secondary Ab (indirect ELISA). When the enzyme (e.g. peroxidase) catalyses the reaction of appropriate substrates, they can change the colour, which is used as the signal. By measuring the amount of colour produced, i.e. the absorption or optical density, it is possible to quantify the amount of Ag in the solution. Specific spectrophotometers have been developed for this purpose that are called ELISA readers. For quantification, a standard curve must be established using known concentrations of the Ag (Fig. A.2b).

The principle of RIA is very similar (Fig. A.3). However, here it is the Abs that are immobilized on a carrier material. The investigator must have access to the Ag of interest, which is radioactively labelled, frequently by incorporation of radioactive ^{125}I (attached to tyrosine). The principle of this technique is the competition for Ab binding between the Ag of interest in the biological sample and the radiolabelled Ag. The amount of radioactivity that is finally associated with the Ab is inversely proportional to the amount of Ag in the sample. The radioactivity is measured with a scintillation counter, and the Ag is quantified using a standard curve. RIA is more sensitive than ELISA, and is particularly useful when the molecules of interest are present in low concentrations (e.g. hormones).

A.2.1.2 Detection and quantification of DNA and RNA by polymerase chain reaction or reverse transcription polymerase chain reaction

Until the end of the 20th century, polymerase chain reaction (PCR) was essentially used to detect nucleic acids but not to quantify them. Other techniques such as Northern blot or Southern blot were applied for the quantification of RNA and DNA, respectively. The principle of these techniques is to immobilize the whole RNA or DNA of a sample on a membrane (nylon or nitrocellulose), and to reveal the presence of the nucleic acid sequence of interest by using DNA probes that are complementary to the sequence and radioactively labelled with ^{32}P, ^{33}P, or ^{35}S. Both techniques are used less and less nowadays and will not be described in detail here.

At present a particular PCR-based technique, called quantitative PCR, allows for precise measurement of the amount of DNA and RNA in biological samples. To understand how this technique works we will go briefly through the principles of conventional PCR (Fig. A.4). PCR requires a certain number of molecular compounds: PCR primers, i.e. oligonucleotides that hybridize with the DNA region of interest, nucleotides (dNTPs), a thermostable DNA polymerase (e.g. Taq-polymerase), and the template DNA [either genomic DNA or complementary DNA (cDNA)]. The primers hybridize with the template, and the DNA polymerase synthesizes the new DNA strands by successive incorporation of dNTPs. The PCR products are separated by gel electrophoresis, stained with intercalating fluorescent dyes (like the frequently used ethidium bromide), and visualized by exposure to UV light. RNA must be converted into DNA before it can be amplified by PCR since the thermostable DNA polymerase applied in PCR cannot use RNA molecules as a template (DNA-dependent DNA polymerase). This initial step, called reverse transcription, relies on a retroviral enzyme, the reverse transcriptase that converts RNA into cDNA. The technique that allows amplification of cDNA fragments is called reverse transcription (RT)-PCR. During a PCR reaction, the number of PCR products increases. The increase is exponential since after each reaction cycle the

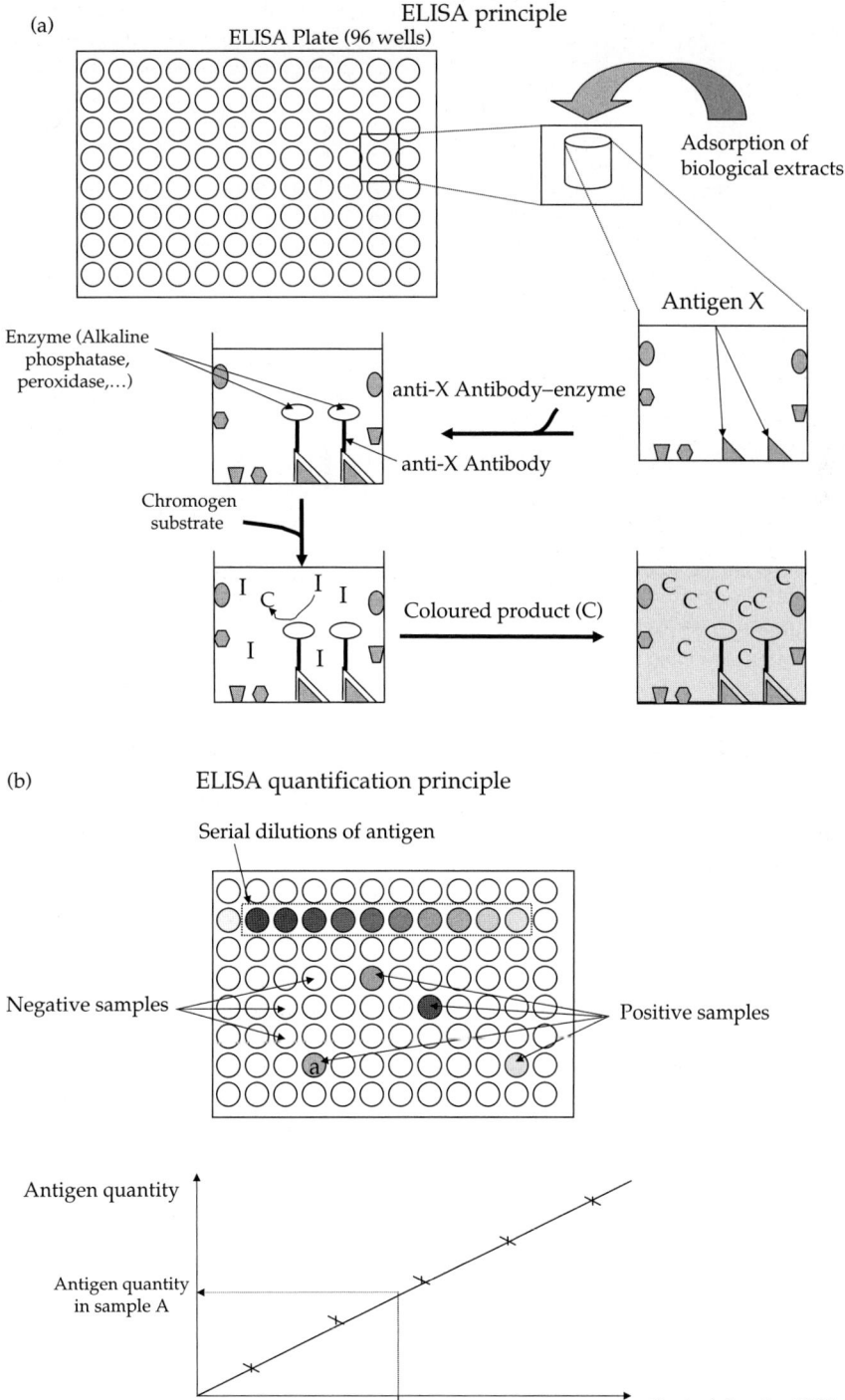

Figure A.2 Principle of enzyme-linked immunosorbent assay (ELISA) and quantification: (a) principle of ELISA; (b) principle of quantification.

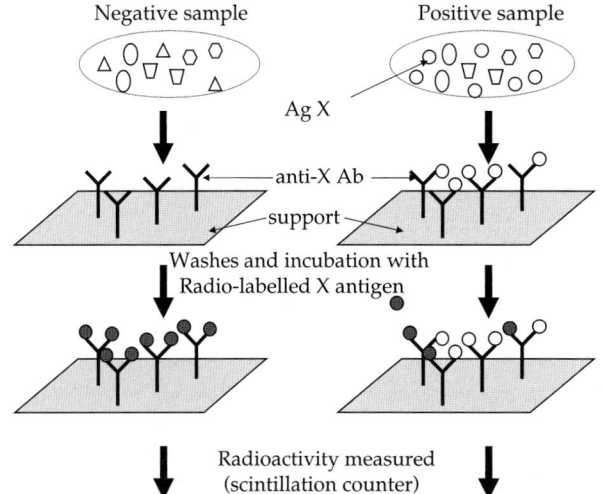

Negative sample   Positive sample

Ag X

anti-X Ab

support

Washes and incubation with
Radio-labelled X antigen

Radioactivity measured
(scintillation counter)

The amount of radioactivity that is finally associated with the Ab for an unknown
sample is inversely proportional to the amount of Ag in the sample. For quantification,
a standard curve must be established using known concentrations of the Ag.

Figure A.3 Principle of radio-immunoassay (RIA). Ag, antigen; Ab, antibody.

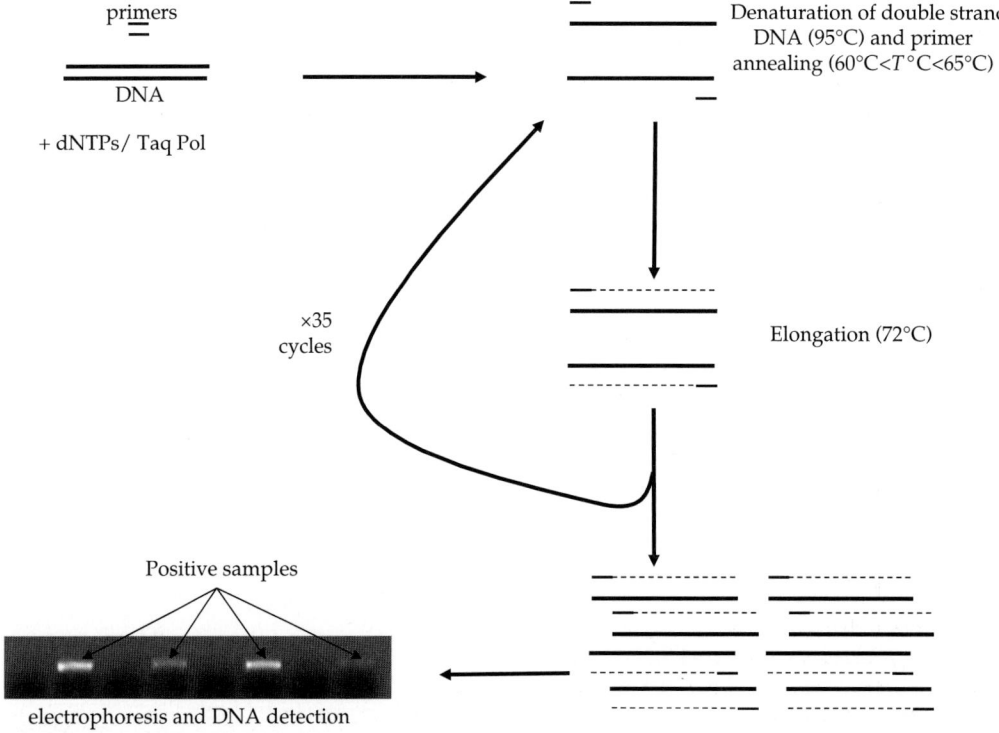

primers

DNA

+ dNTPs/ Taq Pol

Denaturation of double strand
DNA (95°C) and primer
annealing (60°C<T°C<65°C)

×35
cycles

Elongation (72°C)

Positive samples

electrophoresis and DNA detection

Figure A.4 The classic PCR principle. dNTPs, nucleotides; Taq Pol, Taq-polymerase.

newly synthesized strand of a PCR product can serve as a template in the next reaction cycle. In each reaction cycle, the number of PCR products is doubled. The amount of PCR product at the end of a conventional PCR does not necessarily give information about the initial amount of template DNA. As a matter of fact, the enzymatic reaction does not produce an exponential increase of PCR products until the end of the PCR since the resources in the reaction tube are limited. Consequently, a conventional PCR is only quantitative during the initial exponential reaction phase. Quantitative PCR has become widely used thanks to the development of devices that continuously monitor the increase of PCR product number within the reaction tube. For this reason, quantitative PCR is also called real-time PCR. The generation of new PCR products is measured during the entire PCR. As a result, one has access to the exponential phase of the reaction whose shape reflects the initial template concentration. The real-time monitoring is possible due to the direct or indirect labelling of the PCR product with fluorescent dyes. The intensity of the emitted fluorescence corresponds to the number of PCR product molecules and can be measured with a spectrophotometer. Standard curves serve to determine the unknown DNA concentration in the sample. There exist different methods of quantification. (Before these new RT-PCR machines became available, quantitative PCR was performed using an internal DNA standard of known concentration and dilution series of the unknown sample, followed by densitometry of gel-separated PCR products. This was done to ensure measuring in the exponential phase of the PCR.)

One quantification method consists of comparing the exponential amplification curve of the biological sample with an unknown template concentration with those obtained using known concentrations of the template (standard curve). Concerning host–parasite interactions, the quantification of cDNA by RT-PCR is a rapid and current method for evaluating the impact a parasite can have on the gene expression patterns of the host. In addition, the quantification of genomic DNA is also of interest in parasitology because it permits us to measure the **parasite load** in an organism if the parasites are not easily detectable by visual inspection.

A.2.2 Holistic approach

Biotic interactions such as parasitism can be characterized by modulations of the genomic expressions involved. To understand these phenomena, it is necessary to first understand the underlying 'molecular dialogues and conflicts' that take place between the parasite and its host. Given that these molecular dialogues and conflicts involve a great diversity of molecules, previously developed targeted analytical techniques are often inappropriate tools. In this case, it is necessary to use holistic methods of molecular analysis. These methods profit from the formidable progress made during the last decade in terms of genome sequencing. Today, more than a hundred genomes have been sequenced or have their sequencing in progress. This gives access to a list of the proteins potentially coded by genes. Knowing the 'actual possibilities', apart from representing major progress, provides no information about where (e.g. neurons, muscle fibres, liver cells) or when (day/night, before/after digestion, during muscular effort) these genes are expressed. One of the present challenges for biologists is to characterize genomic expression during development, under certain physiological conditions, during abiotic stress, or even under stresses of interest for this work such as those that are induced by parasites. This type of study requires holistic methods of molecular analysis that belong to the field of functional genomics.

In our host–parasite models, these methods will allow us to obtain a 'molecular view' of the infected host that will be compared with that of the healthy host: functional parasite markers will thus be identified. We will mention methods for the analysis of two molecular classes that are involved in the gene expression process, i.e. RNA (in particular, messenger RNAs or mRNAs) and the functional products of genes, i.e. proteins. The genome is the set of genes of an organism, and the transcriptome and proteome are, respectively, the set of mRNAs and the set of proteins expressed within a cell, a tissue, an organ, or an organism (Fig. A.5). Transcriptome and proteome analyses described here allow a holistic study of the qualitative and quantitative expressions of these two gene expression products.

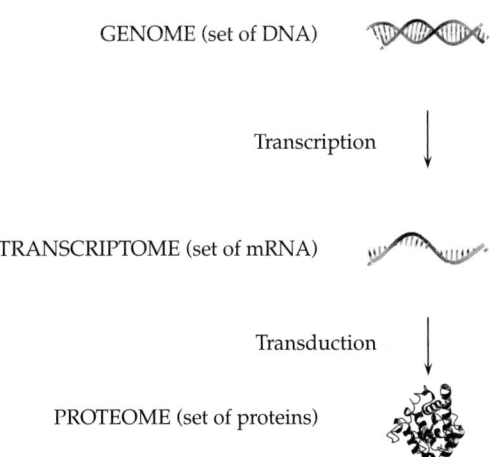

Figure A.5 The products of genetic expression.

A.2.2.1 Transcriptome analysis

Two main strategies exist for the holistic analysis of the transcription activity of a biological specimen:

1. Massive sequencing methods coupled to research within data banks; the most common are the expressed sequence tags (EST), serial analysis of gene expression (SAGE), and RNA display techniques.

2. Methods hybridizing the cDNA resulting from the reverse transcription of the mRNA of a specimen with the DNA of known sequences that are fixed on a solid support; in this case, they are called DNA chip techniques (microarrays, macroarrays, chips).

Only the principles of EST and DNA chip techniques will be addressed here.

Expressed sequence tags

Given that this method implies the identification of the mRNA by sequencing and that the sequencing methods that are routinely used at present only allow sequencing of DNA, it is first necessary to transform mRNA into cDNA. This transformation step, called reverse transcription, has already been described above. This reverse transcription results in a set of cDNAs qualitatively and quantitatively representative of the transcripts present in the specimen. These cDNAs are then individually inserted into bacterial vectors for cloning. A large number of clones (i.e. several thousand) are then randomly sequenced. The resultant sequences, called ESTs, tags, or labels, are compared with known gene sequences within data banks in order to discover the identity of the genes that produce them (Fig. A.6). This gives indirectly a qualitative (the nature of the gene) and quantitative (the number of times the same transcript is identified) overview of the gene's expression within the specimen.

DNA chips

This technique is based on the hybridization of two sorts of DNA: cDNA resulting from reverse transcription of mRNA from the specimen and the DNA of genes attached to a chip. The chip is a miniature solid support, most often a glass plate, on which are attached the sequences of several hundred or thousand known genes. The arrangement of the sequences on the support is very well ordered: each sequence of a given gene is located in a very precise site on the chip, called a well. At the same time, the mRNAs extracted from biological specimens undergo reverse transcription. The cDNAs obtained are fluorescence-marked, by cyanines for instance. By using two different fluorochromes (e.g. Cy3 and Cy5), it is then possible to compare transcript populations from two specimens subjected to different conditions (e.g. healthy/infected individuals, low/high temperature specimens) (Fig. A.7). After hybridization between the marked cDNAs and the chip DNAs is complete, the detection step is performed. By exciting each well with fluorochrome-specific wavelengths (635 nm for Cy5 and 532 nm for Cy3), two images are obtained that reveal the genes expressed within the specimens. An overlay of both images gives a synthesis of the gene expression—in green the genes expressed only in the specimen submitted to the first condition, and in red the genes expressed in the specimen submitted to the second condition, and finally, in yellow the genes expressed by both specimens. The precise measure of the colour shades permits evaluation of the abundance of hybrid cDNA in each well and thus gives indirect information on the quantitative gene expression (Fig. A.7).

A.2.2.2 Proteome analysis

Proteome analysis is based on a protein separation system (most often bidimensional electrophoresis or chromatography) and an identification system (mass spectrometry coupled with research within data bases).

Separation systems

Several strategies exist for the protein separation step: (1) electrophoretic methods, (2) chromatographic methods, and (3) affinity methods. We have chosen here to present bidimensional electrophoresis, the most common separation method used in proteome studies.

Bidimensional electrophoresis combines the isoelectric focusing technique and zone electrophoresis. The proteins extracted from the tissue under study are placed on a gel polyacrylamide strip [of the immobilized pH gradient (IPG) type], in which a static pH gradient is created (Fig. A.8a). An electric field is then applied to this strip to so that the proteins migrate until they reach their isoelectric point (the pH where their charge is equal to zero) (Fig. A.8b). Once the proteins are separated in one dimension, the strip is saturated with sodium dodecyl sulphate (SDS) and placed on a second polyacrylamide gel. SDS is an ionic detergent that denatures proteins (it destroys their three-dimensional structure) and gives them a negative charge that makes the influence of the native charge insignificant (Fig. A.8c). Thus, under the effect of an electric field, proteins are separated in a second dimension based on their molecular mass (Fig. A.8d). The separated proteins, after several hours of migration, are then identified within the gel using stains (silver nitrate, Coomassie Blue, or Sypro ruby). There is a large range of stains, each with its own advantages and disadvantages. One or

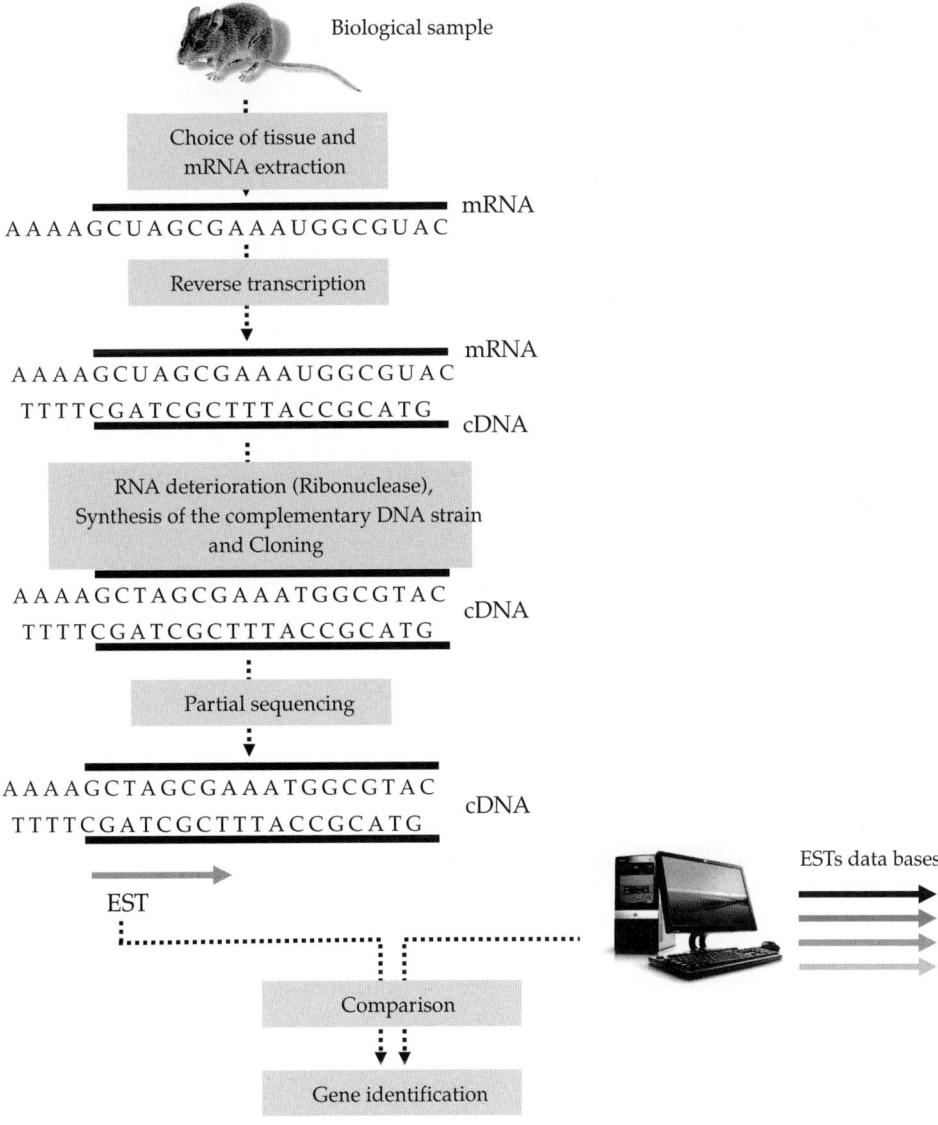

Biological sample

Choice of tissue and
mRNA extraction

mRNA

A A A A G C U A G C G A A A U G G C G U A C

Reverse transcription

mRNA

A A A A G C U A G C G A A A U G G C G U A C
T T T T C G A T C G C T T T A C C G C A T G   cDNA

RNA deterioration (Ribonuclease),
Synthesis of the complementary DNA strain
and Cloning

A A A A G C T A G C G A A A T G G C G T A C   cDNA
T T T T C G A T C G C T T T A C C G C A T G

Partial sequencing

A A A A G C T A G C G A A A T G G C G T A C
T T T T C G A T C G C T T T A C C G C A T G   cDNA

EST

ESTs data bases

Comparison

Gene identification

Figure A.6 Principle of the expressed sequence tag (EST) technique.

the other will be chosen according to the nature of the specimen, the targeted level of sensitivity, the amount of protein placed on the gel, and the technique used in the identification step.

Gels are then scanned to analyse the image. Comparison of the gels obtained from several specimens allows one to see the differences in protein composition between these specimens. Proteins displaying a particular expression pattern within a specimen, for instance a protein that is under- or overrepresented only in infected organisms, are extracted from the gel in order to be identified.

Identification system

Proteins can be identified using one of several techniques. For instance, the Edman method based on the sequencing of the N-terminal end of proteins can be used. However, the N-terminal end of many proteins is blocked (e.g. an amine group with an acetyl or formyl derivative) and these proteins can thus not be sequenced.

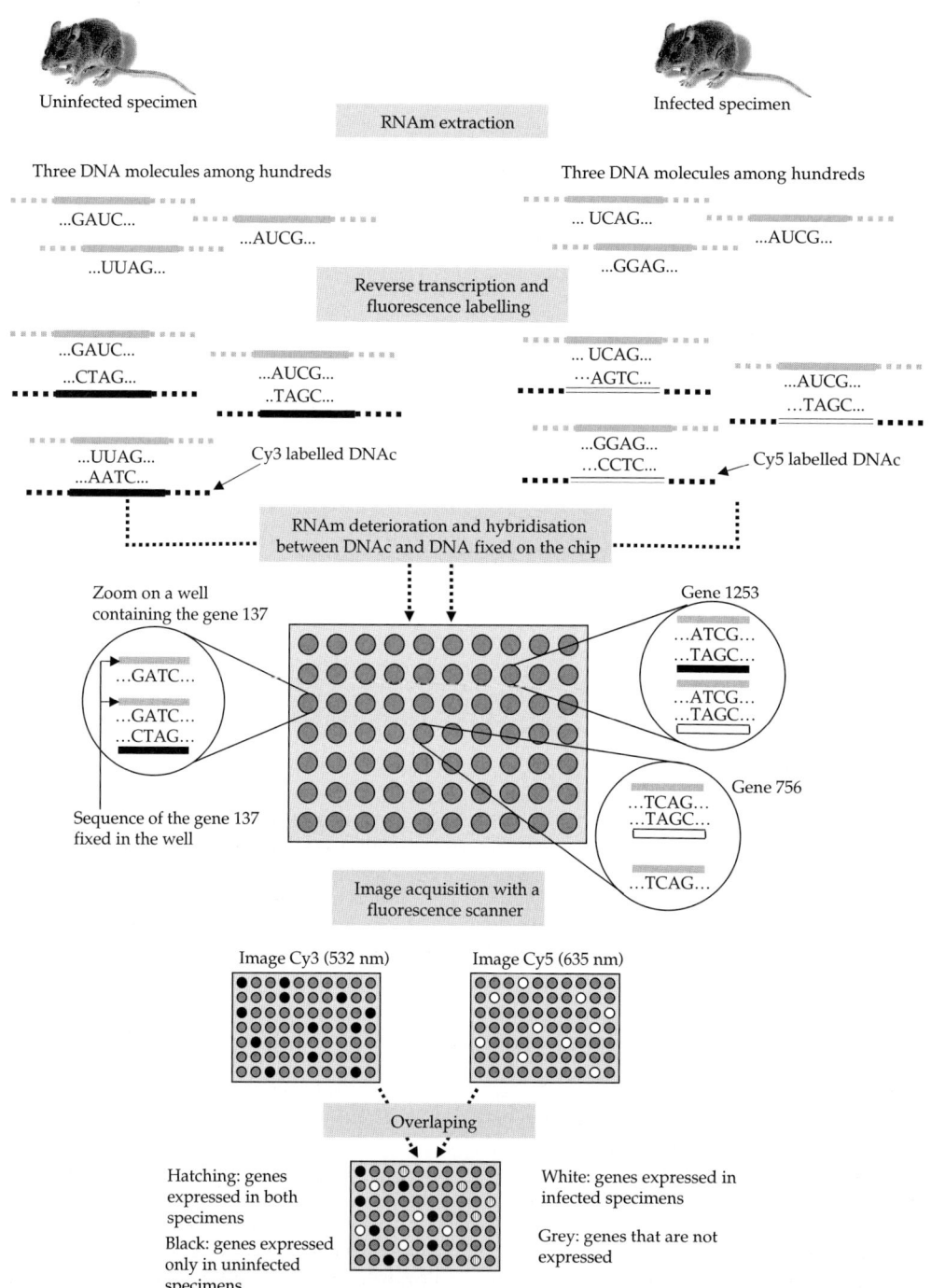

Figure A.7 Principle of the DNA chip technique.

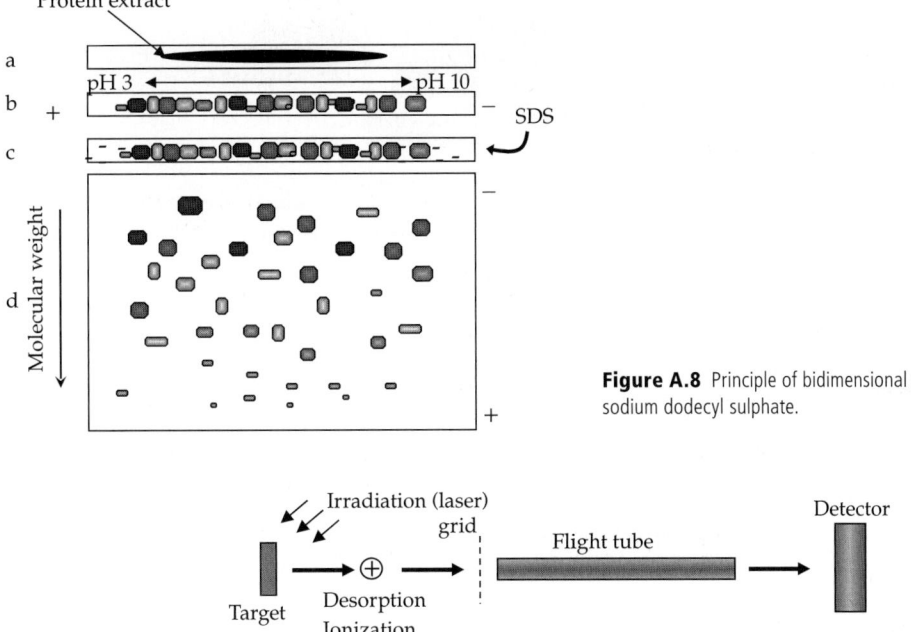

Figure A.8 Principle of bidimensional electrophoresis. SDS, sodium dodecyl sulphate.

Figure A.9 The principle of mass spectrometry. The analyte (fragments resulting from trypsin digestion) is deposited on a target. A laser is fired on the target leading to the desorption of the analytes and their ionization (cation acquisition). The molecules accelerate into the flight tube where they fly towards the detector. Low-mass molecules arrive in a shorter time than heavier ones and molecules of different mass are thus separated.

Furthermore, Edman sequencing is a slow, poorly sensitive process (i.e. the initial amount of protein must be high) and quite expensive. That is why this technique is seldom used for identification in proteomics. The most common method in current use is mass spectrometry (MS). Its application to proteomics has increased tremendously, made possible by (1) the progress in protein ionization and (2) the explosion of gene sequences available within the data bases.

The first step in MS consists of digesting the proteins to be identified using an enzyme, most often trypsin. This enzyme cuts the polypeptides following lysine and arginine when the next amino acid is not proline. A 'peptide fingerprint' is obtained that corresponds to the set of peptide fragments that characterize the digested protein (i.e. the peptide mass fingerprint, PMF). Simple MS allows this fingerprint to be 'materialized' by estimating the precise mass of each fragment (with an accuracy of 0.1 Da) (Fig. A.9). In the end, these peptide fragments are identified as a mass spectrum (Fig. A.10).

If one can precisely measure the mass of the peptides resulting from the digestion of an unknown protein, it is then possible to compare these masses with those obtained by the virtual digestion of the set of proteins currently known and sequenced. These known proteins are indexed in data bases that can be accessed on the Internet (Fig. A.10).

Other MS methods are currently available and are being increasingly used. Among these, tandem mass spectrometry (MS/MS) allows fragmentation of the different peptides resulting from enzyme digestion. The resulting fragmentation spectra are used to partially re-create the primary structure of these peptides. Comparison of the data obtained with those in the protein data bases allows the identification.

A.3 Studying genetic variability

Host and parasite species are usually subdivided into different populations within their areas of distribution. Different evolutionary forces will act on each population and will determine both the maintenance of the intrapopulation genetic variability and the process of differentiation between populations (Hartl, 1994). Very briefly, mutations and migrations tend to increase the local genetic variability and to reduce the differences

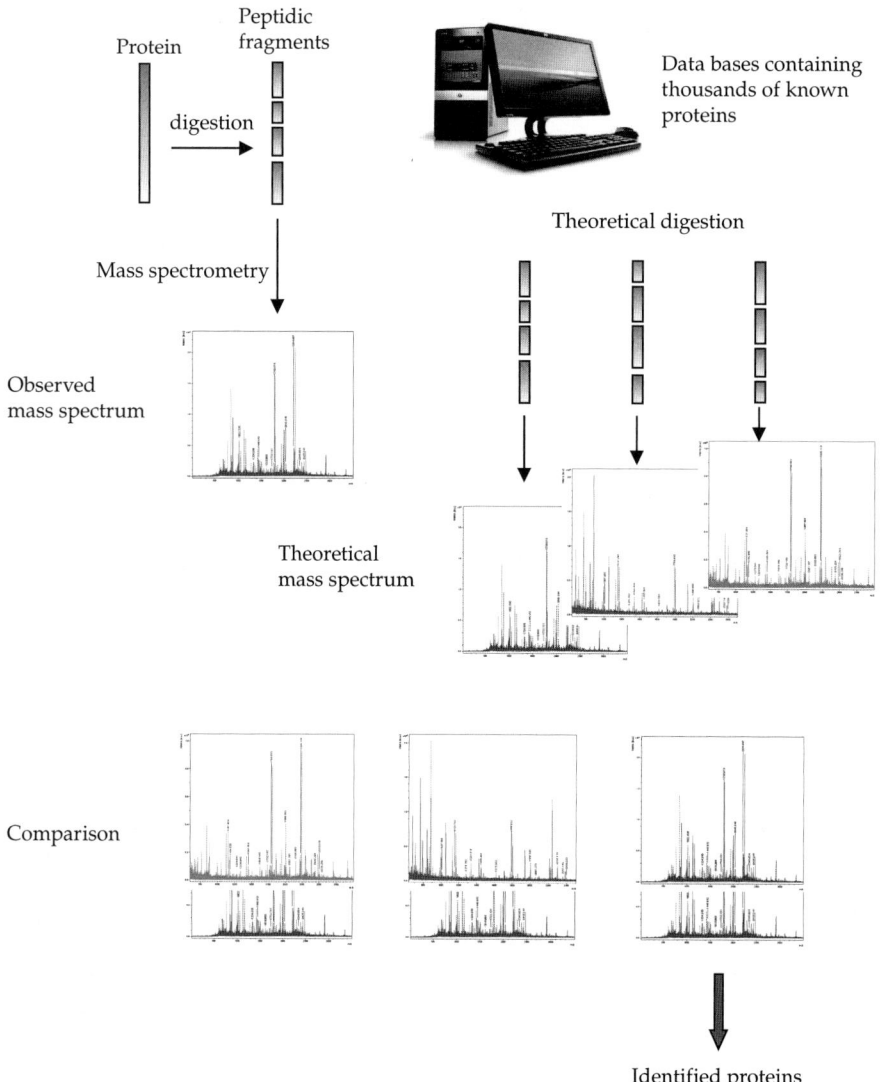

Figure A.10 Mass spectral analysis.

between populations. Conversely, natural selection tends, in numerous cases, to reduce the genetic variability of **fitness** components and favours between-population differentiation if the environment is spatially heterogeneous. Drift also reduces the genetic variability within a population and increases differences between populations; however, it acts on both selected and neutral characters and its consequences are unpredictable and, a priori, non adaptive. Estimating the relative contributions of all these forces is crucial for an understanding of species evolution. In the particular case of host–parasite interactions, the evolutionary processes are even more subtle, since each partner exerts strong selective pressure on the other.

In practice, two kinds of traits can be distinguished: molecular markers and quantitative traits.

A.3.1 Molecular markers

This approach studies the genetic variability using short DNA or protein sequences usually called molecular markers. Knowledge of this variability gives valuable information on many topics in ecology and evolution, such as degree of relatedness, modes of reproduction, population structure, **gene flow** between populations, speciation, hybridization, introgression, or even **phylogeny**. Indeed, molecular markers are interesting from different points of view:

• They are discrete units that can be precisely and unequivocally measured. At the molecular level, it is possible to qualitatively estimate the differences (mutations) in nucleotides of several DNA fragments or the differences in amino acids of several proteins.

• They are supposed to be independent from the environment and the rest of the genome, i.e. neither the environmental conditions nor the expression of an **epistatic** gene affect the expression of the marker.

• They are generally quite polymorphic, i.e. they are present in different forms within the individuals. At the molecular level, this means that several alleles exist for the same locus. Consequently, they are rather discriminant and sometimes allow differentiation of individuals, even if they are closely related.

• In some cases, they are co-dominant which allows one to distinguish heterozygotes from homozygotes.

• Many of them are considered to be neutral, i.e. they do not undergo natural selection. In order for a molecular marker to be neutral, it is necessary that whatever the allele present on the marker locus, the fitness of the individual remains unchanged. This is the case with microsatellites that are often used in population genetics. However, care must be taken because neutrality is not always evident. Some markers can be so close to genes undergoing selection that they cannot be considered as neutral. As a matter of fact, this is the property that is used to establish the genetic maps that helps identify quantitative traits of interest (quantitative trait loci, QTL) and possibly implement genetic improvement programmes (marker assisted selection, MAS).

A.3.1.1 The different molecular marking techniques

Genome sequencing is the only comprehensive method for identifying polymorphisms. Despite the progress made in automation, sequencing remains quite expensive and time-consuming to implement on the large numbers of samples that are necessary for studies of polymorphism. Molecular marking techniques are thus direct, non-comprehensive, and particularly quick methods for identifying polymorphism at the species, population, and individual levels. There are very many molecular marking techniques and their number keeps growing; new types of molecular markers will undoubtedly become available within the next few years. At present none

of the techniques is perfect and a choice must be made according to the targets and experimental possibilities. We have chosen to briefly present here the most common techniques in ecology and evolution; particular attention will be paid to the allozyme and restriction fragment length polymorphism (RFLP) techniques.

The allozyme technique

Allozymes are the different forms of an enzyme coded by different allelic variants of the same locus. In most cases allozymes are cell metabolism enzymes, involved in the Krebs cycle for instance. These forms have different amino acid compositions, but this difference does not affect the enzyme's activity. Allozymes were the first molecular markers to be widely used in ecology. The analysis of allozymes is based on the starch or acrylamide gel electrophoresis technique that allows separation of the protein variants. To begin, a tissue specimen is taken from individuals with the genotype of interest for a given locus. Proteins are extracted by grinding the specimen in an extraction buffer under non-denaturing conditions (i.e. the three-dimensional structure of the protein is maintained) in order to preserve the enzyme's functionalities. The protein fraction is recovered by centrifugation. Then, each protein extract is placed on a gel and an electric field is applied. The proteins migrate through the reticulated molecular network of the gel, more or less quickly depending on their charge, mass, and three-dimensional properties. The different allelic variants of the enzyme (allozymes) will migrate differently according to their amino acid composition. Once the migration is finished, a solution containing the specific substrate and a coloured reagent is applied to the gel. The enzyme then catalyses the substrate to give a product that combines with the reagent to give a coloured band at the site where the enzyme migrated. According to the number and position of the electrophoretic bands, it is possible to infer the genetic composition at the locus that codes for the enzyme (Fig. A.11).

The allozyme neutrality criterion is often discussed because these enzymes are often involved in cellular metabolism, and thus result from DNA coding sequences. Generally speaking, at a given locus, this technique reveals the variability between species and populations, but much less so between individuals of the same population. Consequently, it is very seldom used for paternity tests for instance, and much more to identify hybrids.

The RFLP technique

This technique does not address polymorphism at the level of the gene product (proteins) but at the level of DNA itself. Unlike biochemical markers, it allows access

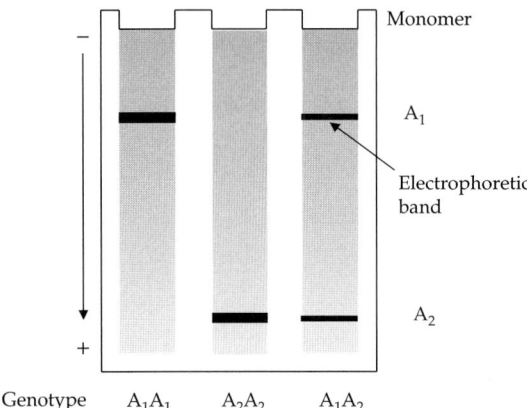

Genotype A_1A_1 A_2A_2 A_1A_2

Figure A.11 Pattern of electrophoretic bands for a monomer allozyme in a diploid organism. The allelic variant A_1 codes for the enzyme A_1 and the allelic variant A_2 for the enzyme A_2. Consequently, in a specimen with an A_1A_1 genotype only A_1 enzymes are produced. In the same way, in a specimen with a A_2A_2 genotype, only A_2 enzymes are produced. Conversely, in a heterozygous specimen having both A_1 and A_2 allelic variants, both enzymes can be produced. That's why the electrophoretic pattern of the heterozygotes shows two distinct bands.

to non-coding parts of DNA that often display much more polymorphism. This technique includes three steps: digestion of DNA by restriction enzymes, electrophoresis, and a hybridization step (Fig. A.12). Restriction enzymes cut the DNA at specific sites (from four to eight pairs of bases according to the enzyme). The specificity is such that the replacement of a base in a site is enough to prevent the nick. On a given portion of DNA, a mutation will lead to the suppression or creation of new restriction sites. This approach allows the differentiation of organisms that carry different alleles by analysing their restriction profile. Following the DNA extraction and digestion by a restriction enzyme, electrophoresis (see the allozymes section above) is carried out which separates the restriction products according to their size and charge (DNA is negatively charged at physiological pHs). At this stage, the problem is to visualize polymorphism. The genome comprises a very large number of nucleotides (several hundred million, even billions of pairs of bases according to the species). There are thus very many DNA nicking sites for a given restriction enzyme. Consequently, very large numbers of restriction fragments are obtained once the enzyme digestion stage is over. It is not possible to separate all of them given the limited resolving power of electrophoresis. In order to overcome this problem, only one locus is targeted which is revealed by a Southern-type molecular hybridization.

The restriction fragments separated by electrophoresis are denatured (i.e. the two strands of DNA are separated by breaking the hydrogen bonds) and transferred from the gel to a nylon or nitrocellulose membrane (Fig. A.12). The membrane is then put in the presence of a marked probe (with radioisotopes, fluorochromes, etc.) that can reveal the locus of interest. This probe is a single-stranded DNA fragment with a sequence that is complementary to only one of the restriction fragments. Thus, when a probe is put in contact with the membrane and all the restriction fragments, a specific hybridization occurs via hydrogen bonds. The resulting hybrid gets marked by the probe. In order to reveal the hybrid, one just has to use an adapted developing system (i.e. the probe plus the associated restriction fragment of interest). With radiolabelling, the technique that allows the hybrid to be revealed is autoradiography (Fig. A.12).

The size and complexity of nuclear DNA can sometimes be a major issue for RFLP analyses, and that is the reason why the ecological applications of this method tend to use mitochondrial DNA, which offers the following advantages: (1) it undergoes less recombination than nuclear DNA and (2) it is only transmitted by the mother. RFLP analysis using mitochondrial DNA is well adapted to show population polymorphism and to establish intra-specific genealogical lineages. For instance, it has allowed the development of phylogeography, which studies the historical reasons (glaciations, mountain range, etc.) for the current distribution of a species.

The microsatellite technique
Microsatellites are short repeating DNA base pair sequences. The repeated unit is generally a di-, tri-, tetra-, or pentanucleotide. For instance, $(CA)n$ (with $8 < n < 50$) is a very common repeat unit. These sequences are often found in non-coding regions of DNA. Microsatellites are good genetic markers because they are highly polymorphic, given the frequent errors that originate in them during DNA replication (added/deleted repeat units). Polymorphism thus resides in the number of repeats; for instance, an individual can have 12 repeats of the CA dinucleotide in one region of its DNA whereas another individual will have 14 repeats in the same region (Fig. A.13). PCR (see above) is used to reveal the differences in the number of repeats. This technique needs complementary primers for the flanking regions (DNA sequences that lie on either side of the repeated sequence). The products obtained with PCR (from 100 to 300 bp) are thereafter separated using electrophoresis. Since alleles only differ by a few nucleotides, the systems used (such as capillary electrophoresis) generally display very high resolutions that allows visualization of the polymorphism.

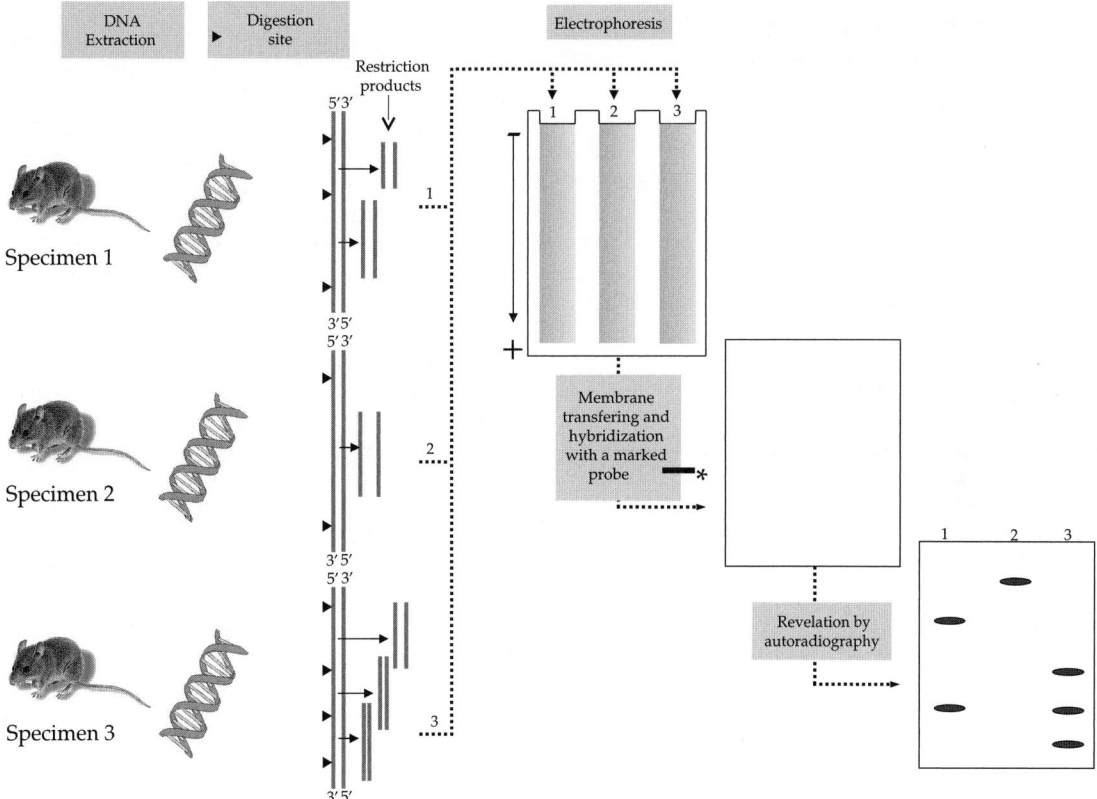

Figure A.12 Principle of the restriction fragment length polymorphism (RFLP) technique.

Individual 1 (allele 1), 12 repeats :

...TGCGTAGGCCT**CACACACACACACACACACACACACA**GTTGCATCGGGTA...

...ACGCATCCGGA**GTGTGTGTGTGTGTGTGTGTGTGT**CAACCGTAGCCCT ...
flanking region   **dinucleotide repeat**   flanking region

Individual 2 (allele 2), 14 repeats:

...TGCGTAGGCCT**CACACACACACACACACACACACACACACA**GTTGCATCGGGTA...

...ACGCATCCGGA**GTGTGTGTGTGTGTGTGTGTGTGTGTGT**CAACCGTAGCCCT...
flanking region   **dinucleotide repeat**   flanking region

Figure A.13 A microsatellite locus showing two allelic variants (individuals 1 and 2).

Given that the mutation rate is fast at the microsatellite locus level, this technique is not often used in phylogeny. On the other hand, it is very useful in population genetics or behavioural ecology to estimate, for example, the extra-pair copulation rate.

Random amplification of polymorphic DNA
This technique is based on PCR, but unlike the microsatellite technique which requires knowledge of the flanking regions of a given locus to visualize polymorphism, the random amplification of polymorphic DNA (RAPD)

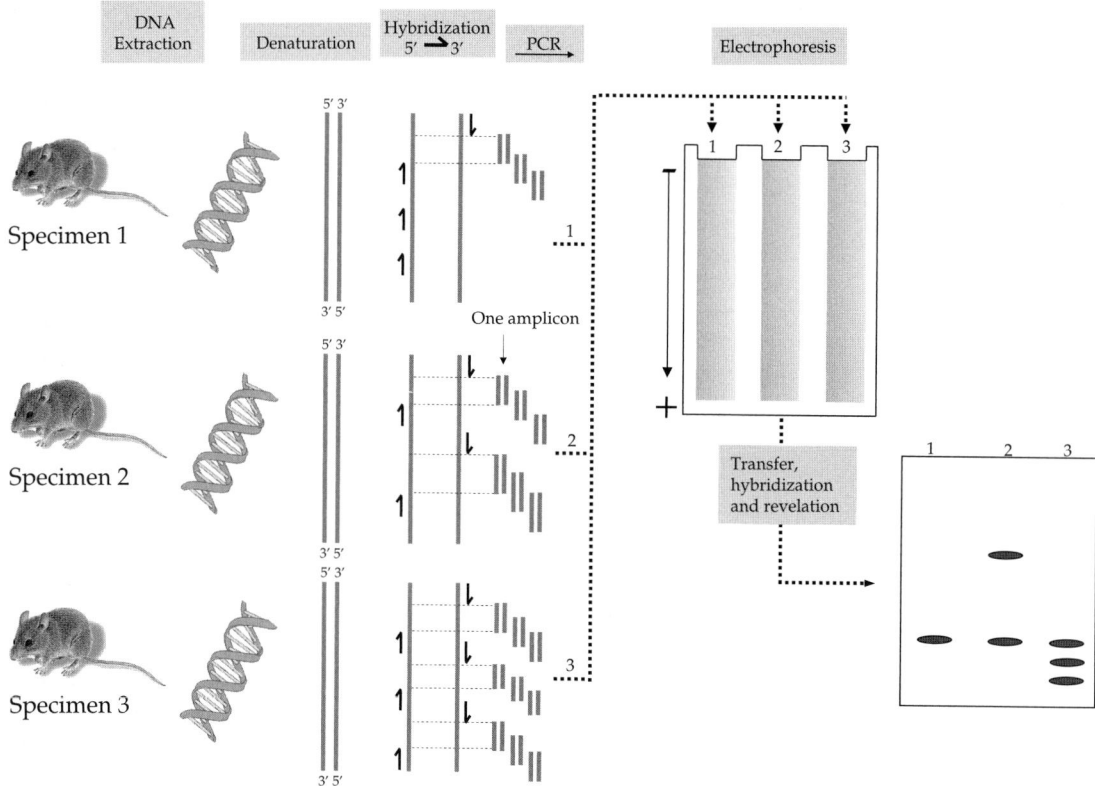

Figure A.14 Principle of the random amplification of polymorphic DNA (RAPD) technique.

technique requires no previous knowledge of the DNA sequence. After DNA extraction, PCR amplification is carried out on nucleotide probes with short and random sequences. When two hybridization sites are near enough and display opposite orientations on the target DNA, the amplification begins (Fig. A.14). The different products generated by PCR (amplicons) are then separated by agarose gel electrophoresis and the polymorphism is visualized through the presence/absence of amplicons (Fig. A.14). The amplicons are visualized thanks to a fluorescent DNA intercalator under UV excitation. Different primers are used to reveal sufficient polymorphic loci for a population analysis.

This technique can be used at all taxonomic levels since it reveals polymorphism at the interspecific, inter-population, and intrapopulation level. However, it lacks repeatability, mainly due to contamination. Given that the technique is very sensitive, it sometime happens that the electrophoretic band patterns come from contaminating DNA (from bacteria for example) and not from the organism under study. On the other hand, the neutrality criterion is not always respected since one does not always know what is amplified: is it part of a gene having undergone selection or conversely a non-coding part of the DNA?

The list of techniques mentioned here is far from comprehensive; among others one can find amplified fragment length polymorphism (AFLP), sequence specific amplified polymorphism (SSAP), single-strand conformation polymorphism (SSCP), minisatellites, sequence tagged sites (STS), single-nucleotide polymorphism (SNP), etc.

A.3.1.2 Analytical methods

The different molecular marking techniques previously described allow one to show polymorphism at the species, population, and individual levels. When one wants to study genetic variability within a population or the differentiation between two populations, these techniques are applied to a large number of individuals and loci. It then becomes necessary to process the data using mathematical methods.

The genotypic and allelic structure of a population

If one again considers the allozyme example (Section A.3.1.1), analysis of the electrophoretic band pattern (Fig. A.11) allows one to identify the genotypes and alleles present in the specimen and to determine their frequency. In the example shown in Fig. A.12, the scientist tested a sample of three individuals for a single locus in a diploid species. Electrophoresis allowed the distinction of three different genotypes for this sample: two homozygotes, one genotype A_1A_1 and one genotype A_2A_2, and one heterozygote A_1A_2. If one now considers a population of N individuals and the following results: n_{11} individuals with genotype, A_1A_1, n_{22} individuals with genotype A_2A_2, and n_{12} individuals with genotype A_1A_2, then the genotypic structure of the population is

$$f(A_1A_1) = n_{11}/N$$

$$f(A_2A_2) = n_{22}/N$$

$$f(A_1A_2) = n_{12}/N$$

where f is the genotypic frequency and $f(A_1A_1) + f(A_2A_2) + f(A_1A_2) = 1$. The allelic structure will be as follows:

$$p = (2n_{11} + n_{12})/2N = f(A_1A_1) + \tfrac{1}{2}f(A_1A_2)$$

and

$$q = 1 - p = (2n_{22} + n_{12})/2N = f(A_2A_2) + \tfrac{1}{2}f(A_1A_2)$$

where p is the frequency of allele A_1 and q is the frequency of allele A_2.

Polymorphism rate, heterozygosity, and allelic diversity

In order to quantify the genetic variability of a population for several loci, it is possible to compute different parameters:

• Polymorphism rate P: this is the proportion of polymorph loci within the set of loci analysed:

$P =$ number of polymorphic loci/number of loci analysed.

For example, if out of 10 allozymes tested within a population, 7 are polymorphic, $P = 7/10 = 0.7$.

However, this parameter is not perfect for quantifying variability since it does not consider the number and frequency of alleles in each of the polymorphic loci. For example, locus B with three alleles B_1, B_2, B_3 brings more variability to a population than locus A with two alleles A_1, A_2. Indeed, the number of combinations of genotypes (G) is higher for locus B than for locus A: $G_{(B)} = 6$ (i.e. B_1B_1, B_2B_2, B_3B_3, B_1B_2, B_1B_3, B_2B_3), whereas $G_{(A)} = 3$ (i.e. A_1A_1,

A_2A_2, A_1A_2). Considering k alleles in a locus $G = k(k+1)/2$. Considering k alleles in two loci, G becomes $k^2(k+1)^2/4$.

• Heterozygosity H: this is the average frequency of heterozygotes observed on each of the loci studied:

$$H = \frac{1}{N}\sum H_i$$

where N is the total number of loci analysed and H_i is the heterozygosity at locus i, i.e. the frequency of heterozygotes at locus i).

• Allelic diversity: another measure of variability is the average number of alleles per locus (total number of alleles/total number of loci analysed).

Note that evolutionary processes such as selection, drift, migration, and mutation affect the parameters presented above. Heterozygosity can also be affected by the mating system of a population. The influence of these processes on genetic variability is not addressed here.

The spatial structure of populations

The spatial subdivision of populations causes a loss of genetic variability within subpopulations under the effect of genetic drift. This is responsible for the decrease in heterozygosity observed with respect to the expected heterozygosity under the assumption of a single large population having reached Hardy–Weinberg equilibrium. In order to estimate the spatial structure of the populations, three fixation indices have been developed:

• F_{IS} measures, within subpopulations, the excess or lack of heterozygotes under the assumption of Hardy–Weinberg equilibrium (the subscript IS means an individual within a subpopulation):

$$F_{IS} = 1 - (H_I / \bar{H}_S)$$

where H_I is the heterozygosity observed on average per individual, for the set of subpopulations and $\bar{H}_S$ is the average heterozygosity under Hardy–Weinberg equilibrium. The value of F_{IS} lies between -1 and 1, negative values indicating an excess of heterozygotes and positive values a lack of heterozygotes compared with the Hardy–Weinberg equilibrium.

• F_{ST} measures the effect of population subdivision, i.e. the lack of heterozygotes resulting from the differentiation between the subpopulations (ST means subpopulation in the whole population):

$$F_{ST} = 1 - (\bar{H}_S / H_T)$$

where H_T is the expected heterozygosity per individual, assuming that the whole population has reached Hardy–Weinberg equilibrium. Its value goes from 0 (no subdivision, no genetic divergence within the population) to 1 (extreme subdivision, complete isolation between subpopulations).

• Finally F_{IT} measures the global reduction in heterozygosity between an individual and the whole population (IT meaning the individual in the whole population): $F_{IT} = 1 - (H_I / H_T)$.

A.3.2 Quantitative traits

Focusing on quantitative traits is often more helpful than the study of molecular markers. Indeed, quantitative traits are usually continuous traits and cannot being sorted by classes of individuals. As a consequence, there exist numerous phenotypes that are slightly differ one from another. Moreover, the expression of phenotypic traits can be modified by environmental factors, either abiotic (e.g. temperature) or biotic (e.g. developmental host). A single genotype can thus produce several phenotypes according to the environmental conditions. This is called phenotypic plasticity. Finally, the genetic determinism of such traits is usually polygenic with numerous possible interactions between alleles (e.g. **dominance**) or between loci (**epistasy**). The intraspecific genetic variability of quantitative traits can be studied at two nested levels: the differentiation between populations and intrapopulation variability. We will briefly present the experimental designs associated with these two levels. We will then widen our reasoning to the study of several quantitative traits.

A.3.2.1 Genetic differentiation between populations

In the fields of ecology and evolution we often want to compare different populations with regard to some quantitative traits supposed to be linked with individual fitness. Such comparisons can help us to understand how the spatial heterogeneity of some environmental factors can lead to different selective pressures and finally to divergent local evolution. Particular processes are concerned in the framework of host–parasite interactions: ecological specialization and host range evolution or co-evolution.

At first sight, the principle of between-population comparison is quite simple since it is necessary to test the populations under the same conditions, this common environment being either controlled (a climate chamber for instance) or natural (*in situ* experiment). However, care must be taken to avoid experimental bias. In particular, the different populations should also be reared in the same conditions for a couple of generations before a comparison can be made. Indeed, it has been shown for numerous organisms that the environment of the parents or grandparents can modify the phenotype of an individual. Moreover, special care must also be taken to maintain the genetic variability of the collected populations. Indeed, a time lag usually exists between the field collection and the test. During this delay, drift events or unintended selection can occur and modify the genetic make-up of the populations. In a mass-rearing culture, the more parents that are used to create the following generation, the less drift there will be. Another method is to create, for each population, numerous isofemale lines. Each of these lines originates from a single pair of parents and only a small number of parents are then used to found the following generations. As a consequence, each isofemale line can be viewed as a small inbred population where heterozygosity will quickly decrease and where alleles will be fixed randomly. The genetic diversity of the population is nevertheless preserved if sufficient isofemale lines have been created (Falconer 1974).

If all these requirement have been followed, statistical differences between populations for some phenotypic traits can only be explained by genetic differences. These differences can then be analysed in terms of local (mal) adaptations, depending on the knowledge on the ecological and evolutionary context.

A.3.2.2 Within-population genetic variability

The within-population genetic variability is an interesting component which determines the capacity of the population to cope with new selective pressures. This is important since numerous habitats often change more or less rapidly or reliably. From an agronomic point of view, the within-population variability also determines the ability to genetically improve the efficiency of useful organisms such as **biological control** agents (Wajnberg 2004).

Quantifying the genetic variability of a quantitative trait is not simple (Falconer 1974). Indeed, the observed phenotypic variance (V_P) is the sum of the variances induced by the environment (V_E) and the actual genetic differences between individuals (V_G). This genetic variance can itself be partitioned into several components: (1) variance associated with the additive effects between alleles (V_A); (2) variance associated with dominance effects (V_D); and (3) variance associated with epistatic differences (V_I). These relationships can be summarized as follows:

$$V_P = V_E + V_G$$

with

$$V_G = V_A + V_D + V_I.$$

A particular parameter, the so-called heritability, can be defined from these equations. The heritability of a quantitative trait measures the contribution of genetic differences to the total phenotypic variance. Broad and narrow heritabilities are respectively estimated as follows:

$$H_b = V_G/V_P$$

and

$$H_n = V_A/V_P.$$

Several experimental designs exist to more or less accurately quantify the different variances and the heritability (Falconer 1974; Wajnberg 2004 for a discussion concerning biological control):

• Isofemale lines analysis. In this case, the variance between several isofemales of a single population is compared with the within-lines variance. This so-called residual variance only estimates the environmental variance (V_E) since most of the alleles should be fixed in each isofemale line. A significant difference between the between-isofemale lines and the residual variance demonstrates a genetic variability for the quantitative traits. Nevertheless, the estimation of the genetic variance is rather rough.

• Parent–offspring regression. In this case, the quantitative trait is measured on randomly paired parents as well as on their offspring. Intuitively, the more the trait variability is genetically determined, the more similar the parents and their offspring should be. Graphically, the value of one of the parents (or the mid-parent) can be then plotted on the x-axis and the value of the offspring on the y-axis. The slope of the regression estimates the broad heritability of the quantitative traits.

• Sib-mating analysis. Several quantitative genetic designs allow one to more accurately partition the genetic variance (V_G). It is worth noting that such experiments are generally quite time-consuming and require a good experimental knowledge of the organisms. In one design N_1 males are taken randomly from a population and each of these males is crossed with N_2 females, producing $N_1 \times N_2$ full-sib families. In each of these families, r individuals are taken randomly and the quantitative trait is measured for each individual. A specific statistical analysis, called nested analysis of variance (ANOVA), allows one to test the effect of the two levels (i.e. the influence of the males and the nested influence of the females) and to estimate the narrow heritability and additive variance. More complicated designs exist and offer the possibility to more precisely estimate other components.

A.3.2.3 Correlations between characters

The two previous sections focused on a single quantitative trait. Of course, numerous quantitative traits usually contribute to the fitness of individuals. Moreover, most of these traits are more less linked one with another. This linkage could be genetic if the same locus simultaneously determines several traits ; this called **pleiotropy**. The linkage can also be physiological if two quantitative traits depend on the same nutrients or on shared metabolic processes. Whatever the origin of the linkage, the traits are not able to evolve independently of with another. '**Trade-off**' appears when the correlation between two traits is negative in term of fitness. In this case, high performance for one character is then associated with low performance for the other and it is not possible to maximize the two traits simultaneously. This trade-off concept is central to numerous fields in ecology and evolution (Roff 1992; Stearns 1992).

From a methodological point of view, the simplest approach is to use phenotypic correlation between two characters as an approximation of their genetic correlation. This approach has nevertheless been criticized, and other designs allow one to more precisely estimate the genetic correlations. In fact, there are some analogies between the estimation of the genetic variance of a single character and the estimation of the genetic co-variance between two characters. That's why sib-mating analysis (see above), for instance, can be also used to estimate the genetic co-variance between two quantitative traits (Roff 1997).

Glossary

Additivity, additive effect (for a gene): A gene is said to have an additive effect on a trait if it modifies trait values by the same amount whatever the identity of the other genes at the same locus (in this case non-additivity means dominance) or at other loci (in this case non-additivity means **epistasy**). Thus within a two-allele locus (with alleles A and a) that codes for a phenotype X additivity means that $X_{AA} - X_{Aa} = X_{aA} - X_{aa}$. With two loci A,a and B,b (and assuming for simplicity haploid genotypes, but this is straightforwardly extended to diploids), additivity means $X_{AB} - X_{Ab} = X_{aB} - X_{ab}$. For any gene (with additive effects or not) the effect on a phenotype can actually be decomposed as the sum of an additive part and a non-additive part. The additive component of the effect of allele A on phenotype X is the difference between $X_{A \cdot}$ (the average X over individuals that have at least one A allele) and $X_{\cdot \cdot}$ (the overall population mean). The non additive component represents all the rest, and varies depending on the other allele (for example in an AB individual, the non-additive effect of A is $X_{AB} - X_{A \cdot}$).

Aetiological agent: The causative agent of a disease condition, often a micro-organism or toxin.

Allee effect: A negative effect for small populations, resulting from a decrease in growth rate with density. The Allee effect can lead small populations to extinction.

Allopatric (speciation): Evolution of new species from geographically isolated populations.

Allozymes: Enzymes encoded by the same gene but different in form or molecular structure from the present alleles. Their function remains the same whatever the allelic form of the gene.

Ankylostoma: A genus of **nematode** intestinal parasites that consists of several species. *Ankylostoma duodenale* is the common hookworm in humans. *Ankylostoma braziliense*, *Ankylostoma ceylonicum*, and *Ankylostoma caninum* occur primarily in cats and dogs, but all have been known to occur in humans.

Antibodies: *see* immunoglobulins.

Antigen: macromolecule recognized by immune cells and that stimulates the immune response. Antigens are generally proteins, polysaccharides, and their lipid derivates.

Bacteriophage (or **phage**): A virus that infects bacteria.

Biological control: the use of living organisms (called **biological control or biocontrol agents**) in order to limit damage caused by noxious crop pests. Insect parasitoids are the main biological control agents used for crop protection.

Biotic capacity: in relation to a given environment, corresponding to the population density obtained in the absence of regulation by extrinsic biotic factors (predation or parasitism).

Biotope: An area that is uniform in environmental conditions and in its distribution of animal and plant life.

Bottleneck: A decrease in population abundance leading to a reduction in **polymorphism**.

Castrating parasites: Parasites that cause the arrest of host reproduction. This arrest may result from a number of mechanisms, ranging from the physical substitution of host gonads by the parasite to the manipulation of host physiology.

Cestodes: Parasitic flatworms ('tapeworms'), the adults of which live in the digestive tract of vertebrates. Cestodes have a flat segmented body, shaped as a ribbon, with a specialized 'head' (scolex) with hooks and suckers enabling them to fix onto the host. The strobila (body segments, called proglottids, from the neck region down) ends with proglottids containing many eggs. With no actual digestive tract, cestodes consume food pre-digested by their host. Their life cycle can be complicated (requiring one or several intermediate hosts). There is practically no free-living phase.

Chromogen substrate: A colourless compound that can be converted into a coloured product by an enzymatic reaction.

Cline: A gradual change in a morphological or molecular character or feature across the distribution of a species or a population, usually related to an environmental or geographical transition.

Colostrum: a type of milk secreted by mammals during the last days of gestation and the very first days after parturition. Colostrum is rich in **antibodies**.

Continuous traits: A trait whose variations are measured on a scale rather than by classification into categories, e.g. height and weight (opposite: discontinuous traits).

Counter-selection: Elimination through the process of natural selection.

Cytoplasmic incompatibility: Post-zygotic reproductive barrier between conspecific individuals harbouring different **endosymbionts** (*Wolbachia* variants for instance). According to the type of cytoplasmic incompatibility, eggs originating from fertilization between incompatible gametes can either abort or develop into males.

Density-dependent increase: Increase in population size inversely proportional to the current number of individuals. Hence, a small population will show a greater increase in the number of individuals than a larger population. Such a mechanism, if it is the only one to determine the size of a population, will lead to limited maximum population size corresponding to the carrying capacity of the environment.

Diagnostic markers: Loci showing obvious differences in sequences from one **taxon** to another. Diagnostic markers allow the definite identification of an individual as belonging to one or another taxon.

Diplo-diploidy: Mechanism of sex determination in which both sexes are diploids.

Diptera: An insect order. Insects in this order have a single pair of wings, the other pair having been reduced to form halteres, that function as balancing organs during flight. This order mainly contains flies and mosquitoes.

Disgenesis: Abnormal development. By extension, a physiological dysfunction arising from a genetic source.

Distorting gene: A 'selfish' gene which lead to the elimination of part of the genome during the meiosis.

Dominance: Non-additive interaction between two alleles of the same genes. For a diploid loci with two alleles A and a and quantitative trait X, dominance occurs if $X_{AA}-X_{Aa} \neq X_{Aa}-X_{aa}$.

Ebola fever: Ebola is the common term for a group of viruses belonging to the genus Ebolavirus, family Filoviridae, which cause Ebola haemorrhagic fever. The disease is deadly and encompasses a range of symptoms including vomiting, diarrhoea, general body pain, internal and external bleeding, and fever.

Ecdysteroids: Hormone implicated in the insect moulting process.

Ectoparasitism: A type of development in which the body of the parasite or **parasitoid** (hence called an ectoparasitoid) is mainly outside the host. In a number of cases only the head of the parasitoid is inside the host in order to allow for nutrition (opposite: **endoparasitism**).

Encapsulation: A defence mechanism used by some hosts leading to the formation of a cell layer around foreign bodies like a **parasitoid** egg. Such a mechanism usually leads the parasitoid to die by asphyxiation.

Endemic: A species living exclusively within a limited area.

Endogenous: A phenomenon or a substance originating within an organism or a tissue (opposite: **exogenous**) (for example hormones are endogenous secretions).

Endoparasitism: A type of development in which the body of the parasite or **parasitoid** (hence called an endoparasitoid) is mainly inside the host. In this case, the development of the parasitoid usually occurs within the haemocoel (where the host's main organs are) or in other different tissues (opposite: **ectoparasitism**).

Endosymbionts: Symbiotic organisms for which all or most of the life cycle occurs within cells (either the cytoplasm or the nucleus) of its host.

Environmental correlation: A component of the **phenotypic correlation** that relies on the influence of environmental conditions on both traits.

Epibiont: Non-parasitic organism that lives on another organism.

Epidemiology: The study of the distribution and determinants of disease in a population. An epidemiological study often compares two groups of people who are alike except for one factor, such as exposure to a chemical or the presence of a health effect. The investigators try to determine which factor is associated with the health effect.

Epistasy (epistatic): Non-additive interaction between two alleles at different loci. For two haploid loci A,a and B,b, and one trait X, epistasy occurs if $X_{AB}-X_{Ab} \neq X_{aB}-X_{ab}$, i.e. substitution of B by b has a different effect depending on whether there is A or a at the other locus.

Epizooty: An epidemic afflicting wild or domestic animals.

Erythema: Redness or inflammation of the skin or mucous membranes.

Exogenous: Originating from outside the organism, caused by external factors (opposite: **endogenous**).

F_1, F_2: Hybrids of the first and second generation. F_1 hybrids result from direct crossing between individuals of parental types. F_2 hybrids result from crossing between F_1 hybrids.

Fitness: (1) [of a gene] The number of copies of a gene that are present in the following generation, or the rate of multiplication of these copies, relative to other genes; the genetic contribution of an individual to the next generation. (2) [**phenotypic fitness**, or **individual fitness**] The capacity of an individual or the average capacity of a phenotype of producing descendants capable of reproducing in turn, relative to other individuals in the same population at the same time. Fitness may depend on the abiotic, biotic, and genomic environment.

Fluctuant asymmetry: Small random deviations instead of the perfect symmetry of a bilateral structure resulting from a genome. Fluctuant asymmetry is regarded as the expression of developmental disturbance. It is the only form of asymmetry which can be considered as a useful indication of the stress to which the organisms are subjected.

Fluorophore: A fluorescent molecule that absorbs light at a certain wavelength (excitation) and rapidly emits light of a different wavelength (emission). When exposed to the excitation wavelength, the fluorophore can emit visible light at the emission wavelength.

Free-living species: Organisms that do not live in an intimate interaction with another living organism (cf. parasites or **symbionts**).

Gametic disequilibrium (also called **linkage disequilibrium**): Non-random association of two alleles at different loci within gametes. Considering two diallelic loci (A,a and B,b) linkage disequilibrium is quantified as the difference between the observed frequency of AB gametes and the frequency expected under random assortment , i.e. $p_A \times p_B$.

Gene flow: Movement of genes between different **taxa** or populations.

Gene-for-gene: Recognition system between hosts and pathogens. According to the classical model, parasite genes produce proteins which are recognized by their hosts. A host recognizing a parasite is resistant. Pathogen genotypes able to overcome resistance are called virulent. Each resistance gene of the host has a counterpart **virulence** gene in the parasite. This interaction model constitutes the dogma of plant pathology and is the first proposed host–parasite recognition model.

Genetic assimilation: Process through which a phenotype, originally produced in response to an environmental stimulus, is expressed constitutively independently of the presence of the stimulus.

Genetic correlation: A component of the **phenotypic correlation** that relies on the influence of the individual genotype on both traits.

Genetic drift: Random fluctuation in allele frequencies from one generation to the next due to sampling effects in a finite population size.

Genetic variance (of a trait): A quantity that measures all interindividual differences that can be inherited through reproduction. When the genetic variance is zero, all differences among individuals for the considered trait must result from differences in the living conditions experienced by individuals (environment) and will not be transmitted to offspring.

Genotype: The entire genetic identity of an individual, including alleles, or gene forms, that do not show as outward characteristics.

Haemolymph: Haemolymph is the arthropod equivalent to blood, particularly in insects.

Hantavirus: Hantaviruses belong to the Bunyaviridae family of viruses and are rodent-borne agents. The word hantavirus is derived from the Hantan River, where the Hantaan virus (the **aetiological** agent of Korean haemorrhagic fever) was first isolated. The disease associated with Hantaan virus is called Korean haemorrhagic fever or haemorrhagic fever with renal syndrome.

Haplo-diploidy: Sex determination mechanism in which females are diploids and males are haploids. Haplo-diploidy is particularly frequent in hymenopteran **parasitoids**. In this case, females originate from fertilized eggs whereas males originate from unfertilized eggs (synonym = arrhenotoky).

Hemiclone: Through the process of **hybridogenesis**, hybrids transmit half of their genome (a **hemigenome**) without any recombination with another parental genome. This half genome is then propagated from generation to generation without alteration. This situation corresponds to the definition of a clone, but as it only concerns half of the hybrid genome it is called a hemiclone.

Hemigenome: In diploids (who inherit half of their genes from each parent), the term hemigenome refers to each parental inheritance: maternal or paternal hemiclone.

Heritability (adj. **heritable**): A trait is considered heritable when (1) it varies among individuals within a population and (2) at least part of this variation can be transmitted from parent to offspring. Heritability in the narrow-sense is the proportion of variation that can be transmitted (compared to total variation). It is defined as $h^2 = V_A/V_P$, where V_A is the additive *genetic variance* and V_P is total *phenotypic variance*.

Heterosis: The superiority of hybrids when compared with their parents, for any measurable character (also known as hybrid vigour).

Heteroxenous life cycle: Life-cycle of a parasite requiring a succession of several hosts according to the stage of development of the parasite: intermediate host(s) for larval stage, definitive host for adult stage.

Heterozygote (heterozygous, heterozygosity): An individual that has two different alleles at a given locus. Heterozygosity, for a locus, is the frequency of heterozygous individuals in the population. Heterozygosity, for an individual, is the frequency of heterozygous loci within its genome.

Hookworm: *see Ankylostoma*.

Hybridogenesis: [unusual] Hemiclonal form of reproduction of hybrids by which F_1 individuals transmit only one of the parental genomes. Examples are found in fish, amphibians, and stick-insects.

Hybridogenetic process: A genetic process observed in hybrids when some genes (*see* **distorting gene**) belonging to one of the parental genomes lead to the elimination of the other parental genome in the germ line.

Hygiene behaviour: Any behaviour helping to maintain or improve an organism's health.

Hymenoptera: An insect order. With the Coleoptera they are the most diverse order of insects. There are more than 120,000 species of Hymenoptera. In this order we find bees, ants, wasps, and most insect **parasitoids**.

Hyperparasitoid: A particular type of **parasitoid** attacking hosts that are themselves parasitoids. Facultative hyperparasitoids can also develop as simple parasitoids.

Hypothetico-deductive approach: An approach consisting of proposing a hypothesis and deducing from it predictions the relevance of which can be tested empirically. Any contradiction between the observed facts and the prediction leads to the hypothesis being rejected.

Hypoxic: Condition linked to a deficiency in the oxygen being provided to tissues in respect of cellular needs.

Idiobiont: A **parasitoid** whose pre-imaginal development is ensured by the host resources present at **oviposition**, leading to the quick death of the host (opposite **koinobiont**).

Immunoglobulins: Proteins found in blood and tissue fluids produced by cells of the immune system to bind to substances in the body that are recognized as foreign **antigens**.

Inbreeding depression: The relative decrease in fitness of inbred individuals (produced by consanguineous matings) compared with outbred individuals.

Incidence: The number of instances of illness commencing, or of persons falling ill, during a given period in a specified population.

Infectious disease: A disease resulting from the presence and activity of a microbial agent.

Infectious dose: The number of infecting organisms generally required to cause disease.

Infra-population: Population of a parasite species harboured in a host individual (in its intestine, for example).

Integrated pest management: Optimal combination of control methods including biological, cultural, mechanical, physical, and/or chemical controls to reduce pest populations to an economically acceptable level with as few harmful effects as possible to the environment.

Intermediate host: A necessary second host needed for the accomplishment of a parasite's life cycle. The parasite can develop in this host but not reproduce sexually; this occurs in the definitive host.

Intertidal zone: Part of the shoreline exposed at low tide.

Introgression index: The percentage of the genome of one of the parental **taxa** (considered as the reference) present in the individuals of a hybrid zone. It is expressed by the average of the allelic frequency of taxon A in diagnostic loci between taxa A and B. For example, in *musculus* and *domesticus* mice, the percentage of *domesticus* alleles in hybrid genomes can be determined through the study of 10 **allozymic** diagnostic loci. It varies roughly from 100% to 0% between the western part and the eastern part of the hybrid zone.

Juvenile hormone: Hormone implicated in the moulting process in insects.

Kairomone: Products emitted by an organism and used by another species leading to a detrimental effect for the emitter. For instance, some odours produced by phytophagous insects are used by **parasitoids** as a cue for host location.

Koinobiont: A **parasitoid** whose pre-imaginal development does not imply the quick death of its host but requires a rather long physiological interaction between the two partners (opposite **idiobiont**).

Latrines: A site specifically reserved for heaping excrement.

Legionnaires' disease (**legionellosis**): A form of pneumonia caused by the *Legionella* bacterium.

Linkage disequilibrium: Statistical association between alleles of different loci. It is defined mathematically as the difference between the frequency of a gamete and its expected frequency based on the frequencies of the individual alleles that compose the gamete. Put differently, it is the difference between the observed frequency of a gamete and its frequency expected under random mating. Linkage disequilibrium may be generated by non-random mating, population subdivision, or selection. It decreases over generations because of recombination.

Long-lasting interaction: Any association of organisms which belong to different **taxa** (heterospecific association) involving the coexistence of partners during part of their life cycle. For example, the lichen (association fungus/alga) or the tapeworm (association tapeworm/human). Such interactions differ from predation by the length in time for which the partners coexist (e.g. a lion kills and immediately eats a gazelle) and the exchanges between partners (e.g. the predator only feeds on its prey, it does not interfere with its behaviour or its habitat).

Lyme disease: A disease that affects the joints, nervous system, and heart that is transmitted by the deer tick and is caused by a parasite known as *Borrelia*.

Lymphocytes: A class of vertebrate immune cells that that respond specifically to foreign antigens. Some lymphocytes enter cells and secrete antibodies into the blood. Others have antibodies exposed on their membranes, and destroy antigen-bearing cells by contact. There are two classes of lymphocytes: B lymphocytes (B cells) and T lymphocytes (T cells).

Macrophages: A class of immune cells that can ingested other cells (or dead cells and cellular debris) by phagocytosis.

Metapopulation: A group of populations, usually of the same species, which exist at the same time but in different places. They are more or less interconnected by dispersal. The term is especially relevant, and different from that of 'subdivided population', when populations are subject to disturbances leading to local extinctions.

Metazoa: Multicellular animals.

Migrations: Genes flow (gametes, individuals) between populations.

Morbidity: Effect of an infectious disease on the general condition of an infected host individual. State of being ill or diseased. Morbidity is the occurrence of a disease or condition that alters health and quality of life.

Mutation: Any change in the DNA sequence. There are several kinds of mutation: e.g. point mutation, deletion, insertion, chromosomal rearrangement.

Natural selection: The process by which individuals that are well adapted to their environment contribute more than others to gene transmission.

Nematodes: Roundworms having bilateral symmetry, a non-segmented body, and a complete digestive tract. The body is covered with a thick cuticle. Their growth is discontinuous. They are dioecious and have four juvenile stages. Genetic studies indicate that arthropods are very closely related to nematodes. Depending on the species, they are free-living or parasitic, with one or several hosts.

Overdominance: Situation in which the phenotypic value of a heterozygous genotype (AB) is greater than that of both corresponding homozygotes (AA and BB).

Oviposition: The action of laying an egg by a female **parasitoid** inside or outside a host.

Parapatric (species): Species whose areas of distribution do not overlap (opposite of sympatric species) but are adjacent (cf. **allopatric**).

Parasite abundance: Mean number of parasites per host individual, including uninfected hosts.

Parasite intensity: Mean number of parasites per infected host individual, excluding uninfected hosts.

Parasite load: Number of parasites belonging to the same species per host. A term mostly used for macroparasites and ectoparasites.

Parasitoid: An organism that develops within or on another organism (its host), that feeds on this host, and acts directly or indirectly on its development in such a way that it ends up killing it. Adult parasitoids are free-living.

Parental effects: Environmental effects transmitted over generations. For example, 'good' quality mothers may produce 'good' quality offspring because they transmit more resources in their eggs. Parental effects are often limited to maternal effects, because in general fathers do not influence the phenotypic values of their offspring (this is of course not the case when fathers contribute to parental care).

Pathogenicity: The ability of an organism (a pathogen) to cause disease in another organism.

Phenotypic correlation (between two traits): A statistical association between two individual traits (A, B) within a population. It can be positive (in this case individuals with high A tend to have high B), or negative (individuals with high A have low B). Mathematically, it is defined as $r(A,B) = Cov(A,B)\ V(A)^{-1/2}\ V(B)^{-1/2}$, where Cov means covariance and V means variance.

Phenotypic plasticity: The ability of an organism to show one phenotype among others given environmental conditions. The value of the trait of a given genotype varies across different environments.

Phylogeny (adj. **phylogenetic**): The genealogical relationships among living species or **taxa**, represented in general as tree. Branch tips represent the studied living taxa and the root is their common ancestor. Reconstruction of the evolutionary history of organisms and their relatedness.

Pleiotropy (adj. **pleiotropic**): The phenomenon whereby a single gene or mutation affects several apparently unrelated aspects of the phenotype. For example, the S allele of a human haemoglobin gene (sickle-cell anaemia allele) impairs oxygen storage in the blood and increases resistance to malaria.

Pleiotropy (antagonist): A single allele can induce several apparently independent phenotypic effects. When pleiotropic effects negatively affect a function or a phenotype, they are said to be antagonistic. For example, in Siamese cats, the allele responsible for colour gradation (light coloured where the body is warm and dark color at the cold extremities) also causes the convergent strabismus.

PolyDNAvirus: Viral symbionts that persist as stably integrated proviruses in the genome of some **parasitoids** (Braconidae and Ichneumonidae). Their replication occurs within the ovaries of the wasp and they are injected during **oviposition** within the host where their expression modifies the host physiology and enhances the development of the wasp (and also their own transmission).

Polygenic trait: Phenotypic trait coded by several genes with several alleles.

Polymorphism: Genetic variation (for a trait or at a given locus) within a population.

Post-ovipositional steps: Steps of the parasitic process occurring after **oviposition**. The main post-ovipositional steps include host immune response.

Post-zygotic (reproductive isolation): Mating and fertilization are possible between **taxa**, but the hybrids cannot develop and produce viable and fertile progeny.

Pre-ovipositional steps: Steps of the parasitic process occurring before **oviposition**. The main pre-ovipositional steps include the location of a favourable habitat and a suitable host.

Pre-zygotic (reproductive isolation): Mating or fertilization between two **taxa** are prevented in different ways: the taxa live in different habitats (ecological isolation); their periods of reproduction are different (temporal isolation); they display special courtship rituals; their genitalia are anatomically incompatible; their gametes do not merge (gametic isolation).

Prevalence: The percentage of a host population infected by a given parasite.

Promiscuity: Exaggerated proximity among individuals of the same species, resulting from high population density; the term 'sexual promiscuity' refers to the fact that males and females within a species copulate with several individuals, often within the same social group.

Pro-ovigeny: The possession by adult female wasps of a complete stock of mature oocytes when they emerge (cf. **synovigeny**).

Prophylactic (behaviour, drug): **Prophylactic** behaviour or drugs reduce the likelihood of illness or infection.

Red Queen hypothesis: A hypothesis proposed by L. Van Valen, stating that there is a continuous arms race between host and parasite, which maintains a continuous selection pressure for each species to match to the evolutionary changes of the other. The reference to the Red Queen comes from *Through the Looking-glass* by Lewis Carroll: Alice meets the Red Queen of an imaginary country where the landscape is in perpetual movement. People have to run forward to stay in the same place, just as hosts and parasites must always change to keep each other in check.

Reservoir: Anything (a person, animal, plant, substance) in which an infectious agent normally lives and multiplies: an infectious agent depends on a reservoir for its survival.

Resistance to parasites: The ability to avoid infection, eliminate parasites, or decrease parasite loads for a host in contact with a given parasite.

Retro-crossings: Crossing between a F_1 hybrid and one of the parents ('backcrossing').

Septicaemia: A disease affecting many organ systems due to toxins in the blood which are released by bacteria or other micro-organisms. Signs include fever, pinpoint bruises on mucous membranes, and lesions in the joints, heart valves, eyes, or other organs.

Sexual dimorphism: Phenotypic difference between males and females of a species. Sexual dimorphism can be linked to the intensity of sexual selection.

Sexual selection: Competition among members of one sex (generally males) for fertilization opportunities with the other sex.

Sociality: The tendency of organisms to live and organize in groups with their conspecifics.

Species (in the biological sense): A group of organisms who can interbreed and produce fertile offspring.

Subspecies: **Taxa** having reached some level of differentiation (genetic divergence without morphological divergence), but which are not reproductively isolated.

Superparasitism: **Oviposition** in a host already opposite: parasitized by the female **parasitoid** itself (self-superparasitism) or by a conspecific (conspecific superparasitism).

Sympatric (speciation): Evolution of new species from populations having the same geographical distribution.

Sympatry: Spatial overlap between the areas of distribution of two species or populations (opposite: allopatry).

Synomone: Chemical compounds whose production benefits both the emitter and receiver.

Synovigeny: The ability of a female to produce or mature oocytes throughout its adult life (cf. **pro-ovigeny**).

Systematics: The branch of biology that classifies living organisms.

Taxa (singular **taxon**, or **taxonomic unit**): Categories or taxonomic groups like phylum, order, family, genus, or species. 'Taxon' must be preferentially used when the taxonomic level of the group is unknown.

Taxis: Directed movement of an organism moving freely in space in response to a stimulus from its environment. A taxis is positive if the organism's movement takes it closer to a stimulus and negative if it moves it away from it.

Territoriality: Characteristic of a species whose individuals, alone or in family groups, defend a limited area within their habitat or are excluded from it by conspecifics.

Thelytoky (or **thelytokous parthenogenesis**): Mode of reproduction whereby females can produce daughters in the absence of males, mating, and fertilization, through parthenogenesis.

Therapeutics (behaviour, food supplement): For curing or treating a disease.

Tolerance: The ability of a host to maintain a high probability of survival or high fecundity even with high parasite loads. Contrary to the notion of resistance, tolerance does not imply the induction of defence mechanisms or specific physiological adaptations for the host to stand the parasitic load. The physiological characteristics allowing the host to be tolerant exist apart from the parasitic infection.

Trade-off: A trade-off is a situation that involves losing one quality or aspect of something in return for gaining another quality or aspect. Often this means that both traits compete for a common resource. A simple example is the trade-off between size and number of eggs. Species that lay more eggs tend to have smaller eggs. It is assumed that with a given amount of energy, one cannot increase the size of eggs without decreasing their number. Trade offs may occur at different levels: among individuals within a population, among species etc.

Transmission (horizontal or vertical): A parasite is vertically transmitted when infected parents transmit it to their offspring. In practice, vertical transmission occurs maternally. When transmission does not occur from parent to offspring it is called horizontal.

Vector: A carrier, especially the animal (usually an arthropod) that transfers an infective agent from one host to another. In molecular biology, a vector is a mean (virus or plasmides for instance) to transfer more or less durably a DNA sequence of interest into a cell.

Virulence: In the context of plant pathology and **host-parasitoid** interactions, virulence corresponds to the capacity of a pathogenic or **parasitoid** genotype to circumvent the resistance of host genotype (*see* '**gene-for-gene**'). Also, more broadly, the ability of a pathogen to cause damage to its host.

Zoonosis: Diseases caused by agents transmitted from animals to humans.

References

Abdel-Moneim, A.S., Abdel-Gawad, M.M.A. (2006). Genetic variations in maternal transfer and immune responsiveness to infectious bursal disease virus. *Veterinary Microbiology* **14**: 16–24.

Able, D.J. (1996). The contagion indicator hypothesis for parasite-mediated sexual selection. *Proceedings of the National Academy of Sciences USA* **93**: 2229–2233.

Adamo, S.A. (1999). Evidence for adaptive changes in egg laying in crickets exposed to bacteria and parasites. *Animal Behaviour* **57**: 117–124.

Adamo, S.A. (2002). Modulating the modulators: parasites, neuromodulators and host behavioral change. *Brain, Behavior and Evolution* **60**: 370–377.

Adamo, S.A. (2004). How should behavioural ecologists interpret measurements of immunity? *Animal Behaviour* **68**: 1443–1449.

Agnew, P., Bedhomme, S., Haussy, C., Michalakis, Y. (1999). Age and size at maturity of the mosquito *Culex pipiens* infected by the microsporidian parasite *Vavraia culicis*. *Proceedings of Royal Society B: Biological Sciences* **266**: 947–952.

Ågren, G.I., Bosatta, E. (1996). *Theoretical ecosystem ecology*. Cambridge University Press, Cambridge.

Aguirre, A.A., Ostfeld, R.S., Tabor, G.M., House, C., Pearl, M.C. (2002). *Conservation medicine. Ecological health in practice*. Oxford University Press, Oxford.

Albon, S., Stien, A., Irvine, R.J., Ropstad, R., Halvorsen, O. (2002). The roles of parasites in the dynamics of a reindeer population. *Proceedings of Royal Society B: Biological Sciences* **269**: 1625–1632.

Alibert, P., Renaud, S., Dod, B., Bonhomme, F., Auffray, J.C. (1994). Fluctuating asymmetry in the Mus musculus hybrid zone: a heterotic effect in disrupted co-adapted genomes. *Proceedings of Royal Society B: Biological Sciences* **258**: 53–59.

Allison, A.C. (1954). Notes on sickle-cell polymorphism. *Annals of Human Genetics* **19**: 39–57.

Alonso-Alvarez, C., Bertrand, S., Devevey, G., Prost, J., Faivre, B., Sorci, G. (2004). Are carotenoids limiting resources? An experimental test of dose-dependent effect of carotenoids and immune activation on sexual signals and antioxidant activity. *The American Naturalist* **164**: 651–659.

van Alphen, J.J.M., Visser, M.E. (1990). Superparasitism as an adaptive strategy for insect parasitoids. *Annual Review of Entomology* **35**: 59–79.

Altizer, S., Nunn, C.L., Thrall, P.H., Gittleman, J.L., Antonovics, J., Cunningham, A.A., Dobson, A.P., Ezenwa, V., Jones, K.E., Pedersen, A.B., Poss, M., Pulliam, J. (2003). Social organization and parasite risk in mammals: integrating theory and empirical studies. *Annual Review of Ecology and Systematics* **34**: 517–547.

Amat, I., Castello, M. Desouhant, D., Bernstein, C. (2006). The influence of temperature and host availability on the host exploitation strategies of sexual and asexual parasitic wasp of the same species. *Oecologia* **148**: 153–161.

Amundsen, T. (2000). Why are female birds ornamented? *Trends in Ecology and Evolution* **15**: 149–155.

Amundsen, T., Forsgren, E. (2001). Male mate choice selects for female coloration in a fish. *Proceedings of the National Academy of Sciences USA* **98**: 13155–13160.

Anderson, E. (1948). Hybridization of the habitat. *Evolution* **2**: 1–9.

Anderson, R.C. (1972). The ecological relationships of meningeal worm and native cervids in North America. *Journal of Wildlife Diseases* **8**: 304–310.

Anderson, R.M., Gordon, D.M. (1982). Processes influencing the distribution of parasite numbers within host populations with special emphasis on parasite-induced host mortalities. *Parasitology* **85**: 373–398.

Anderson, R.M., May, R.M. (1978). Regulation, stability of host-parasite population interactions. I. Regulatory processes. *Journal of Animal Ecology* **47**: 219–247.

Anderson, R.M., May, R.M. (1982). Co-evolution of hosts and parasites. *Parasitology* **85**: 411–426.

Anderson, R.M., May, R.M. (1985). Helminth infections of humans: mathematical models, population dynamics, control. *Advances in Parasitology* **24**: 1–101.

Anderson, R.M., May, R.M. (1991). *Infectious diseases of humans: dynamics and control*. Oxford: University Press, Oxford.

Andersson, M. (1994). *Sexual selection*. Princeton University Press, Princeton, NJ.

Apanius, V. (1998). Stress and immune defense. In: *Advances in the study of behavior* (ed. A.P. Møller), pp. 133–153. Academic Press, New York.

Apanius, V., Penn, D., Slev, P.R., Ruff, L.R., Potts, W.K. (1997). The nature of selection on the major histocompatibility complex. *Critical Reviews in Immunology* **17**: 179–224.

Arnold M.L. (1997). *Natural hybridization and evolution*. Oxford University Press, Oxford.

Arnold, M.L. (2004). Transfer and origin of adaptations through natural hybridization: were Anderson and Stebbins right? *Plant Cell* **16**: 562–570.

Arnold, M.L., Hodges, S.A. (1995). Are natural hybrids fit or unfit relative to their parents? *Trends in Ecology and Evolution* **10**: 67–71.

Arnott, S.A., Barber, I., Huntingford, F.A. (2000). Parasite-associated growth enhancement in a fish–cestode system. *Proceedings of the Royal Society B: Biological Sciences* **267**: 657–663.

Aron, J.L., Patz, J.A. (2001). *Ecosystem change and public health. A global perspective*. The Johns Hopkins University Press, Baltimore, MD.

Askew, R.R., Shaw, M.R. (1986). Parasitoid communities: their size, structure and development. In: *Insect parasitoids* (ed. J.K. Waage, D. Greathead), pp. 225–264. Academic Press, London.

Aufreiter, S., Mahaney, W.C., Milner, M.W., Huffman, M.A., Hancock, R.G.V., Wink, M., Reich, M. (2001). Mineralogical and chemical interactions of soils eaten by chimpanzees of the Mahale Mountains and Gombe Stream National Parks, Tanzania. *Journal of Chemical Ecology* **27**: 285–308.

Ayling, A.M. (1981). The role of biological disturbance in temperate subtidal encrusting communities. *Ecology* **62**: 830–847.

Babendreier, D., Bigler, F., Kuhlmann, K. (2005). Methods used to assess non-target effects of invertebrate biological control agents of arthropod pests. *BioControl* **50**: 820–870.

Baggen, L., Gurr, G. (1998). The influence of food on *Copidosoma koehleri* (Hymenoptera: Encyrtidae), and the use of flowering plants as a habitat management tool to enhance biological control of potato moth, *Phthorimaea operculella* (Lepidoptera: Gelechiidae). *Biological Control* **11**: 9–17.

Baggen, L., Gurr, G.M., Meats, A. (1999). Flowers in tri-trophic systems: mechanisms allowing selective exploitation by insect natural enemies for conservation biological control. *Entomologia Experimentalis et Applicata* **91**: 155–161.

Bakker, T.C.M., Mazzi, D., Zala, S. (1997). Parasite-induced changes in behaviour and color make *Gammarus pulex* more prone to fish predation. *Ecology* **78**: 1098–1104.

Ballabeni, P. (1995). Parasite-induced gigantism in a snail: a host adaptation? *Functional Ecology* **9**: 887–893.

Barbault, R. (1981). *Écologie des populations et des peuplements*. Masson, Paris.

Barbault, R. (1995). *Écologie des peuplements, structure et dynamique de la biodiversité*. Masson, Paris.

Barton, D.P. (1997). Introduced animals and their parasites: the cane toad, *Bufo marinus*, in Australia. *Australian Journal of Ecology* **22**: 316–324.

Barton, N. (2001). The role of hybridization in evolution. *Molecular Ecology* **10**: 551–568.

Barton, N.H., Hewitt, G.M. (1985). Analysis of hybrid zones. *Annual Review of Ecology and Systematics* **16**: 113–148.

Beckage, N.E., Gelman, D.B. (2004). Wasp parasitoid disruption of host development: implications for new biologically based strategies for insect control. *Annual Review of Entomology* **49**: 299–330.

Beddington, J.R., Free, C.A., Lawton, J.H. (1975). Dynamic complexity in predator–prey models framed in difference equations. *Nature* **255**: 58–60.

Begon, M., Mortimer, M. (1986). *Population ecology: a unified study of animals and plants*, 2nd edn. Blackwell Scientific Publications, Oxford.

Begon, M., Bowers, R.G., Kadianakis, N., Hodgkinson, D.E. (1992). Disease and community structure: the importance of host self-regulation in a host–pathogen model. *The American Naturalist* **139**: 1131–1150.

Bendich, A. (1989). Carotenoids and the immune response. *Journal of Nutrition* **119**: 112–115.

Bernstein, C. (2000). Host–parasitoid models: the story of a successful failure. In: *Parasitoid population biology* (ed. M.E. Hochberg, A.R. Ives), pp. 41–57. Princeton University Press, Princeton, NJ.

Bert, T.M., Hesselman, D.M., Arnold, W.S., Moore, W.S., Cruz-Lopez, H., Marelli, D.C. (1993). High frequency of gonadal neoplasia in a hard clam (*Mercenaria spp.*) hybrid zone. *Marine Biology* **117**: 97–104.

Bethel, W.M., Holmes, J.C. (1974). Correlation of development of altered evasive behavior in *Gamarus lacustris* (Amphipoda) harboring cystacanths of *Polymorphus paradoxus* (Acanthocephala) with the infectivity to the definitive host. *Journal of Parasitology* **60**: 272–274.

Bethel, W.M., Holmes, J.C. (1977). Increased vulnerability of amphipods to predation owing to altered behavior induced by larval acanthocephalans. *Canadian Journal of Zoology* **55**: 110–115.

Biron, D., Joly, C., Galeotti, N., Ponton, F., Marché, L. (2005). The proteomics: a new prospect for studying parasitic manipulation. *Behavioural Processes* **68**: 249–253.

Bize, P., Roulin, A., Richner, H. (2003). Adoption as an offspring strategy to reduce ectoparasite exposure. *Proceedings of the Royal Society B: Biological Sciences* **270**: S114–S116.

Blondel, J. (1995). *Biogéographie. Approche écologique et évolutive*. Masson, Paris.

Blount, J.D., Metcalfe, N.B., Birkhead, T.R., Surai, P.F. (2003). Carotenoid modulation of immune function and sexual attractiveness in zebra finches. *Science* **300**: 125–127.

Boag, B., Lello, J., Fenton, A., Tompkins, D.M., Hudson, P.J. (2001). Patterns of parasite aggregation in the wild European rabbit *Oryctolagus cuniculus*. *International Journal for Parasitology* **31**: 1421–1428.

Bollache, L., Gambade, G., Cézilly, F. (2001). The effects of two acanthocephalan parasites, *Pomphorhynchus laevis* and *Polymorphus minutus*, on pairing success in male *Gammarus pulex* (Crustacea: Amphipoda). *Behavioral Ecology and Sociobiology* **49**: 296–303.

Bollache, L., Cézilly, F., Rigaud, T. (2002). Effects of two acanthocephalan parasites on the fecundity and pairing status of female *Gammarus pulex* (Crustacea: Amphipoda). *Journal of Invertebrate Pathology* **79**: 102–110.

Bonneaud, C., Mazuc, J., Gonzalez, G., Haussy, C., Chastel, O., Faivre, B., Sorci, G. (2003). Assessing the cost of mounting an immune response. *The American Naturalist* **161**: 367–379.

Bonneaud, C., Mazuc, J., Chastel, O., Westerdahl, H., Sorci, G. (2004). Terminal investment induced by immune challenge and MHC associated fitness traits in the house sparrow. *Evolution* **58**: 2823–2830.

Bonneaud, C., Perez-Tris, J., Federici, P., Chastel, O., Sorci, G. (2006). MHC alleles confer local resistance to malaria in a wild passerine. *Evolution* **60**: 383–389.

Bonner, J.T. (1988). *The evolution of complexity*. Princeton University Press, Princeton, NJ.

Borgia, G. (1986). Satin bowerbird parasites: a test of the bright male hypothesis. *Behavioral Ecology and Sociobiology* **19**: 355–358.

Borgia, G., Collis, K. (1989). Female choice for parasite-free male satin bowerbirds and the evolution of bright male plumage. *Behavioral Ecology and Sociobiology* **25**: 445–454.

Boulinier, T., Staszewski, V. (2008). Maternal transfer of antibodies: raising immuno-ecology issues. *Trends in Ecology and Evolution* **23**: 282–288.

Boulinier, T., Sorci, G., Monnat, J.-Y., Danchin, E. (1997). Parent–offspring regression suggests heritable susceptibility to ectoparasites in a natural population of kittiwake *Rissa tridactyla*. *Journal of Evolutionary Biology* **10**: 77–85.

Boulétreau, M., David, J.R. (1980). Sexually dimorphic response to host habitat toxicity in *Drosophila* parasitic wasps. *Evolution* **35**: 395–399.

Boulétreau, T., Fouillet, P., Allemand, R. (1991). Parasitoids affect competitive interactions between the sibling species, *Drosophila melanogaster* and *D. simulans*. *Redia* **84**: 171–177.

Boursot, P., Bonhomme, F., Britton-Davidian, J., Catalan, J., Yonekawa, H., Orsini, P., Guerasimov, S., Thaler, L.

(1984). Introgression différentielle des génomes nucléaires et mitochondriaux chez deux semi-espèces européennes de souris. *Comptes Rendus de l'Académie des Sciences de Paris* **299**: 365–370.

Boursot, P., Auffray, J.C., Britton-Davidian, J., Bonhomme, F. (1993). The evolution of house mice. *Annual Review of Ecology and Systematics* **24**: 119–152.

Brinkhof, M.W.G., Heeb, P., Köllinker, M., Richner, H. (1999). Immunocompetence of nestling great tits in relation to rearing environment and parentage. *Proceedings of the Royal Society B: Biological Sciences* **266**: 2315–2322.

Broutin, H., Elguero, E., Simondon, F., Guégan, J.-F. (2004). Spatial dynamics of pertussis in a small region of Senegal. *Proceedings of the Royal Society B: Biological Sciences* **27**: 2091–2098.

Buchanan, K.L., Evans, M.R., Goldsmith, A.R., Bryant, D.M., Rowe, L.V. (2001). Testosterone influences basal metabolic rate in male house sparrows: a new cost of dominance signalling? *Proceedings of the Royal Society B: Biological Sciences* **268**: 1337–1344.

Buchholz, R. (1995). Female choice, parasite load and male ornamentation in wild turkeys. *Animal Behaviour* **50**: 929–943.

Buchholz, R. (1997). Male dominance and variation in fleshy ornamentation in wild turkeys. *Journal of Avian Biology* **28**: 223–230.

Buckling, A., Rainey, P.B. (2002). The role of parasites in sympatric and allopatric host diversification. *Nature* **420**: 496–499.

Burke, J.M., Arnold, M.L. (2001). Genetic and the fitness of hybrids. *Annual Review of Genetics* **35**: 31–52.

Camp, J.W., Huizinga, H.W. (1979). Altered color, behavior and predation susceptibility of the isopod *Asellus intermedius* infected with *Acanthocephalus dirus*. *Journal of Parasitology* **65**: 667–669.

Campan, R., Scapini, F. (2002). *Ethologie. Approche systémique du comportement*. DeBoeck, Brussels.

Carton, Y., Nappi, A.J. (1997). *Drosophila* cellular immunity against parasitoids. *Parasitology Today* **13**: 218–227.

de Castro, F., Bolker, B. (2005). Mechanisms of disease-induced extinction. *Ecology Letters* **8**: 117–126.

Catchpole, C.K., Slater, P.J.B. (1995). *Bird song: biological themes and variations*. Cambridge University Press, Cambridge.

Catchpole, E.A., Morgan, B.J.T., Coulson, T.N., Freeman, S.N., Albon, S.D. (2000). Factors influencing Soay sheep survival. *Applied Statistics Journal of the Royal Statistical Society Series C* **49**: 453–472.

Cattadori, I.M., Haukisalmi, V., Henttonen, H.H., Hudson, P.J. (2006). Transmission ecology and the structure of parasite communities in small mammals. In: *Micromammals and macroparasites: from evolutionary biology to management* (ed. S. Morand, B. Krasnov, R. Poulin). Springer, Tokyo: 349–399.

Cézilly, F. (2006). *Le paradoxe de l'hippocampe. Une histoire naturelle de la monogamie.* Buchet-Chastel, Paris.

Cézilly, F., Grégoire, A., Bertin, A. (2000). Conflict between co-occurring manipulative parasites? An experimental study of the joint influence of two acanthocephalan parasites on the behaviour of *Gammarus pulex*. *Parasitology* **120**: 625–630.

Cézilly, F., Danchin, E., Giraldeau, L.-A. (2008). Research methods in behavioural ecology. In: *Behavioural ecology* (ed. E. Danchin, L.-A. Giraldeau, F. Cézilly), pp. 55–95. Oxford University Press, Oxford.

Chandra, R.K., Newberne, P.M. (1977). *Nutrition, immunity and infection.* Plenum Press, New York.

Chew, B.P., Park, J.S. (2004). Carotenoid action on the immune response. *Journal of Nutrition* **134**: 257S–261S.

Chippindale, A.K., Gibson, J.R., Rice, W.R. (2001). Negative genetic correlation for adult fitness between sexes reveals ontogenetic conflict in *Drosophila*. *Proceedings of the National Academy of Sciences USA* **98**: 1671–1675.

Chown, S.L., Sinclair, B.J., Leinaas, H.P., Gaston, K.J. (2004). Hemispheric asymmetries in biodiversity—a serious matter for ecology. *PLoS Biology* **2**: 1701–1707.

Christe, P., Richner, H., Oppliger, A. (1996a). Begging, food provisioning, and nestling competition in great tit broods infested with ectoparasites. *Behavioral Ecology* **7**: 127–131.

Christe, P., Richner, H., Oppliger, A. (1996b). Of great tits and fleas: sleep baby sleep.... *Animal Behaviour* **52**: 1087–1092.

Cichon, M., Sendecka, J., Gustafsson, L. (2003). Age-related decline in humoral immune function in collared flycatchers. *Journal of Evolutionary Biology* **16**: 1205–1210.

Clayton, D.H. (1991). Avian grooming and ectoparasite avoidance. In: Bird–parasite interactions: ecology, evolution and behavior (ed. J.E. Loye, M. Zuk), pp. 258–289. Oxford University Press, Oxford.

Clayton, D.H., Pruett-Jones, S.G., Lande, R. (1992). Reappraisal of the interspecific prediction of parasite-mediated sexual selection: opportunity knocks. *Journal of Theoretical Biology* **157**: 95–108.

Cleaveland, S., Laurenson, M.K., Taylor, L.H. (2001). Diseases of humans and their domestic mammals: pathogen characteristics, host range and the risk of emergence. *Philosophical Transactions of the Royal Society B. Biological Sciences* **356**: 991–999.

Cleaveland, S., Hess, G.R., Dobson, A.P., Laurenson, M.K., McCallum, H.I., Roberts, M.G., Woodroffe, R. (2002). The role of pathogens in biological conservation. In: *The ecology of wildlife diseases* (ed. P.J. Hudson, A. Rizzoli, B.T. Grenfell, H. Heesterbeek, A.P. Dobson), pp. 139–150. Oxford University Press, Oxford.

Clutton-Brock, T.H., Price, O.F., Albon, S.D., Jewell, P.A. (1991). Persistent instability and population regulation in Soay sheep. *Journal of Animal Ecology* **60**: 593–608.

Clutton-Brock, T.H., Illius, A., Wilson, K., Grenfell, B.T., MacColl, A.D.C., Albon, S.D. (1997). Stability and instability in ungulate populations: an empirical analysis. *The American Naturalist* **149**: 195–219.

Colautti, R.I., Ricciardi, A., Grigorovitch, I.A., Mac Isaac, H.J. (2004). Is invasion success explained by the enemy release hypothesis? *Ecology Letters* **7**: 721–733.

Collyer, M.L., Stockwell, C.A. (2004). Experimental evidence for cost of parasitism for a threatened species, white sands pupfish (*Cyprinodon tularosa*). *Journal of Animal Ecology* **73**: 821–830.

Colon, L. (2004). Dispersion de *Rana ridibunda* dans la vallée du Rhône et relations génétiques avec le complexe d'hybridation *esculenta*. PhD Thesis, Lyon1 University.

Coltman, D.W., Wilson, K., Pilkington, J.G., Stear, M.J., Pemberton, J.M. (2001a). A microsatellite polymorphism in the gamma interferon gene is associated with resistance to gastrointestinal nematodes in a naturally-parasitized population of Soay sheep. *Parasitology* **122**: 571–582.

Coltman, D.W., Pilkington, J., Kruuk, L.E.B., Wilson, K., Pemberton, J.M. (2001b). Positive genetic correlation between parasite resistance and body size in a free-living ungulate population. *Evolution* **55**: 2116–2125.

Colwell, R.R. (1996). Global climate and infectious disease: the cholera paradigm. *Science* **274**: 2025–2031.

Combes, C. (1991). Ethological aspects of parasite transmission. *The American Naturalist* **138**: 866–880.

Combes, C. (1995). *Interactions durables. Écologie et évolution du parasitisme.* Masson, Paris.

Combes, C. (1996). Parasites, biodiversity and ecosystem stability. *Biodiversity and Conservation* **5**: 953–962.

Combes, C. (2001). *Parasitism. The ecology and evolution of intimate interactions.* University of Chicago Press, Chicago, IL.

Conn, J.A., Wilkerson, R.C., Segura, M.N.O., de Souza, R.T.L., Schlichting, C.D., Wirtz, R.A., Povoa, M.M. (2002). Emergence of a new neotropical malaria vector facilitated by human migration and changes in land use. *American Journal of Tropical Medicine and Hygiene* **66**: 18–22.

Coté I., Poulin, R. (1995). Parasitism and group size in social animals: a meta-analysis. *Behavioral Ecology* **6**: 159–165.

Coustau, C., Renaud, F., Maillard, C., Pasteur, N., Delay, B. (1991). Differential susceptibility to a trematode parasite among genotypes of the *Mytilus edulis/galloprovincialis* complex. *Genetical Research* **57**: 207–212.

Crofton, H.D. (1971). A quantitative approach to parasitism. *Parasitology* **62**: 179–193.

Crowl, T.A., Covich, A.P. (1990). Predator-induced life-history shifts in a fresh-water snail. *Science* **247**: 949–951.

Cunningham, A.A. (1996). Disease risks of wildlife translocations. *Conservation Biology* **10**: 349–353.

Danchin, E., Cézilly, F. (2008). Sexual selection: another evolutionary process. In: *Behavioural ecology* (ed. E. Danchin, L.-A. Giraldeau, F. Cézilly), pp. 363–426. Oxford University Press, Oxford.

Danchin, E, Cézilly, F., Giraldeau, L.-A. (2008). Fundamental concepts in behavioural ecology. In: *Behavioural ecology* (ed. E. Danchin, L.-A. Giraldeau, F. Cézilly), pp. 29–53. Oxford University Press, Oxford.

Darwin, C. (1859). *On the origin of species.* John Murray, London.

Darwin, C. (1871). *The descent of man, and selection in relation to sex.* John Murray, London.

Daszak, P., Cunningham, A.A. (1999). Extinction by infection. *Trends in Ecology and Evolution* **14**: 279.

Daszak, P., Cunningham, A.A., Hyatt, A.D. (2000). Emerging infectious diseases of wildlife-threats to biodiversity and human health. *Science* **287**: 443–449.

Dawkins, R. (1982). *The extended phenotype.* Oxford University Press, Oxford.

Deerenberg, C., Apanius, V., Daan, S., Bos, N. (1997). Reproductive effort decreases antibody responsiveness. *Proceedings of the Royal Society B: Biological Sciences* **264**: 1021–1029.

Demarais, B.D., Dowling, T.E., Douglas, M.E., Minckley, W.L., Marsh, P.C. (1992). Origin of *Gila seminuda* (Teleostei: Cyprinidae) through introgressive hybridization: implications for evolution and conservation. *Proceedings of the National Academy of Sciences USA* **89**: 2747–2751.

Demas, G.E., Chefer, V., Talan, M.I., Nelson, R. (1997). Metabolic costs of mounting an antigen-stimulated immune response in adult and aged C57BL/6J mice. *American Journal of Physiology* **42**: R1631–R1637.

Denoth, M., Frid, L., Myers, J. H. (2002). Multiple agents in biological control: improving the odds? *Biological Control* **24**: 20–30.

Deredec, A., Courchamp, F. (2003). Extinction thresholds in host-parasite dynamics. *Annales Zoologi Fennici* **40**: 115–130.

Derothe, J.M., Loubes, C., Perriat-Sanguinet, M., Orth, A., Moulia, C. (1999). Experimental trypanosomiasis of natural hybrids between house mouse subspecies. *International Journal for Parasitology* **29**: 1011–1016.

Derothe, J.M., Le Brun, N., Loubes, C., Perriat-Sanguinet, M., Moulia, C. (2001). Susceptibility of natural hybrids between house mouse subspecies to *Sarcocystis muris*. *International Journal for Parasitology* **31**: 15–19.

Derothe, J.M., Porcherie, A., Perriat-Sanguinet, M., Loubes, C., Moulia, C. (2004). Recombination does not generate pinworm susceptibility during experimental crosses between two mouse subspecies. *Parasitology Research* **93**: 356–363.

Di Giulio, D.B., Eckburg, P.B. (2004). Human monkeypox: an emerging zoonosis. *The Lancet* **4**: 15–25.

Diamond, J. (2000). *De l'inégalité parmi les sociétés. Essai sur l'homme et l'environnement dans l'histoire.* Gallimard, Paris.

Dickson, L.L., Whitham, T.G. (1996). Genetically-based plant resistance traits affect arthropods, fungi, and birds. *Oecologia* **106**: 400–406.

Ditchkoff, S.S., Lochmiller, R.L., Masters, R.E., Hoofer, S.R, Van Den Bussche, R.A. (2001). Major-histocompatibility-complex-association variation in secondary sexual traits of white-tailed deer (*Odocoileus virginianus*): evidence for good-genes advertisement. *Evolution* **55**: 616–625.

Dobson, A.P. (1995a). The ecology and epidemiology of rinderpest virus in Serengeti and Ngorongoro crater conservation area. In: *Serengeti II: research, management and conservation of an ecosystem* (ed. A.R.E. Sinclair, P. Arcese), pp. 485–505. The University of Chicago Press, Chicago, IL.

Dobson, A.P. (1995b). Rinderpest in the Serengeti ecosystem: the ecology and control of a keystone virus. In: *Proceedings of a Joint Conference of the American Association of Zoo Veterinarians, Wildlife Disease Association, and American Association of Wildlife Veterinarians* (ed. R.E. Junge), pp. 518–519. AAZV, East Lansing, Michigan.

Dobson, A.P., Hudson, P.J. (1992). Regulation and stability of a free-living host–parasite system: *Trichostrongylus tenuis* in red grouse. II. Population models. *Journal of Animal Ecology* **61**: 487–498.

Dobson, A.P., Roberts, M. (1994). The population dynamics of parasitic helminth communities. *Parasitology* **109**: S97–S108.

Dobzhansky, T. (1936). Studies on hybrid sterility. II. Localization of sterility factors in *Drosophilia pseudoobscura* hybrids. *Genetics* **21**: 113–135.

Dorchies, P., Bergeaud, J.P., Nguyen Van Khanh, N., Morand, S. (1997). Reduced egg counts in mixed infections with *Oestrus ovis* and *Haemonchus contortus*: influence of eosinophils? *Parasitology Research* **83**: 727–730.

Doutt, R.L. (1959). The biology of parasitic Hymenoptera. *Annual Review of Entomology* **4**: 161–182.

Duclaux, E. (1902). *L'hygiène sociale.* Félix Alcan, Paris.

Dupont, F., Crivelli, A.J. (1988). Do parasites confer a disadvantage to hybrids? A case study of *Alburnus alburnus × Rutilus rubilio*, a natural hybrid of Lake Mikri Prespa, Northern Greece. *Oecologia* **75**: 587–592.

Eberhard, W.G. (2000). Spider manipulation by a wasp larva. *Nature* **406**: 255–256.

Ebi, K.L., Exuzides, K.A., Lau, E., Kelsh, M., Barnston, A. (2001). Association of normal weather periods and El Nino events with hospitalization for viral pneumonia in females: California, 1983–1998. *American Journal of Public Health* **91**: 1200–1208.

Edelaar, P., Drent, J., De Goeij, P. (2003). A double test of the parasite manipulation hypothesis in a burrowing bivalve. *Oecologia* **134**: 66–71.

Eggleton, P., Belshaw, R. (1992). Insect parasitoid: an overview. *Philosophical Transactions of the Royal Society B: Biological Sciences* **337**: 1–20.

Ehman, K.D., Scott, M.E. (2001). Urinary odour preferences of MHC congenic female mice, *Mus domesticus*: implications for kin recognition and detection of parasitized males. *Animal Behaviour* **62**: 781–789.

Ehman, K.D., Scott, M.E. (2002). Female mice mate preferentially with non-parasitized males. *Parasitology* **125**: 461–466.

Eilenberg, J., Hajek, A., Lomer, C. (2001). Suggestions for unifying the terminology in biological control. *BioControl* **46**: 387–400.

Ekblom, R., Saether, S.A., Grahn, M., Fiske, P., Kalas, J.A., Hoglund, J. (2004). Major histocompatibility complex variation and mate choice in a lekking bird, the great snipe (*Gallinago media*). *Molecular Ecology* **13**: 3821–3828.

Emlen, S., Wrege, T. (1989). Experimental induction of infanticide in female wattled jacanas. *The Auk* **106**: 1–7.

Emlen, S., Wrege, T. (2004). Size dimorphism, intrasexual competition, and sexual selection in wattled jacana (*Jacana jacana*), a sex-role reversed shorebird in Panama. *The Auk* **121**: 391–403.

Endler, J.A. (1980). Natural selection on color patterns in *Poecilia reticulata*. *Evolution* **34**: 76–91.

Enserink, M. (2002). West Nile's surprisingly swift continental sweep. *Science* **297**: 1988–1989.

Eraud C., Duriez O., Chastel O., Faivre B. (2005). The energetic cost of humoral immunity in the collared dove, *Streptopelia decaocto*: is the magnitude sufficient to force energy-based trade-offs? *Functional Ecology* **19**: 110–118.

Ezenwa,V.O. (2004a). Selective defecation and selective foraging: antiparasite behavior in wild ungulates? *Ethology* **110**: 851–862.

Ezenwa,V.O. (2004b). Host social behavior and parasitic infection: a multifactorial approach. *Behavioral Ecology* **15**: 446–454.

Faivre, B., Préault, M., Salvadori, F., Théry, M., Gaillard, M., Cézilly, F. (2003a). Bill colour and immunocompetence in the blackbird. *Animal Behaviour* **65**: 1125–1131.

Faivre, B., Grégoire, A., Préault, M., Cézilly, F., Sorci, G. (2003b). Immune activation rapidly mirrored in a secondary sexual trait. *Science* **300**: 103.

Falconer, D.S., Mackay, T.F.C. (1996). Introduction to quantitative genetics (4th Edition). Addison Wesley Longman, Harlow, Essex, UK.

Faruque, S.M., Bin Naser, I., Islam, M.J., Faruque, A.S.G., Ghosh, A.N., Nair, G.B., Sack, D.A., Mekalanos, J.J. (2005a). Seasonal epidemics of cholera inversely correlate with the prevalence of environmental cholera phages. *Proceedings of the National Academy of Sciences USA* **102**: 1702–1707.

Faruque, S.M., Islam, M.J., Ahmad, Q.S., Faruque, A.S.G., Sack, D.A., Nair, G.B., Mekalanos, J.J. (2005b). Self-limiting nature of seasonal cholera epidemics: role of host-mediated amplification of phage. *Proceedings of the National Academy of Sciences USA* **102**: 6119–6124.

Feener, D.H.J., Brown, B.V. (1997). Diptera as parasitoid. *Annual Review of Entomology* **42**: 73–97.

Fellowes, M.D.E., Kraaijeveld, A.R., Godfray, H.C.J. (1999). Cross-resistance following artificial selection for increased defence against parasitoids in *Drosophila melanogaster*. *Evolution* **53**: 966–972.

Fenner, F., Cairns, J. (1959). Variation in virulence in relation to adaptation to new hosts. In: *The viruses: biochemical and biophysical properties* (ed. F.M. Burnet, W.M. Stanley), pp. 225–249. Academic Press, New York.

Fenner, F., Ratcliffe, F.N. (1965). *Myxomatosis*. Cambridge University Press, Cambridge.

Fenner, F., Day, M.F., Woodroofe, G.M. (1956). Epidemiological consequences of the mechanical transmission of myxoma by mosquitoes. *Journal of Hygiene* **54**: 284–303.

Ferguson, N.M., Cummings, D.A.T., Cauchemez, S., Frasen, C., Riley, S., Meeyai, A., Iamsirithaworn, S., Burke, D.S. (2005). Strategies for containing an emerging influenza pandemic in Southeast Asia. *Nature* **437**: 209–214.

Fernandez, V., Corley, J.C. (2003). The functional response of parasitoids and its implications for biological control. *Biocontrol Science and Technology* **13**: 403–413.

Finlay, B.J. (2002). Global dispersal of free-living microbial eukaryote species. *Science* **296**: 1061–1063.

Fisher, R.A. (1958). *The genetical theory of natural selection*, 2nd edn. Dover, New York.

Flatt, T. (2005). The evolutionary genetics of canalization. *Quarterly Review of Biology* **80**: 287–316.

Floate, K.D., Whitham, T.G. (1993). The 'hybrid bridge' hypothesis: host shifting via plant hybrid swarms. *The American Naturalist* **141**: 651–662.

Folstad, I., Karter, A.J. (1992). Parasites, bright males and the immunocompetence handicap. *The American Naturalist* **139**: 603–622.

Folstad, I., Hope, A.M., Karter, A., Skorping, A. (1994). Sexually-selected color in male sticklebacks—a signal of both parasite exposure and parasite resistance. *Oikos* **69**: 511–515.

Forbes, M.R.L. (1993). Parasitism and host reproductive effort. *Oikos* **67**: 444–450.

Forbes, M.R.L. (1991). Ectoparasites and mating success of male *Enallagma erbium* damselflies (Odonata: Coenagrionidae). *Oikos* **60**: 336–342.

Fox, L.R., Morrow, P.A. (1981). Specialization: species property and local phenomenon? *Science* **211**: 887–893.

Franceschi, N., Bauer, A., Bollache, L., Rigaud, T. (2008). The effects of parasite age and intensity on variability in acanthocephalan-induced behavioural manipulation. *International Journal for Parasitology* **38**: 1161–1170.

Frandon, J., Kabiri, F. (1999). La lutte biologique contre la pyrale du maïs avec les trichogrammes. *Dossier de l'Environnement* **19**: 107–112.

Frank, S.A. (2002). *Immunology and evolution of infectious disease*. Princeton University Press, Princeton, NJ.

Fredensborg, B.L., Poulin, R. (2006). Parasitism shaping host life-history evolution: adaptive responses in a marine gastropod to infection by trematodes. *Journal of Animal Ecology* **75**: 44–53.

Freeland, W.J. (1976). Pathogens and the evolution of primate sociality. *Biotropica* **8**: 12–24.

Freeland, W.J. (1983). Parasite and the coexistence of animal species. *The American Naturalist* **121**: 223–236.

Fritz, R.S. (1999). Resistance of hybrid plants to herbivores: genes, environment, or both? *Ecology* **80**: 382–391.

Fritz, R.S., Nichols-Orians, C.M., Brunsfeld, S.J. (1994). Interspecific hybridization of plants and resistance to herbivores: hypotheses, genetics, and variable responses in a diverse herbivore community. *Oecologia* **97**: 106–117.

Fritz, R.S., Roche, B.M., Brunsfeld, S.J., Orians, C.M. (1996). Interspecific and temporal variation in herbivore responses to hybrid willows. *Oecologia* **108**: 121–129.

Fritz, R.S., Moulia, C., Newcombe, G. (1999). Resistance of hybrid plants and animals to herbivores, pathogens, and parasites. *Annual Review of Ecology and Systematics* **30**: 565–591.

Galvani, A.P. (2003a). Immunity, antigenic heterogeneity, and aggregation of helminth parasites. *Journal of Parasitology* **89**: 232–241.

Galvani, A.P. (2003b). Epidemiology meets evolutionary ecology. *Trends in Ecology and Evolution* **18**: 132–139.

Gandon, S., Mackinnon, M.J., Nee, S., Read, A.F. (2001). Imperfect vaccines and the evolution of parasite virulence. *Nature* **414**: 751–755.

Gandon, S., Agnew, P.A., Michalakis, Y. (2002). Coevolution between parasite virulence and host life-history traits. *The American Naturalist* **160**: 374–388.

Garamszegi, L.S. (2005). Bird song and parasites. *Behavioral Ecology and Sociobiology* **59**: 167–180.

Garamszegi, L.S., Møller, A.P., Török, J., Michl, G., Péczely, P., Richard, M. (2004). Immune challenge mediates vocal communication in a passerine bird: an experiment. *Behavioral Ecology* **15**: 148–157.

Gardon, J., Heraud, J.M., Laventure, S., Ladam, A., Capot, P., Fouquet, E., Favre, J., Weber, S., Hommel, D., Hulin, A., Couratte, Y., Talarmin, A. (2001). Suburban transmission of Q fever in French Guiana: evidence of a wild reservoir. *Journal of Infectious Diseases* **184**: 278–284.

Gasparini, J., McCoy, K.D., Haussy, C., Tveraa, T., Boulinier, T. (2001). Induced maternal response to the Lyme disease spirochete *Borrelia burgdorferi sensu lato* in a colonial seabird, the kittiwake *Rissa tridactyla*. *Proceedings of the Royal Society B: Biological Sciences* **1467**: 647–650.

Gasparini, J., McCoy, K.D., Tveraa, T., Boulinier, T. (2002). Related concentrations of specific immunoglobulins against the Lyme disease agent *Borrelia burgdorferi sensu lato* in eggs, young and adults of the kittiwake (*Rissa tridactyla*). *Ecology Letters* **5**: 519–524.

Gasparini, J., Roulin, A., Gill, V., Hatch, S.A., Boulinier, T. (2006a). In kittiwakes food availability partially explains the seasonal decline in humoral immunocompetence. *Functional Ecology* **20**: 457–463.

Gasparini, J., McCoy, K.D., Staszewski, V., Haussy, C., Boulinier, T. (2006b). Dynamics of anti-*Borrelia* antibodies in kittiwake chicks (*Rissa tridactyla*) suggests a maternal educational effect. *Canadian Journal of Zoology* **84**: 623–627.

Gegear, R.J., Otterstatter, M.C., Thomson, J.D. (2006). Bumble-bee foragers infected by a gut parasite have an impaired ability to utilize floral information. *Proceedings of the Royal Society B: Biological Sciences* **273**: 1073–1078.

Gérard, C., Théron, A. (1997). Age/size- and time-specific effects of *Schistosoma mansoni* on energy allocation patterns of its snail host *Biomphalaria glabrata*. *Oecologia* **112**: 447–452.

Gershwin, M.E., Beach, R.S., Hurley, L.S. (1985). *Nutrition and immunity*. Academic Press, New York.

Ghani, A.C., Ferguson, N.M., Donnelly, C.A., Anderson, R.M. (2000). Predicted vCJD mortality in Great Britain. *Nature* **406**: 583–584.

Gilbert, L., Norman, R., Laurenson, K.M., Reid, H.W., Hudson, P.J. (2001). Disease persistence and apparent competition in a three-host community: an empirical and analytical study of large-scale, wild populations. *Journal of Animal Ecology* **70**: 1053–1061.

Giorgi, M., Arlettaz, R., Christe, P., Vogel, P. (2001). The energetic grooming costs imposed by a parasitic mite (*Spinturnix myoti*) upon its bat host (*Myotis myotis*). *Proceedings of the Royal Society B: Biological Sciences* **268**: 2071–2075.

Giraldeau, L.-A. (2008). Social foraging. In: *Behavioural ecology* (ed. E. Danchin, L.-A. Giraldeau, F. Cézilly), pp. 257–283. Oxford University Press, Oxford.

Godfray, H.C.J. (1994). *Parasitoids*. Princeton University Press, Princeton, NJ.

Gog, J., Woodroffe, R., Swinton, J. (2002). Disease in endangered metapopulations: the importance of alternative hosts. *Proceedings of the Royal Society B: Biological Sciences* **269**: 671–676.

Gonzalez, G., Sorci, G., De Lope, F. (1999). Seasonal variation in the relationship between cellular immune response and badge size in male house sparrows (*Passer domesticus*). *Behavioral Ecology and Sociobiology* **46**: 117–122.

Gonzalez, G., Sorci, G., Smith, L.C., De Lope, F. (2001). Testosterone and sexual signalling in male house sparrows (*Passer domesticus*). *Behavioral Ecology and Sociobiology* **50**: 557–562.

Graf, J.D., Polls-Pelaz, M. (1989). Evolutionary genetics of the *Rana esculenta* complex. In: *Evolution and ecology of unisexual vertebrates* (ed. R.M. Dawley, J.P. Bogart), Bulletin 466, pp. 289–301. New York State Museum, Albany, NY.

Graham, A.L., Allen, J.E., Read, A.F. (2005). Evolutionary causes and consequences of immunopathology. *Annual Reviews in Ecology and Systematics* **36**: 373–397.

Grant, B.R., Grant, P.R. (1996). High survival of Darwin's finch hybrids: effects of beak morphology and diets. *Ecology* **77**: 500–509.

Grassly, N.C., Fraser, C., Garnett, G.P. (2005). Host immunity and synchronized epidemics of syphilis across the United States. *Nature* **433**: 417–421.

Greathead, D.J. (1995). Benefits and risks of classical biological control. In: *Biological control. Benefits and risks* (ed. H.M.T. Hokhanen, J.M. Lynch), pp. 53–63. Cambridge University Press, Cambridge.

Greathead, D.J., Greathead, A.J. (1992). Biological control of insect pests by insect parasitoids and predators: the BIOCAT database. *Biocontrol News Information* **13**: 61–68.

Green, J.L., Holmes, A.J., Westoby, M., Oliver, I., Briscoe, D., Dangerfield, M., Gillings, M., Beattie A.J. (2004). Spatial scaling of microbial eukaryote diversity. *Nature* **432**: 747–750.

Greenman, J.V., Hudson, P.J. (1997). Infected coexistence instability with and without density-dependent regulation. *Journal of Theoretical Biology* **185**: 345–356.

Grenfell, B.T., Dobson, A.P. (eds) (1995). *Ecology of infectious diseases in natural populations*. Cambridge University Press, Cambridge.

Grenfell, B.T., Harwood, J. (1997). (Meta)population dynamics of infectious diseases. *Trends in Ecology and Evolution* **12**: 395–399.

Grenfell, B.T., Price, O.F., Albon, S.D., Clutton-Brock, T.H. (1992). Overcompensation and population cycles in an ungulate. *Nature* **355**: 823–826.

Grenfell, B.T., Dietz, K., Roberts, M.G. (1995). Modelling the immuno-epidemiology of macroparasites in wildlife host populations In: *Ecology of infectious diseases in natural populations* (ed. B.T. Grenfell, A.P. Dobson), pp. 362–383. Cambridge University Press, Cambridge.

Grenfell, B.T., Bjørnstad, O.N., Kappey, J. (2001). Travelling waves and spatial hierarchies in measles epidemics. *Nature* **414**: 716–723.

Grindstaff, J.L., Brodie, E.D., Ketterson, E.D. (2003). Immune function across generations: integrating mechanism and evolutionary process in maternal antibody transmission. *Proceedings of the Royal Society B: Biological Sciences* **270**: 2309–2319.

Grindstaff, J.L., Hasselquist, D., Nilsson, J.-A., Sandell, M., Smith, H.G., Stjernman, M. (2006). Transgenerational priming of immunity: maternal exposure to a bacterial antigen enhances offspring humoral immunity. *Proceedings of the Royal Society B: Biological Sciences* **273**: 2551–2557.

Grosholtz, E.D. (1994). The effect of host genotype and spatial distribution on trematode parasitism in a bivalve population. *Evolution* **48**: 1514–1524.

Guan, Y., Zheng, B.J., He, Y.Q., Liu, X.L., Zhuang, Z.X., Cheung, C.L. *et al.* (2003). Isolation and characterization of viruses related to the SARS coronavirus from animals in southern China. *Science* **302**: 276–278.

Guernier, V., Hochberg, M.E., Guégan, J.-F. (2004). Ecology drives the worldwide distribution of human diseases. *PLoS Biology* **2**: 740–746.

Guiler, E.R. (1961). The former distribution and decline of the thylacine. *Australian Journal of Sciences* **23**: 207–210.

Gulland, F.M.D. (1992). Role of nematode parasites in Soay sheep mortality during a population crash. *Parasitology* **105**: 493–503.

Gulland, F.M.D., Fox, M. (1992). Epidemiology of nematode infections in Soay sheep (*Ovis aries* L.) on St. Kilda. *Parasitology* **105**: 481–492.

Guégan, J.-F., Broutin, H. (2008). Microbial diversity: patterns and processes. In: *Biodiversity change and human health: from ecosystem services to spread of disease* (ed. O. Sala, L.A. Skevington, C. Parmesan). Island Press–SCOPE–DIVERSITAS, New York (in press).

Guégan, J.-F., Morand, S., Poulin, R. (2004). Are there general laws in parasite community ecology? The emergence of spatial parasitology and epidemiology. In: *Parasitism and ecosystems* (ed. F. Thomas, F. Renaud, J.F. Guégan), pp. 22–42. Oxford University Press, Oxford.

Gylfe, A., Bergström, S., Lundström, J., Olsen, B. (2000). Reactivation of *Borrelia* infection in birds. *Nature* **403**: 724–725.

Hamilton, W.D. (1971). Geometry for the selfish herd. *Journal of Theoretical Biology* **31**: 295–311.

Hamilton, W.D. (1980). Sex versus non-sex versus parasites. *Oikos* **35**: 282–290.

Hamilton, W.D., Zuk, M. (1982). Heritable true fitness and bright birds: a role for parasites? *Science* **218**: 384–387.

Harris-Warrick, R.M., Marder, E. (1991). Modulation of neural networks for behaviour. *Annual Review of Neuroscience* **14**: 39–57.

Harrison, R.G. (1993). Hybrid zones and the evolutionary process. In: *Hybrids and hybrid zones: historical perspective* (ed. R.G. Harrison), pp. 3–12. Oxford University Press, Oxford.

Hart, B.L. (1990). Behavioral adaptations to pathogens and parasites: five strategies. *Neuroscience and Biobehavioral Reviews* **14**: 273–294.

Hart, B.L. (2005). The evolution of herbal medicine: behavioural perspectives. *Animal Behaviour* **70**: 975–989.

Hartl, D.L. (1994). *A primer of population genetics*, 2nd edn. Sinauer Associates Inc., Sunderland, MA.

Harvell, C.D., Mitchell, C.E., Ward, J.R., Altizer, S., Dobson, A.P., Ostfeld, R.S., Samuel, M.D. (2002). Climate warming and disease risks for terrestrial and marine biota. *Science* **296**: 2158–2162.

Harvey, P.H., Pagel, M.D. (1991). *The comparative method in evolutionary biology*. Oxford University Press, Oxford.

Hassell, M.P. (1978). *The dynamics of arthropod predator–prey systems*. Princeton University Press, Princeton, NJ.

Hassell, M.P., May, R.M. (1973). Stability in insect host–parasite models. *Journal of Animal Ecology* **42**: 693–726.

Hassell, M.P., Pacala, S.W. (1990). Heterogeneity of host-parasitoid interactions. *Philosophical Transactions of the Royal Society B: Biological Sciences* **330**: 203–220.

Hassell, M.P., Varley, G.C. (1969). New inductive population model for insect parasites and its bearing on biological control. *Nature* **223**: 1133–1136.

Hasselquist, D., Marsh, J.A., Sherman, P.W., Wingfield, J.C. (1999). Is avian humoral immunocompetence suppressed by testosterone? *Behavioral Ecology and Sociobiology* **45**: 167–175.

Hawkins, B.A., Field, R., Cornell, H.V., Currie, D.J., Guégan, J.-F. *et al.* (2003). Energy, water, and broad-scale geographic patterns of species richness. *Ecology* **84**: 3105–3117.

Haydon, D.T., Shaw, D.J., Cattadori, I.M., Hudson, P.J., Thirgood, S.J. (2002). Analysing noisy time-series: describing regional variation in the cyclic dynamics of red grouse. *Philosophical Transactions Royal Society B: Biological Sciences* **269**: 1609–1617.

Heaney, L.R., Timm, R.M. (1985). Morphology, genetics, and ecology of pocket gophers (genus *Geomys*) in a narrow hybrid zone. *Biological Journal of the Linnean Society* **25**: 301–317.

Hechtel, L.J., Johnson, C.L., Juliano, S.A. (1993). Modification of antipredator behaviour of *Caecidotea intermedius* by its parasite *Acanthocephalus dirus*. *Ecology* **74**: 710–713.

Heeb, P., Werner, I., Kölliker, M., Richner, H. (1998). Benefits of induced host responses against an ectoparasite. *Proceedings of the Royal Society of London Series B: Biological Sciences* **265**: 51–56.

Heeb, P., Werner, I., Mateman, A.C., Kolliker, M., Brinkhof, M.W.G., Lessels, C.M., Richner, H. (1999). Ectoparasites infestation and sex-biased local recruitment of hosts. *Nature* **400**: 63–65.

Heimpel, G.E., Collier, T.R. (1996). The evolution of host-feeding behaviour in insect parasitoids. *Biological Review* **71**: 373–400.

Helluy, S. (1983) Un mode de favorisation de la transmission parasitaire: *la manipulation du comportement de l'hôte intermediaire*. *Revue d' Ecologie* **38**: 211–223.

Helluy, S. (1983). Relations hôtes–parasites du trematode *Microphallus papillorobustus* (Rankin 1940). II. Modifications du comportement des *Gammarus* hôtes intermédiaires et localisation des métacercaires. *Annales de Parasitologie Humaine et Comparée* **58**: 1–17.

Helluy, S. (1984). Relations hôtes–parasites du trematode *Microphallus papillorobustus* (Rankin 1940). III. Facteurs impliqués dans les modifications du comportement des *Gammarus* hôtes intermédiaires et tests de prédation. *Annales de Parasitologie Humaine et Comparée* **59**: 41–56.

Helluy, S., Holmes, J.C. (1990). Serotonin, octopamine, and the clinging behavior induced by the parasite *Polymorphus paradoxus* (Acanthocephala) in *Gammarus lacustris* (Crustacea). *Canadian Journal of Zoology* **68**: 1214–1220.

Helluy, S., Thomas, F. (2003). Effects of *Microphallus papillorobustus* (Platyhelminthes: Trematoda) on serotonergic immunoreactivity and neuronal architecture in the brain of *Gammarus insensibilis* (Crustacea: Amphipoda). *Proceedings of the Royal Society B: Biological Sciences* **270**: 563–568.

Hemmes R.B., Alvarado, A., Hart, B.L. (2002). Use of California bay foliage by wood rats for possible fumigation of nest-borne ectoparasites. *Behavioral Ecology* **13**: 381–385.

Hewitt, G.M. (1988). Hybrid zones—natural laboratories for evolutionary studies. *Trends in Ecology and Evolution* **3**: 158–167.

Heymer, H. (1977). *Vocabulaire éthologique*. Presses Universitaires de France, Paris.

Hill, G.E. (1999). Is there an immunological cost to carotenoid-based ornamental coloration? *The American Naturalist* **154**: 589–595.

Hill, G.E., Inouye, C.Y., Montgomerie, R. (2002). Dietary carotenoids predict plumage coloration in wild house finches. *Proceedings of the Royal Society B: Biological Sciences* **269**: 119–1124.

Hillebrand, H. (2004). On the generality of the latitudinal diversity gradient. *The American Naturalist* **163**: 192–211.

Hillgarth, N. (1990). Parasites and female choice in the ring-necked pheasant. *American Zoologist* **30**: 227–233.

Hochberg, M.E. (1991). Intra-host interactions between a braconid endoparasitoid, *Apanteles glomeratus*, and a baculovirus for larvae of *Pieris brassicae*. *Journal of Animal Ecology* **60**: 51–63.

Hochberg, M.E., Michalakis, Y., de Meeus, T. (1992). Parasitism as a constraint on the rate of life-history evolution. *Journal of Evolutionary Biology* **5**: 491–504.

Hochberg, M.E., Gomulkiewicz, R., Holt, R.D., Thompson, J.N. (2000). Weak sinks could cradle symbioses—strong sources should harbor parasitic symbiosis. *Journal of Evolutionary Biology* **13**: 213–222.

Holmes, J.C. (1996). Parasites as threats to biodiversity in shrinking ecosystems. *Biodiversity and Conservation* **5**: 975–983.

Holmes, J.C., Price, P.W. (1986). Communities of parasites. In: *Community ecology: patterns and processes* (ed. J. Kikkawa, D.J. Anderson), pp. 187–213. Blackwell, London.

Holt, R.D., Pickering, J. (1985). Infectious disease and species coexistence: a model of Lotka–Volterra form. *The American Naturalist* **126**: 196–211.

Holt, R.D., Dobson, A.P., Begon, M., Bowers, R.G., Schauber, E.M. (2003). Parasite establishment in host communities. *Ecology Letters* **6**: 837–842.

Horak, P., Ots, I., Vellau, H., Spottiswoode, C., Moller, A.P. (2001). Carotenoid-based plumage coloration reflects hemoparasite infection and local survival in breeding great tits. *Oecologia* **126**: 166–173.

Hõrak, P., Saks, L., Karu, U., Ots, I. (2006). Host resistance and parasite virulence in greenfinch coccidiosis. *Journal of Evolutionary Biology* **19**: 277–288.

Hotz, H., Uzzell, T., Berger, L. (1997). Linkage groups of protein-coding genes in Western Palearctic water frogs reveal extensive evolutionary conservation. *Genetics* **147**: 255–270.

Houde, A.E., Torio, A.J. (1992). Effect of parasitic infection on male color pattern and female choice in guppies. *Behavioral Ecology* **3**: 346–351.

Houle, D. (1992). Comparing evolvability and variability of quantitative traits. *Genetics* **130**: 195–204.

Howard, D.J. (1993). Reinforcement: origin, dynamics and fate of an evolutionary hypothesis. In: *Hybrid zones and the evolutionary process* (ed. R.G. Harrisson), pp. 46–69. Oxford University Press, New York.

Howard, R.D., Minchella, D.J (1990). Parasitism and mate competition. *Oikos* **58**: 120–122.

Hudson, P.J., Newborn, D., Dobson, A.P. (1992). Regulation and stability of a freeliving host-parasite system: *Trichostrongylus tenuis* in red grouse. Monitoring and parasite reduction experiments. *Journal of Animal Ecology* **61**: 477–486.

Hudson, P.J., Dobson, A.P., Newborn, D. (1998). Prevention of population cycles by parasite removal. *Science* **282**: 2256–2258.

Hudson, P.J., Dobson, A.P., Newborn, D. (2002). Parasitic worms and population cycles of red grouse. In: *Population cycles* (ed. A. A. Berryman). pp. 109–129. Oxford University Press, Oxford.

Huffaker, C.B., Simmons, F.J., Laing, J.E. (1976). The theoretical and empirical basis of biological control. In: *Theory and practice of biological control* (ed. C.B. Huffaker, P.S. Messenger), pp. 41–78. Academic Press, New York.

Hurd, H. (2001). Host fecundity reduction: a strategy for damage limitation? *Trends in Parasitology* **17**: 363–368.

Hurd, H., Warr, E., Polwart, A. (2001). A parasite that increases host lifespan. *Proceedings of the Royal Society B: Biological Sciences* **268**: 1749–1753.

Hurtrez-Boussès, S., Blondel, J., Perret, P., Fabreguettes, J., Renaud, F. (1998). Chick parasitism by blowflies affects feeding rates in a Mediterranean population of blue tits. *Ecology Letters* **1**: 17–20.

Hutchings, M.R., Gordon, I.J., Kyriazakis, I. (2001). Sheep avoidance of faeces-contaminated patches leads to a trade-off between intake rate of forage and parasitism in subsequent foraging decisions. *Animal Behaviour* **62**: 955–964.

Idris, A.B., Grafius, E. (1995). Wildflowers as nectar sources for *Diadegma insulare* (Hymenoptera: Ichneumonidae), a parasitoid of diamondback moth (Lepidoptera: Yponomeutidae). *Environmental Entomology* **24**: 1726–1735.

Irwin M.T., Samonds, K.E., Raharison, J.-L., Wright, P.C. (2004). Lemur latrines: observations of latrine behavior in wild primates and possible ecological significance. *Journal of Mammalogy* **85**: 420–427.

Ives, A.R., Murray, D.L. (1997). Can sublethal parasitism destabilize predator-prey population dynamics? A model of snowshoe hares, predators and parasites. *Journal of Animal Ecology* **66**: 265–278.

Iwasa, Y., Pomiankowski, A., Nee, S. (1991). The evolution of costly mate preferences II. The 'handicap' principle. *Evolution* **45**: 1431–1442.

Jaenike, J. (1990). Host specialization in phytophagous insects. *Annual Review of Ecology and Systematics* **21**: 243–273.

Jaenike, J. (1992). Mycophagous *Drosophila* and their nematode parasites. *The American Naturalist* **139**: 893–906.

Jaenike, J. (1995). Interactions between mycophagous *Drosophila* and their nematode parasites: from physiological to community ecology. *Oikos* **72**: 235–244.

Jervis, M.A., Kidd, N.A.C. (1986). Host-feeding strategies in hymenopteran parasitoids. *Biological Review* **61**: 395–434.

Jervis, M.A., Hawkins, B.A., Kidd, N.A.C. (1996). The usefulness of destructive host feeding parasitoids in classical biological control: theory and observation conflict. *Ecological Entomology* **21**: 41–46.

Jervis, M.A., Heimpel, G.E., Ferns, P.N., Harvey, J.A., Kidd, N.A.C. (2001). Life-history strategies in parasitoid wasps: a comparative analysis of 'ovigeny'. *Journal of Animal Ecology* **70**: 442–458.

Jiggins, C.D., Mallet, J. (2000). Bimodal hybrid zones and speciation. *Trends in Ecology and Evolution* **15**: 250–255.

Jokela, J., Lively, C.M. (1995). Parasites, sex and early reproduction in a mixed population of freshwater snails. *Evolution* **49**: 1268–1271.

Joly, P. (2001). The future of the selfish hemiclone: a neodarwinian approach of waterfrog evolution. *Mitteilungen aus dem Museum für Naturkunde in Berlin* **77**: 31–38.

Joly, P., Guesdon, V., Fromont, E., Plenet, S., Grolet, O., Guegan, J.F., Hurtrez-Bousses, S., Thomas, F., Renaud, F. (2007). Heterozygosity and parasite intensity: lung parasites in the water frog hybridization complex. *Parasitology* **135**: 95–104.

Jones, C.G., Lawton, J.H., Shachak, M. (1994). Organisms as ecosystem engineers. *Oikos* **69**: 373–386.

Jones, C.G., Lawton, J.H., Shachak, M. (1997). Positive and negative effects of organisms as physical ecosystem engineers. *Ecology* **78**: 1946–1957.

Jones, H.R. (1990). *Population geography*. Paul Chapman Publishing, London.

Kaldonski, N., Perrot-Minnot, M.-J., Cézilly, F. (2007). Differential influence of two acanthocephalan parasites on the antipredator behaviour of their common intermediate host *Animal Behaviour* **74**: 1311–1317.

Kavaliers, M., Colwell, D.D. (1995). Reduced spatial learning in mice infected with the nematode *Heligmosomoides polygyrus*. *Parasitology* **110**: 591–597.

Kavaliers, M., Choleris, E., Pfaff, D.W. (2005). Genes, odours and the recognition of parasitized individuals by rodents. *Trends in Parasitology* **21**: 423–429.

Kennedy, C.R., Bush, A.O. (1994). The relationship between pattern and scale in parasite communities: a stranger in a strange land. *Parasitology* **109**: 187–196.

Kiesecker, J.M., Skelly, D.K., Beard, K.H., Preisser, E. (1999). Behavioral reduction of infection risk. *Proceedings of the National Academy of Sciences USA* **96**: 9165–9168.

King, G.M., Kirchman, D., Salyers, A.A., Schlesinger, W., Tiedje, J.M. (2001). *Global environmental change. Microbial contributions, microbial solutions*. American Society for Microbiology, Washington, DC.

Kirchner, J.W., Roy, B.A. (1999). The evolutionary advantages of dying young: epidemiological implications of longevity in metapopulations. *The American Naturalist* **154**: 140–159.

Kirchner, J.W., Roy, B.A. (2001). Evolutionary implications of host–pathogen specificity: the fitness consequences of host life history traits. *Evolutionary Ecology* **14**: 665–692.

Kirkpatrick, M. (1982). Sexual selection and the evolution of female choice. *Evolution* **36**: 1–12.

Kirkpatrick, M. (1996). Good genes and direct selection in the evolution of mating preferences. *Evolution* **50**: 2125–2140.

Kirkpatrick, M., Ryan, M.J. (1991). The evolution of mating preferences and the paradox of the lek. *Nature* **350**: 33–38.

Klasing, K.C. (1998). Nutritional modulation of resistance to infectious diseases. *Poultry Science* **77**: 1119–1125.

Klein, J., Tichy, H., Figueroa, F. (1987). On the origin of mice. *Annal Universidad de Chile* **5**: 91–120.

Knudsen, R., Gabler, H.-M., Kuris, A. M., Amudsen, P.-A. (2001). Selective predation on parasitized prey—a comparison between two helminth species with different life-history strategies. *Journal of Parasitology* **87**: 941–945.

Kodric-Brown, A. (1985). Female preference and sexual selection for male coloration in the guppy (*Poecilia reticulata*). *Behavioral Ecology and Sociobiology* **17**: 199–206.

Kodric-Brown, A. (1989). Dietary carotenoids and male mating success in the guppy: an environmental component to female choice. *Behavioral Ecology and Sociobiology* **25**: 393–401.

Koella, J.C. (2000). Coevolution of parasite life cycles and host life-histories. In: *Evolutionary biology of host-parasite relationships. Theory meets reality* (ed. R. Poulin, S. Morand, A. Skorping), pp. 185–200. Elsevier, London.

Koella, J.C., Restif, O. (2001). Coevolution of host virulence and host life history. *Ecology Letters* **4**: 207–214.

Kotiaho, J.S., Simmons, L.W., Thompkins, J.L. (2001). Towards a resolution of the lek paradox. *Nature* **410**: 684–686.

Kraaijeveld, A.R., Godfray, H.C.J. (1999). Geographic patterns in the evolution of resistance and virulence in *Drosophila* and its parasitoids. *The American Naturalist* **153**: S61–S74.

Krist, A.C. (2001). Variation in fecundity among populations of snails is predicted by prevalence of castrating parasites. *Evolutionary Ecology Research* **3**: 191–197.

Krist, A.C. (2006). Prevalence of parasites does not predict age at first reproduction or reproductive output in the freshwater snail, *Helisoma anceps*. *Evolutionary Ecology Research* **8**: 753–763.

Krist, A.C., Lively, C.M. (1998). Experimental exposure of juvenile snails (*Potamopyrgus antipodarum*) to infection by trematode larvae (*Microphallus* sp.): infectivity, fecundity compensation and growth. *Oecologia* **116**: 575–582.

Ksiazek, A., Konarzewski, M., Chadzinska, M., Chicon, M. (2003). Costs of immune response in cold-stressed laboratory mice selected for high and low basal metabolism rates. *Proceedings of the Royal Society B: Biological Sciences* **270**: 2025–2031.

Kulkarni, S. and Heeb, P. (2007). Social and sexual behaviours aid transmission of bacteria in birds. *Behavioural Processes* **74**: 88–92.

Lafferty, K.D. (1993). The marine snail, *Cerithidea californica*, matures at smaller sizes where parasitism is high. *Oikos* **68**: 3–11.

Lafferty, K.D., Kuris, A.M. (2005). Parasitism and environmental disturbances. In: *Parasitism and ecosystems* (ed. F. Thomas, F. Renaud, J.F. Guégan), pp. 113–123. Oxford University Press, Oxford.

Lafferty, K.D., Morris, K. (1996). Altered behavior of parasitized killifish increases susceptibility to predation by bird final hosts. *Ecology* **77**: 1390–1397.

Lafuma, L., Lambrechts, M.M., Raymond, M. (2001). Aromatic plants in bird nests as a protection against blood-sucking flying insects? *Behavioral Processes* **56**: 113–120.

Lagrue, C., Kaldonski, N., Perrot-Minnot, M.-J., Motreuil, S., Bollache, L. (2007). Modification of hosts' behavior by a parasite: field evidence for adaptive manipulation. *Ecology* **88**: 2839–2847.

Lambin, X., Krebs, C.J., Moss, R., Stenseth, N.C., Nigel, G. (1999). Population cycles and parasitism. *Science* **286**: 2425.

Landis, D., Wratten, S., Gurr, G.M. (2000). Habitat management to conserve natural enemies of arthropod pests in agriculture. *Annual Review of Entomology* **45**: 175–201.

Landry, C., Garant, D., Duchesne, P., Bernatchez, L. (2001). 'Good genes as heterozygosity': the major histocompatibility complex and mate choice in Atlantic salmon (*Salmo salar*). *Proceedings of the Royal Society B: Biological Sciences* **268**: 1279–1285.

Lane, S.D., Mills, N.J., Getz, W.M. (1999). The effects of parasitoid fecundity and host taxon on the biological control of insect pests: the relationship between theory and data. *Ecological Entomology* **24**: 181–190.

Lawrence, P.O. (1986). Host–parasite hormonal interactions: an overview. *Journal of Insect Physiology* **13**: 295–298.

Lawton, J.H. (2000). *Community ecology in a changing world*. International Ecology Institute, Luhe, Germany.

Lebel, J. (2003). *In-focus: health. An ecosystem approach*. IDRC Publications, Ottawa.

Le Brun, N., Renaud, F., Berrebi, P., Lambert, A. (1992). Hybrid zones and host–parasite relationships: effect on the evolution of parasitic specificity. *Evolution* **46**: 56–61.

Lederberg, J. (2000). Infectious history. *Science* **288**: 287–293.

Lefcort, H., Blaustein, A.R. (1991). Parasite load and brightness in lizards: an interspecific test of the Hamilton–Zuk hypothesis. *Journal of Zoology* **224**: 491–499.

Lefèvre, T., Koella, J.C., Renaud, F., Hurd, H., Biron, D.G., Thomas, F. (2006). New prospects for research on manipulation of insect vectors by pathogens. *PLoS Pathogens* **2**: 633–635.

Lefèvre, T., Thomas, F., Ravel, S., Patrel, D., Renault, L., Le Bourligu, L., Cuny, G., Biron, D.G. (2008). *Trypanosoma brucei brucei* induces alteration in the head proteome of the tsetse fly vector *Glossina palpalis gambiensis*. *Insect Molecular Biology* **16**: 651–660.

Lengagne, T., Grolet, O., Joly, P. (2006). Male mating speed promote hybridization in the *Rana lessonae–Rana esculenta* waterfrog system. *Behavioural Ecology and Sociobiology* **60**: 123–130.

van Lenteren, J.C. (1997). From *Homo economicus* to *Homo ecologicus*: towards environmentally safe pest control. In: *Modern agriculture and the environment* (ed. D. Rosen, E. Tel-or, Y. Hadar, Y. Chen), pp. 17–31. Kluwer Academic Publishers, Dordrecht.

van Lenteren, J.C. (2000). A greenhouse without pesticides: fact or fantasy? *Crop Protection* **19**: 375–384.

van Lenteren, J.C. (2003). *Quality control and production of biological control agents: theory and testing procedures*. CABI Publishing, Wallingford.

van Lenteren, J.C., Bueno, V.H.P. (2003). Augmentative biological control of arthropods in Latin America. *BioControl* **48**: 123–139.

van Lenteren, J.C., Bale, J., Bigler, F., Hokkanen, H.T.M., Loomans, A.J.M. (2006). Assessing risks of releasing exotic biological control agents of arthropods pests. *Annual Review of Entomology* **51**: 609–634.

Leon-Vizcaino, L., de Ybanez, M.R.R., Cubero, M.J., Ortiz, J.M., Espinosa, J., Perez, L. *et al.* (1999). Sarcoptic mange in Spanish ibex from Spain. *Journal of Wildlife Diseases* **35**: 647–659.

Li, K.S., Guan, Y., Wan, J., Smith, G.J., Xu, K.M., Duan, L., Rahardjo, A.P. *et al.* (2004). Genesis of a highly pathogenic and potentially pandemic H5N1 influenza virus in eastern Asia. *Nature* **430**: 209–213.

Lilley, B., Lammie, P., Dickerson, J., Eberhard, M. (1997). An increase in hookworm infection temporally associated with ecologic change. *Emerging Infectious Diseases* **3**: 391–393.

Lindström K., Lundström J. (2000). Male greenfinches (*Carduelis chloris*) with brighter ornaments have higher virus infection clearance rate. *Behavioral Ecology and Sociobiology* **48**: 44–51.

Lion, S., van Baalen, M., Wilson, W.G. (2006). The evolution of parasite manipulation of host dispersal. *Proceedings of the Royal Society B: Biological Sciences* **273**: 1063–1071.

Lipatov, A.S., Govorkova, E.A., Webby, R.J., Ozaki, H., Peiris, M., Guan, Y., Poon, L., Webster, R.G. (2004). Influenza: emergence and control. *Journal of Virology* **78**: 8951–8959.

Li-Ying, L. (1994). Worldwide use of *Trichogramma* for biological control on different crops: a survey. In: *Biological control with egg parasitoids* (ed. E. Wajnberg, S.A. Hassan). pp. 37–53. CABI Publishing, Wallingford.

Lochmiller, R.L., Deerenberg, C. (2000). Trade-offs in evolutionary immunology: just what is the cost of immunity? *Oikos* **88**: 87–98.

Lochmiller R.L., Vestey M.R., Boren J.C. (1993). Relationship between protein nutritional status and immunocompetence in northern bobwhite chicks. *The Auk* **110**: 503–510.

Loehle, C. (1995). Social barriers to pathogen transmission in wild animal populations. *Ecology* **76**: 326–335.

Loiseau, C., Zoorob, R., Garnier, S., Birard, J., Federici, P., Julliard, R., Sorci, G. (2008). Antagonistic effects of a MHC class I allele on malaria-infected house sparrows. *Ecology Letters* **11**: 258–265.

Louda, S.M., Pemberton, R.W., Johnson, M.T., Follett, P.A. (2003). Nontarget effects: the Achilles' heel of biological control? *Annual Review of Entomology* **48**: 365–396.

Loyau, A., Saint Jalme, M., Cagniant, C., Sorci, G. (2005). Multiple sexual advertisements honestly reflect health status in peacocks (*Pavo cristatus*). *Behavioral Ecology and Sociobiology* **58**: 552–557.

Loye, J.E., Caroll, S.P. (1991). Nest ectoparasite abundance and cliff swallow colony site selection, nestling development, and departure time. In: *Bird–parasite interactions. Ecology, behaviour and evolution* (ed. J.E. Loye, M. Zuk), pp. 222–241. Oxford University Press, Oxford.

Lozano G.A. (1994). Carotenoids, parasites, and sexual selection. *Oikos* **70**: 309–311.

Lozano, G.A. (1998). Parasitic stress and self-medication in wild animals. *Advances in the Study of Behaviour* **27**: 291–317.

Lozano G.A. (2001). Carotenoids, immunity, and sexual selection: comparing apples and oranges? *The American Naturalist* **158**: 200–203.

Lunt, N., Hulley, P.E., Craig, A.J.F.K. (2004). Active anting in captive Cape white-eyes *Zosterops pallidus*. *Ibis* **146**: 360–362.

Lynch, M. (1984). Destabilizing hybridization, general-purpose genotypes and geographic parthenogenesis. *Quarterly Review of Biology* **59**: 257–290.

McCallum, H., Dobson, A. (1995) Detecting disease and parasite threats to endangered species and ecosystems. *Trends in Ecology and Evolution* **10**: 190–194.

McCallum, H.I., Singleton, G.R. (1989). Models to assess the potential of *Capillaria hepatica* to control outbreaks of house mice. *Parasitology* **98**: 425–437.

McCarthy, H.O., Fitzpatrick, S., Irwin, S.W.B. (2000). A transmissible trematode affects the direction and rhythm of movement in a marine gastropod. *Animal Behaviour* **59**: 1161–1166.

McCurdy, D.G., Shutler, D., Mullie, A., Forbes, M.R. (1998). Sex-biased parasitism of avian hosts: relations to blood parasite taxon and mating system. *Oikos* **82**: 303–312.

McCurdy, D.G., Boates, J.S., Forbes, M.R. (2001). An empirical model of the optimal timing of reproduction for female amphipods infected by trematodes. *Journal of Parasitology* **87**: 24–30.

McDevitt, H.O. (1998). The role of MHC class II molecules in susceptibility and resistance to autoimmunity. *Current Opinion in Immunology* **10**: 677–681.

McGraw, K.J., Ardia, D.R. (2003). Carotenoids, immunocompetence, and the information content of sexual colors: an experimental test. *The American Naturalist* **162**: 704–712.

McMichael, A.J. (2004). Environmental and social influences on emerging infectious diseases: past, present and future. *Philosophical Transactions of the Royal Society B: Biological Sciences* **359**: 1049–1058.

Markusson, E., Folstad, I. (1997). Reindeer antlers: visual indicators of individual quality? *Oecologia* **110**: 501–507.

Martin, L.B., Scheuerlein, A., Wikelski M. (2002). Immune activity elevates energy expenditure of house sparrows: a link between direct and indirect costs? *Proceedings of the Royal Society B: Biological Sciences* **270**: 153–158.

Martin, L.B., Han, P., Lewittes, J., Kuhlman, J.R., Klasing, K.C., Wikelski, M. (2006). Phytohemagglutinin-induced skin swelling in birds: histological support for a classic immunoecological technique. *Functional Ecology* **20**: 290–299.

Martin, L.B., Weil, Z.M., Nelson, R.J. (2008). Seasonal trade-offs between reproduction and immune activity. *Philosophical Transactions of the Royal Society B: Biological Sciences* **363**: 321–339.

Martins, E.P. (1996). *Phylogenies and the comparative method in animal behaviour*. Oxford University Press, Oxford.

Mason, J.R., Clark, L. (1990). Sarcosporidiosis observed more frequently in hybrids of mallards and American black ducks. *The Wilson Bulletin* **102**: 160–162.

Mauck R.A., Matson, K.D., Philipsborn, J., Ricklefs, R.E. (2005). Increase in the constitutive innate humoral immune system in Leach's storm-petrel (*Oceanodroma leucorhoa*) chicks is negatively correlated with growth rate. *Functional Ecology* **19**: 1001–1007.

May, R.M. (1978). Host-parasitoid systems in patchy environments: a phenomenological model. *Journal of Animal Ecology* **47**: 833–843.

May, R.M., Anderson, R.M. (1978). Regulation and stability of host–parasite population interactions II. Destabilizing processes. *Journal of Animal Ecology* **47**: 249–267.

Mayhew, P.J., Blackburn, T.M. (1999). Does development mode organize life-history traits in the parasitoid Hymenoptera? *Journal of Animal Ecology* **68**: 906–916.

Maynard, B.J., DeMartini, L., Wright, W.G. (1996). *Gammarus lacustris* harboring *Polymorphus paradoxus* show altered patterns of serotonin-like immunoreactivity. *Journal of Parasitology* **82**: 663–666.

Maynard-Smith, J. (1982). *Evolution and the theory of games*. Cambridge University Press, Cambridge.

Maynard-Smith, J. (1985). Sexual selection, handicaps and true fitness. *Journal of Theoretical Biology* **115**: 1–8.

Mayr, E. (1963). *Animal species and evolution*. The Belknap Press of Harvard University Press, Cambridge, MA.

Mazzi, D. (2004). Parasites make male pipefish careless. *Journal of Evolutionary Biology* **17**: 519–527.

Mead-Briggs, A.R., Vaughan, J.A. (1975). The differential transmissibility of myxoma virus strains of differing virulence grades by the rabbit flea *Spilopsyllus cuniculi* (Dale). *Journal of Hygiene* **75**: 237–247.

de Meeûs, T., Renaud, F. (2002). Parasites within the new phylogeny of eukaryotes. *Trends in Parasitology* **18**: 247–251.

Michalakis, Y., Olivieri, I., Renaud, F., Raymond, M. (1992). Pleiotropic action of parasites: how to be good for the host. *Trends in Ecology and Evolution* **7**: 59–62.

Milinski, M. (1990). Parasites and host decision-making. In: *Parasitism and host behaviour* (ed. C.J. Barnard, J.M. Behnke), pp. 95–116. Taylor and Francis, London.

Milinski, M. (2003). The function of mate choice in sticklebacks: optimizing MHC genetics. *Journal of Fish Biology* **63** (Suppl. A): 1–16.

Milinski, M., Bakker, T.C.M. (1990). Female sticklebacks use male coloration in mate choice and hence avoid parasitized males. *Nature* **344**: 330–333.

Mills, N.J. (1997). Techniques to evaluate the efficiency of natural enemies. In: *Methods in ecological and agricultural entomology* (ed. D.R. Dent, M.P. Walton), pp. 271–291. CAB International, Wallingford.

Mills, N.J., Babendreier, D., Loomans, A.J.M. (2006) Methods for monitoring the dispersal of natural enemies from point source releases associated with augmentative biological control. In: *Environmental impact of arthropod biological control: methods and risk assessment* (ed. F. Bigler, D. Babendreier, U. Kuhlmann). CABI Bioscience, Wallingford, 114–131.

Minchella, D.J. (1985). Host life-history variation in response to parasitism. *Parasitology* **90**: 205–216.

Minchella, D.J., Loverde, P.T. (1981). A cost of increased early reproductive effort in the snail *Biomphalaria glabrata*. *The American Naturalist* **118**: 876–881.

Møller, A.P. (1990a). Effects of hematophagous mite on the barn swallow (*Hirundo rustica*): a test of the Hamilton and Zuk hypothesis. *Evolution* **44**: 771–784.

Møller, A.P. (1990b). Parasites and sexual selection : current status of the Hamilton and Zuk hypothesis. *Journal of Evolutionary Biology* **3**: 319–328.

Møller, A.P. (1994). *Sexual selection and the barn swallow*. Oxford University Press, Oxford.

Møller, A.P., Martinelli, R., Saino, N. (2004). Genetic variation in infestation with a directly transmitted ectoparasite. *Journal of Evolutionary Biology* **17**: 41–47.

Moore, J. (1983). Responses of an avian predator and its isopod prey to an acanthocephalan parasite. *Ecology* **64**: 1000–1015.

Moore, J. (2002). *Parasites and the behavior of animals*. Oxford University Press, Oxford.

Moore, J., Gotelli, N.J. (1996). Evolutionary patterns of altered behavior and susceptibility in parasitized hosts. *Evolution* **50**: 807–819.

Moore, S.L., Wilson, K. (2002). Parasites as a viability cost of sexual selection in natural populations of mammals. *Science* **297**: 2015–2018.

Moore, W.S. (1977). An evaluation of narrow hybrid zones in vertebrates. *Quarterly Review of Biology* **52**: 263–278.

Moore, W.S. (1984). Evolutionary ecology of unisexual fishes. In: *Evolutionary genetics of fishes* (ed. B.J. Turner), pp. 329–398. Plenum Press, New York.

Mooring, M.S., Hart, B.L. (1992). Animal grouping for protection from parasites: selfish herd and encounter-dilution effects. *Behaviour* **123**: 173–193.

Mooring, M.S., Blumstein, D.T., Stoner, C.J. (2004). The evolution of parasite-defense grooming in ungulates. *Biological Journal of the Linnean Society* **81**: 17–37.

Morand, S., Poulin, R. (2000). Optimal time to patency in parasitic nematodes: host mortality matters. *Ecology Letters* **3**: 186–190.

Morange, M. (2005). *Les secrets du vivant. Contre la pensée unique en biologie*. Editions de la Découverte, Paris.

Moret, Y., Schmid-Hempel, P. (2000). Survival for immunity: the price of activation for bumblebee workers. *Science* **290**: 1166–1167.

Morse, S.S. (2004). Factors and determinants of disease emergence. *Revue Scientifique et Technique de l'Office International des Epizooties* **23**: 443–451.

Mouchet, J. (1984). L'onchocercose, ou cécité des rivières. *Revue du Palais de la Découverte* **4**: 15–31.

Mougeot, F., Redpath, S.M., Piertney, S.B., Hudson, P.J. (2005). Separating behavioural and physiological mechanisms in testosterone-mediated trade-offs. *The American Naturalist* **166**: 158–168.

Moulia, C. (1999). Parasitism of plant and animal hybrids—are facts and fates the same? *Ecology* **80**: 392–406.

Moulia, C., Aussel, J.P., Bonhomme, F., Boursot, P., Nielsen, J.T., Renaud, F. (1991). Wormy mice in a hybrid zone, a genetic control of susceptibility to parasite infection. *Journal of Evolutionary Biology* **4**: 679–687.

Moulia, C., Le Brun, N., Dallas, J., Orth, A., Renaud, F. (1993). Experimental evidence of genetic determinism in high susceptibility to intestinal pinworm infection in mice : a hybrid zone model. *Parasitology* **106**: 387–393.

Mouritsen, K.N., Poulin, R. (2002). Parasitism, community structure and biodiversity in intertidal ecosystems. *Parasitology* **124**: S101–S117.

Mouritsen, K.N., Poulin, R. (2003). Parasite-induced trophic facilitation exploited by a non-host predator: a manipulator's nightmare. *International Journal for Parasitology* **33**: 1043–1050.

Mouritsen, K.N., Poulin, R. (2005). Parasites boosts biodiversity and changes animal community structure by trait-mediated indirect effects. *Oikos* **108**: 344–350.

Müller, C.B. (1994). Parasitoid induced digging behaviour in bumblebee workers. *Animal Behaviour* **48**: 961–966.

Muller, H.J. (1942). Isolating mechanism, evolution and temperature. *Biological Symposium* **6**: 71–125.

Murray, D.L. (2002). Differential body condition and vulnerability to predation in snowshoe hares. *Journal of Animal Ecology* **71**: 614–625.

Murray, J.F. (1998). Tuberculosis and HIV infection. A global perspective. *Respiration* **65**: 335–342.

Nacher, M. (2002). Worms and malaria: noisy nuisances and silent benefits. *Parasite Immunology* **24**: 391–393.

Nicholson, A.J., Bailey, V.A. (1935). The balance of animal population. *Proceedings of the Zoological Society of London* **3**: 551–598.

Nisbet, R.M., Gurnet, W.S.C. (1982). *Modelling fluctuating populations*. John Wiley and Sons, Chichester.

Noldus, L.P.J.J. (1989). *Chemical espionage by parasitic wasps*. PhD, University of Wageningen, The Netherlands.

Van Noordwijk, A.J., De Jong, G. (1986). Acquisition and allocation of resources : their influence on variation in life-history tactics. *The American Naturalist* **128**: 137–142.

Normak, B.J. (2003). The evolution of alternative genetic systems in insects. *Annual Review of Entomology* **48**: 397–423.

Norris K., Evans M.R. (2000). Ecological immunology: life-history trade-offs and immune defense in birds. *Behavioral Ecology* **11**: 19–26.

Nosil, P. (2002). Transition rates between specialization and generalization in phytophagous insects. *Evolution* **56**: 1701–1706.

Nowak, M.A., Tarczy-Hornoch, K., Austyn, J.M. (1992). The optimal number of major histocompatibility complex molecules in an individual. *Proceedings of the National Academy of Sciences USA* **89**: 10896–10899.

Nunn, C.L. (2003). Behavioural defences against sexually transmitted diseases in primates. *Animal Behaviour* **66**: 37–48.

Nunn, C.L., Altizer, S. (2006). *Infectious diseases in primates. behavior, ecology and evolution.* Oxford University Press, Oxford.

Nunn, C.L., Heymann, E. W. (2005). Malaria infection and host behavior: a comparative study of neotropical primates. *Behavioural Ecology and Sociobiology* **59**: 30–37.

Nunn, C.L., Altizer, S., Jones, K.E., Sechrest, W. (2003). comparative tests of parasite species richness in primates. *The American Naturalist* **162**: 597–614.

Nürnberger, B., Barton, N., Maccallum, C., Gilchrist, J., Appleby, M. (1995). Natural selection on quantitative traits in the *Bombina* hybrid zone. *Evolution* **49**: 1224–1238.

O'Brien, S.J., Roelke, M.E., Marker, L., Newman, A., Winkler, C.A., Meltzer, D. *et al.* (1985). Genetic basis for species vulnerability in the cheetah. *Science* **227**: 1428–1434.

Ogielska, M. (1994). Nucleus-like bodies in gonial cells of *Rana esculenta* (Amphibia, Anura) tadpoles—a putative way of chromosome elimination. *Zoologica Poloniae* **39**: 461–474.

Ogutu-Ohwayo, R. (1990). The impact of native fishes of Lake Victoria and Kyoga (east Africa) and the impact of introduced species, especially the Nile perch., *Lates niloticus* and the Nile tilapia, *Oreochromis niloticus*. *Environmental Biology Fisheries* **27**: 81–86.

OILB-SROP (1973). Statuts. *Bulletin SROP* **1973/1**: 1–25.

Olson, V.A., Owens, I.P.F. (1998). Costly sexual signals: are carotenoids rare, risky or required? *Trends in Ecology and Evolution* **13**: 510–514.

Oppliger, A., Richner, H., Christe, P. (1994). Effect of an ectoparasite on lay date, nest-site choice, desertion, and hatching success in the great tit (*Parus major*). *Behavioral Ecology* **5**: 130–134.

Orr, H.A. (1995). The population genetics of speciation: the evolution of hybrid incompatibilities. *Genetics* **139**: 1805–1813.

Ostfeld, R.S., Keesing, F. (2000a). Biodiversity and disease risk: the case of Lyme disease. *Conservation Biology* **14**: 722–728.

Ostfeld, R.S., Keesing, F. (2000b). The function of biodiversity in the ecology of vector-borne zoonotic diseases. *Canadian Journal of Zoology* **78**: 2061–2078.

Ots, I., Kerimov, A.B., Ivankine, E.V., Ilyina, T.A., Hõrak, P. (2001). Immune challenge affects basal metabolic activity in wintering great tits. *Proceedings of the Royal Society B: Biological Sciences* **268**: 1175–1181.

Ouedraogo, R.M., Goettel, M.S., Brodeur, J. (2004). Behavioral thermoregulation in the migratory locust: a therapy to overcome fungal infection. *Oecologia* **138**: 312–319.

Overli, O., Pall, M., Borg, B., Jobling, M., Winberg, S. (2001). Effects of *Schistocephalus solidus* infection on brain monoaminergic activity in female three-spined sticklebacks *Gasterosteus aculeatus*. *Proceedings of the Royal Society B: Biological Sciences* **268**: 1411–1415.

Owens I.P.F. (2002). Sex differences in mortality rate. *Science* **297**: 2008–2009.

Owens I.P.F., Wilson K. (1999). Immunocompetence: a neglected life history trait or conspicuous red herring? *Trends in Ecology and Evolution* **14**: 170–172.

Pablos-Mendez A., Raviglione M.C., Lazlo A., Binkin, N., Rieder, H.L., Bustreo, F., Cohn, D.L., Lambregts-van Weezenbeek, C.S.B., Jae Kim, S., Chaulet, P., Nunn, P. (1998). Global surveillance for antituberculosis resistance 1994–1997. *New England Journal of Medicine* **338**: 1641–1649.

Packer, C., Holt, R.D., Hudson, P.J., Lafferty, K.D., Dobson, A.P. (2003). Keeping the herds healthy and alert: implication of predator control for infectious disease. *Ecology Letters* **6**: 797–802.

Palmer, C.A., Edmands, S. (2000). Mate choice in the face of both inbreeding and outbreeding depression in the intertidal copepod *Tigriopus californicus*. *Marine Biology* **136**: 693–698.

Park, T. (1948). Experimental studies of interspecies competition. I. Competition between populations of the flour beetles, *Tribolium confusum* Duval and *Tribolium castaneum* Herbst. *Ecological Monographs* **18**: 265–308.

Parmentier, H.K., Abuzeid, S.Y., Reilingh, G.D., Nieuwland, M.G.B., Graat, E.A.M. (2001). Immune responses and resistance to *Eimeria acervulina* of chickens divergently selected for antibody responses to sheep red blood cells. *Poultry Science* **80**: 894–900.

Patz, J., Confalonieri, U. (2005). Human health: infectious and parasitic diseases. In: *Millennium ecosystem assessment: conditions and trends*. Island Press, Washington DC.

Patz J., Confalonieri, U. (coordinators) *et al.* (2005). Human health: ecosystem regulation of infectious diseases. In: *Ecosystems and human well-being: current state and trends: findings of the Condition and Trends Working Group*, Millennium Ecosystem Assessment Series, pp. 393–415. Island Press, Washington, DC.

Pennacchio, F., Strand, M.R. (2006). Evolution of developmental strategies in parasitic Hymenoptera. *Annual Review of Entomology* **51**: 233–258.

Perrin, N., Christe, P., Richner, H. (1996). On host life-history response to parasitism. *Oikos* **75**: 317–321.

Perrot-Minnot, M.-J., Kaldonski, N., Cézilly, F. (2007). Susceptibility to predation and anti-predator behaviour in an acanthocephalan-infected amphipod: a test for the manipulation hypothesis. *International Journal for Parasitology* **37**: 645–651.

Peters, A., Delhey, K., Denk, A.G., Kempenaers, B. (2004). Trade-offs between immune investment and sexual signaling in male mallards. *The American Naturalist* **164**: 51–59.

Plantegenest, M., Outreman, Y., Goubault, M., Wajnberg, E. (2004). Parasitoids flip a coin before deciding to superparasitize. *Journal of Animal Ecology* **73**: 802–806.

Plénet, S., Pagano, A., Joly, P., Fouillet, P. (2000). Variation of plastic responses to oxygen availability within the hybridogenetic *Rana esculenta* complex. *Journal of Evolutionary Biology* **13**: 20–29.

Plénet, S., Joly, P., Hervant, F., Fromont, E., Grolet, O. (2005). Are hybridogenetic zones structured by environmental gradients? *In situ* experiments in the water-frog hybridization complex. *Journal of Evolutionary Biology* **18**: 1575–1586.

Plowright, W. (1982). The effect of rinderpest and rinderpest control on wildlife in Africa. *Symposia of the Zoological Society of London* **50**: 1–28.

Polak, M., Starmer, W.T. (1998). Parasite-induced risk of mortality elevates reproductive effort in male *Drosophila*. *Proceedings of the Royal Society B: Biological Sciences* **265**: 2197–2201.

Pomiankowski, A., Iwasa, Y., Nee, S. (1991). The evolution of costly mate preferences I. Fisher and biased mutation. *Evolution* **45**: 1422–1430.

Pontier, D., Fromont, E., Courchamp, F., Artois, M., Yoccoz, N.G. (1998). Retroviruses and sexual size dimorphism in domestic cats (*Felis catus* L.). *Proceedings of the Royal Society B: Biological Sciences* **265**: 167–173.

Ponton, F., Biron, D.G., Joly, C., Helluy, S., Duneau, D., Thomas, F. (2005). Ecology of parasitically modified populations: a case study from a gammarid–trematode system. *Marine Ecology Progress Series* **299**: 205–215.

Ponton, F., Lefèvre, T., Lebarbenchon, C., Thomas, F., Loxdale, H. D., Marche, L., Renault, L., Perrot-Minnot, M.J., Biron, D.G. (2006). Do distantly related parasites rely on the same proximate factors to alter the behaviour of their hosts? *Proceedings of the Royal Society B: Biological Sciences* **273**: 2869–2877.

Porcherie, A., Boumiza, R., Loubes, C., Dornand, J., Moulia, C. (2003). Modalités des interactions hote-parasite chez la souris domestique: première analyse de l'aspect immunologique de la résistance/sensibilité à *Aspiculuris tetraptera*. *Les Actes du BRG* **4**: 505–518.

Potts, W.K., Manning, C.J., Wakeland, E.K. (1991). Mating patterns in seminatural populations of mice influenced by MHC genotype. *Nature* **352**: 619–621.

Poulin, R. (1994). Mate choice decisions by parasitized female upland bullies, *Gobiomorphus breviceps*. *Proceedings of Royal Society B: Biological Sciences* **256**: 183–187.

Poulin, R. (1995). 'Adaptive' change in the behaviour of parasitized animals: a critical review. *International Journal for Parasitology* **25**: 1371–1383.

Poulin, R. (1996). Helminth growth in vertebrate hosts: does host sex matter? *International Journal for Parasitology* **26**: 1311–1315.

Poulin, R. (1998). *Evolutionary ecology of parasites: from individuals to communities*. Chapman and Hall, London.

Poulin, R. (1999). The functional importance of parasites in animal communities: many roles at many levels? *International Journal for Parasitology* **29**: 903–914.

Poulin, R., Fitzgerald, G.J. (1989). Shoaling as an anti-ectoparasite mechanism in juvenile stickelbacks (*Gasterosteus* spp.). *Behavioral Ecology and Sociobiology* **24**: 251–255.

Poulin, R., Morand, S. (2004). *Parasite biodiversity*. Smithsonian Books, Washington, DC.

Poulin, R., Mouillot, D. (2003). Host introductions and the geography of parasite taxonomic diversity. *Journal of Biogeography* **30**: 837–845.

Poulin, R., Vickery, W.L. (1993) Parasite-mediated sexual selection: just how choosy are parasitized females? *Behavioral Ecology and Sociobiology* **38**: 43–49.

Poulin, R., Nichol, K., Latham, A.D. (2003). Host sharing and host manipulation by larval helminths in shore crabs: cooperation or conflict? *International Journal for Parasitology* **33**: 425–433.

Pounds, J.A., Fogden, P.L., Savage, J.M., Gorman, G.C. (1997). Test of null models for amphibian declines on a tropical mountain. *Conservation Biology* **11**: 1307–1322.

Prince, J.S., Leblanc, W.G., Macia, S. (2004). Design and analysis of multiple choice feeding preference data. *Oecologia* **138**: 1–4.

Pruett-Jones, S.G., Pruett-Jones, M.A., Jones, H.I. (1990). Parasites and sexual selection in birds of parasites. *American Zoologist* **30**: 287–298.

Prugnolle, F., Manica, A., Charpentier, M., Guégan, J-F., Guernier, V., Balloux, F. (2005). Pathogen-driven selection and worldwide HLA class I diversity. *Current Biology* **15**: 1022–1027.

Pusey, A., Wolf, M. (1996). Inbreeding avoidance in animals. *Trends in Ecology and Evolution* **11**: 201–206.

Quicke, D.L.J. (1997). *Parasitic wasps*. Chapman and Hall, London.

Quinnell, R.J., Grafen, A., Woolhouse, M.E.J. (1995). Changes in parasite aggregation with age: A discrete infection model. *Parasitology* **111**: 635–644.

Råberg, L., Nilsson, J.A., Ilmonen, P., Stjernman, M., Hasselquist, D. (2000). The cost of an immune response: vaccination reduces parental effort. *Ecology Letters* **3**: 382–386.

Råberg, L., Vestberg, M., Hasselquist, D., Holdahl, R., Svensson, E., Nilsson, J.A. (2002). Basal metabolic rate and the evolution of the adaptive immune system. *Proceedings of the Royal Society B: Biological Sciences* **269**: 817–821.

Råberg, L., Stjernman, M., Hasselquist, D. (2003). Immune responsiveness in adult blue tits: heritability and effects of nutritional status during ontogeny. *Oecologia* **136**: 360–364.

Ralley, W.E., Gallaway, T.D., Crow, G.H. (1993). Individual and group behaviour of pastured cattle in response to attack by biting flies. *Canadian Journal of Zoology* **71**: 725–734.

Rantala, M.J., Koskimaki, J., Taskinen, J., Tynkkynen, K., Suhonen, J. (2000). Immunocompetence, developmental stability and wingspot size in the damselfly *Calopteryx splendens* L. *Proceedings of the Royal Society B: Biological Sciences* **267**: 2453–2457.

Rapport, D.J., Lee, V. (2003). Ecosystem approaches to human health: some observations on North/South experiences. *Ecosystem Health* **3**: 26–39.

Raufaste, N., Orth, A., Belkhir, K., Senet, D., Smadja, C., Baird, S.J.E., Bonhomme, F., Dod, B., Boursot, P. (2005) Inferences of selection and migration in the Danish house mouse hybrid zone. *Biological Journal of the Linnean Society* **84**: 447–459.

Rauque, C.A., Viozzi, G.P., Semenas, L.G. (2003). Component population study of *Acanthocephalus tumescens* (Acanthocephala) in fishes from Lake Moreno, Argentina. *Folia Parasitologica* **50**: 72–78.

Read, A.F. (1987). Comparative evidence supports the Hamilton and Zuk hypothesis on parasites and sexual selection. *Nature* **328**: 68–70.

Read, A.F., Harvey, P.H. (1989). Reassessment of the comparative evidence for the Hamilton and Zuk theory on the evolution of secondary sexual characters. *Nature* **339**: 618–619.

Renaud, F., de Meeüs, T., Read, A.F. (2005). Parasitism in man-made ecosystems. In: *Parasitism and ecosystems* (ed. F. Thomas, F. Renaud, J.-F. Guégan), pp. 155–176. Oxford University Press, Oxford.

Restif, O., Hochberg, M.E., Koella, J.C. (2001). Virulence and age at reproduction: new insights into host–parasite coevolution. *Journal of Evolutionary Biology* **14**: 967–979.

Reusch, T.B.H., Haberli, M.A., Aeschlimann, P.B., Milinski, M. (2001). Female sticklebacks count alleles in a strategy of sexual selection explaining MHC polymorphism. *Nature* **414**: 300–302.

Reynolds, D.G. (1970). Laboratory studies of the microsporidian *Plistophora culicis* (Weiser) infecting *Culex pipiens fatigans* Wied. *Bulletin of Entomological Research* **60**: 339–349.

Reznick, D.A., Bryga, H., Endler, J.A. (1990). Experimentally induced life-history evolution in a natural population. *Nature* **346**: 357–359.

Richner, H. (1998). Host-parasite interactions and life-history evolution. *Zoology—Analysis of Complex Systems* **101**: 333–344.

Richner, H., Tripet, F. (1999). Ectoparasitism and the trade-off between current and future reproduction. *Oikos* **86**: 535–538.

Richner, H., Christe, P., Oppliger, A. (1995). Paternal investment affects prevalence of malaria. *Proceedings of the National Academy of Sciences USA* **92**: 1192–1194.

Rieseberg, L.H. (1997). Hybrid origins of plant species. *Annual Review of Ecology and Systematics* **28**: 359–389.

Rieseberg, L.H., Archer, M.A., Wayne, R.K. (1999). Transgressive segregation, adaptation and speciation. *Heredity* **83**: 363–372.

van Riper, C., van Riper, S.G., Goff, M.L., Laird, M. (1986). The epizootiology and ecological significance of malaria in Hawaiian land birds. *Ecological Monographs* **56**: 327–344.

van Riper, C., van Riper, S.G., Hansen, W.R. (2002). Epizootiology and effect of avian pox on Hawaiian forest birds. *The Auk* **119**: 929–942.

Ris, N., Allemand, R., Fouillet, P., Fleury, F. (2004). The joint effect of temperature and host species induce complex genotype-by-environment interactions in the larval parasitoid of *Drosophila*, *Leptopilina heterotoma* (Hymenoptera: Figitidae). *Oikos* **106**: 451–456.

Roberts, M.I., Buchanan, K.L., Evans, M.R. (2004). Testing the immunocompetence handicap hypothesis: a review of the evidence. *Animal Behaviour* **68**: 227–239.

Rodó, X., Pascual, M., Fuchs, G., Faruque, A.S.G. (2002). ENSO and cholera: a nonstationary link related to climate change? *Proceedings of the National Academy of Sciences USA* **99**: 12901–12906.

Roff, D.A. (1992). *The evolution of life histories: theory and analysis*. Chapman and Hall, London.

Roff, D.A. (1997). *Evolutionary quantitative genetics*. Chapman and Hall, London.

Rohani, P., Earn, D.J., Grenfell, B.T. (1999). Opposite patterns of synchrony in sympatric disease metapopulations. *Science* **286**: 968–971.

Rohde, K. (2005). *Nonequilibrium theory*. Cambridge University Press, Cambridge.

Roitt, I., Brostoff, J., Male, D (1998). *Immunology*, 4th edn. Mosby, London.

Rolff, J., Siva-Jothy, M.T. (2003). Invertebrate ecological immunology. *Science* **301**: 472–475.

Ros, A.F.H, Groothuis, T.G.G., Apanius, V. (1997). The relation among gonadal steroids, immunocompetence, body mass and behavior in young black-headed gulls (*Larus ridibundus*). *The American Naturalist* **150**: 201–219.

Rosa, R., Pugliese, A. (2002). Aggregation, stability, and oscillations in different models for host–macroparasite interactions. *Theoretical Population Biology* **61**: 319–334.

de Rosnay, J. (1975). *Le macroscope. Vers une vision globale*. Points, Seuil, Paris.

Rotschild, M. (1973). Remarks on carotenoids in the evolution of signals. In: *Coevolution in animals and plants* (ed. L.E. Gilberts, P.H Raven), pp. 20–51. University of Texas Press, Austin, TX.

Roulin, A., Heeb, P. (1999). The immunological function of allosuckling. *Ecology Letters* **2**: 319–324.

Rousset, F., Thomas, F., de Meeüs, T., Renaud, F. (1996). Inference of parasite-induced host mortality from distributions of parasite loads. *Ecology* **77**: 2203–2211.

Rowe, L., Houle, D. (1996). The lek paradox and the capture of genetic variance by condition dependent traits. *Proceedings of the Royal Society B: Biological Sciences* **263**: 1415–1421.

Rumi, A., Hamann, M.I. (1990). Potential schistosome-vector snails and associated trematodes in ricefields of Corrientes province, Argentina. Preliminary results. *Memórias do Instituto Oswaldo Cruz* **85**: 321–328.

Rushton, S.P., Lurz, P.W.W., Gurnell, J., Fuller, R. (2000). Modelling the spatial dynamics of parapoxvirus disease in red and grey squirrels: a possible cause of the decline in the red squirrel in the UK? *Journal of Applied Ecology* **37**: 997–1012.

Ryder, J.J., Siva-Jothy, M.T. (2000). Male calling song provides a reliable signal of immune function in a cricket. *Proceedings of the Royal Society B: Biological Sciences* **267**: 1171–1175.

Sage, R.D., Heyneman, D., Lim, K.C., Wilson, A.C. (1986). Wormy mice in a hybrid zone. *Nature* **324**: 60–63.

Saino, N., Stradi, R., Ninni, P., Pini, E., Møller, A.P. (1999). Carotenoid plasma concentration, immune profile, and plumage ornamentation of male barn swallows (*Hirundo rustica*). *The American Naturalist* **154**: 441–448.

Salvador, A., Veiga, J.P., Martin, J., Lopez, P. (1997). Testosterone supplementation in subordinate, small male lizards: consequences for aggressiveness, color development, and parasite load. *Behavioral Ecology* **8**: 135–139.

Sasal, P., Durand, P., Faliex, E., Morand, S. (2000). Experimental approach to the importance of parasitism in biological conservation. *Marine Ecology Progress Series* **198**: 293–302.

Schade, R., Calzado, E.G., Sarmiento, R., Chacana, P.A., Porankiewicz-Asplund, J., Terzolo, R. (2005). Chicken egg yolk antibodies (IgY-technology): a review of progress in production and use in research and human and veterinary medicine. *ATLA* **33**: 1–26.

Schall, J.J., Dearing, M.D. (1987). Malarial parasitism and male competition for mates in the western fence lizard, *Scelopus occidentalis*. *Oecologia* **73**: 389–392.

Schallig, H.D., Sassen, M.J., Hordijk, P.L., De Jong-Brink, M. (1991). *Trichobilharzia ocellata*: influence of infection on the fecundity of its intermediate snail host *Lymnaea stagnalis* and cercarial induction of the release of schistosomin, a snail neuropeptide antagonizing female gonadotropic hormones. *Parasitology* **102**: 85–91.

Schmid-Hempel, P. (2003). Variation in immune defense as a question of evolutionary ecology. *Proceedings of the Royal Society B: Biological Sciences* **270**: 357–366.

Schmidt, J.M., Smith, J.J.B. (1986). Correlations between body angles and substrate curvature on the parasitoid wasp *Trichogramma minutum*: a possible mechanism of host radius measurements. *Journal of Experimental Biology* **125**: 271–285.

Schmitz, O.J., Nudds, T.D. (1994). Parasite-mediated competition in deer and moose: how strong is the effect of meningeal worm on moose? *Ecological Applications* **4**: 91–103.

Schultz, R.J. (1969). Hybridization, unisexuality, and polyploidy in the teleost *Poeciliopis* (Poeciliidae) and other vertebrates. *The American Naturalist* **103**: 605–619.

Scott, M.E. (1987). Temporal changes in aggregation: a laboratory study. *Parasitology* **94**: 583–595.

Scott, M.E., Dobson, A.P. (1989). The role of parasites in regulating host abundance. *Parasitology Today* **5**: 176–183.

Seivwright, L.J., Redpath, S.M., Mougeot, F., Leckie, F., Hudson P.J. (2005). Interactions between intrinsic and extrinsic mechanisms in a cyclic species: testosterone increases parasite infection in red grouse. *Proceedings of the Royal Society B: Biological Sciences* **272**: 2299–2304.

Semlitsch, R.D. (1993). Asymmetric competition in mixed populations of tadpoles of the hybridogenetic *Rana esculenta* complex. *Evolution* **47**: 510–519.

Semlitsch, R.D., Reyer, H.-U. (1992). Performance of tadpoles from the hybridogenetic *Rana esculenta* complex;
interactions with pond drying and interspecific competition. *Evolution* **46**: 665–676.

Semlitsch, R.D., Hotz, H., Guex, G.D. (1997). Competition among tadpoles of coexisting hemiclones of hybridogenetic *Rana esculenta*: support for the frozen niche variation model. *Evolution* **51**: 1249–1261.

Seppälä, A., Karvonen, A., Valtonen, E.T. (2004). Parasite-induced change in host behaviour and susceptibility to predation in an eye fluke–fish interaction. *Animal Behaviour* **68**: 257–263.

Seppälä, A., Karvonen, A., Valtonen, E.T. (2006). Susceptibility of eye fluke-infected fish to predation by bird hosts. *Parasitology* **132**: 575–579.

Settle, W.H., Wilson, L.T. (1990). Invasion by the variegated leafhopper and biotic interactions, parasitism, competition and apparent competition. *Ecology* **71**: 1461–1470.

Shaw, D.J., Dobson, A.P. (1995). Patterns of macroparasite abundance and aggregation in wildlife populations: a quantitative review. *Parasitology* **111**: S111–S133.

Sheldon, B.C., Verhulst, S. (1996). Ecological immunology: costly parasite defences and trade-offs in evolutionary ecology. *Trends in Ecology and Evolution* **11**: 317–321.

Shykoff, J.A., Widmer, A. (1996). Parasites and carotenoid based signal intensity: how general should the relationship be? *Naturwissenschaften* **83**: 113–121.

Singleton, G.R., Chambers, L.K. (1996). A manipulative field experiment to examine the effect of *Capillaria hepatica* (Nematoda) on wild mouse populations in southern Australia. *International Journal for Parasitology* **26**: 383–398.

Singleton, G.R., McCallum, H.I. (1990). The potential of *Capillaria hepatica* to control mouse plagues. *Parasitology Today* **6**: 190–193.

Singleton, G.R., Spratt, D.M. (1986). The effects of *Capillaria hepatica* (Nematoda) on natality and survival to weaning in BALB/c mice. *Australian Journal of Zoology* **34**: 687–681.

Siva-Jothy, M.T. (2000). A mechanistic link between parasite resistance and expression of a sexually selected trait in a damselfly. *Proceedings of the Royal Society B: Biological Sciences* **267**: 2523–2527.

Skarstein, F., Folstad, I. (1996). Sexual dichromatism and the immunocomptence handicap: an observational approach. *Oikos* **76**: 359–367.

Smadja, C., Ganem, G. (2002). Subspecies recognition in the house mouse: a study of two populations from the border of a hybrid zone. *Behavioral Ecology* **13**: 312–320.

Smith, K.F., Sax, D.F., Gaines, S.D., Guernier, V., Guégan, J.-F. (2007). Global homogenization of human infectious disease. *Ecology* **88**: 1903–1910.

Smith Trail, D.R. (1980). Behavioral interactions between parasites and hosts: host suicide and the evolution of complex life cycles. *The American Naturalist* **116**: 77–91.

Solberg, E.J., Jordhøy, P., Strand, O., Aanes, R., Loison, A., Sæther B.-E., Linnell, J.D.C. (2001). Effects of density-dependence and climate on the dynamics of a Svalbard reindeer population. *Ecography* **24**: 441–451.

Soler, J.J., de Neve, L., Perez-Contreras, T., Soler, M., Sorci, G. (2003a). Trade-off between immunocompetence and growth in magpies: an experimental study. *Proceedings of the Royal Society B: Biological Sciences* **270**: 241–248.

Soler, J.J., Moreno, J., Potti, J. (2003b). Environmental, genetic and maternal components of immunocompetence of nestling pied flycatchers from a cross-fostering study. *Evolutionary Ecology Research* **5**: 259–272.

Sorci, G., Massot, M., Clobert, J. (1994). Maternal parasite load increases sprint speed and philopatry in female offspring of the common lizard. *The American Naturalist* **144**: 153–164.

Sorci, G., Clobert, J., Michalakis, Y. (1996). Cost of reproduction and cost of parasitism in the common lizard, *Lacerta vivipara*. *Oikos* **76**: 121–130.

Sorci, G., Møller, A.P., Boulinier, T. (1997a). Genetics of host–parasite interactions. *Trends in Ecology and Evolution* **12**: 196–199.

Sorci, G., Soler, J.J., Moller, A.P. (1997b). Reduced immunocompetence of nestlings in replacement clutches of the European magpie (*Pica pica*). *Proceedings of the Royal Society B: Biological Sciences* **264**: 1593–1598.

Sorensen, R.E., Minchella, D.J. (1998). Parasite influences on host life history: *Echinostoma revolutum* parasitism of *Lymnea elodes* snails. *Oecologia* **115**: 188–195.

Speelmon, E.C., Checkley, W., Gilman, R.H., Patz, J., Calderon, M., Manga, S. (2000). Cholera incidence and El Niño-related higher ambient temperature. *Journal of the American Medical Association* **283**: 3072–3074.

Spencer, K.A., Buchanan, K.L., Leitner, S., Goldsmith, A.R., Catchpole, C.K. (2005). Parasites affect song complexity and neural development in a songbird. *Proceedings of the Royal Society B: Biological Sciences* **272**: 2037–2043.

Spielman, A., Andreadis, T.G., Apperson, C.S. *et al.* (2004). Outbreak of West Nile virus in North America. *Science* **306**: 1473–1475.

Spurrier, M.F., Boyce, M.S., Manly, B.F.J. (1991). Effects of parasites on mate choice by captive sage grouse, *Centrocercus urophasianus*. In: *Bird–parasite interactions: ecology, evolution and behavior* (ed. J.E. Loye, M. Zuk), pp. 389–398. Oxford University Press, Oxford.

Staszewski, V., Boulinier, T. (2004). Vaccination: a way to address questions in behavioural and population ecology? *Trends in Parasitology* **20**: 17–22.

Staszewski, V., Gasparini, J., McCoy, K., Tveraa, T., Boulinier, T. (2007). Evidence of an interannual effect of maternal immunization on the immune response of juveniles in a long-lived colonial bird. *Journal of Animal Ecology* **76**: 1215–1223.

Stearns, S.C. (1992). *The evolution of life histories*. Oxford University Press, Oxford.

Stien, A., Irvine, R.J., Ropstad, R., Halvorsen, O., Langvatn, R., Albon, S.D. (2002). The impact of gastrointestinal nematodes on wild reindeer: experimental and cross-sectional studies. *Journal of Animal Ecology* **71**: 937–945.

Stiling, P., Cornelissen, T. (2005). What makes a successful biocontrol agent? A meta-analysis of biological control agent performance. *Biological Control* **34**: 236–246.

Stireman, J.O., O'Hara, J.E., Wood, M.D. (2006). Tachinidae: evolution, behavior, and ecology. *Annual Review of Entomology* **51**: 525–555.

Stow, A.J., Sunnucks, P. (2004). Inbreeding avoidance in Cunningham's skinks (*Egernia cunninghami*) in natural and fragmented habitat. *Molecular Ecology* **13**: 631–639.

Strand, M.R., Pech, L.L. (1995). Immunological basis for compatibility in parasitoid-host relationships. *Annual Review of Entomology* **40**: 31–56.

Strauss, S.Y. (1994). Levels of herbivory and parasitism in host hybrid zones. *Trends in Ecology and Evolution* **9**: 209–214.

Suarez, O.V., Cueto, G.R., Cavia, R., Gomez Villafane, I.E., Bilenca, D.N., Edelstein, A., Martinez, P., Miguel, S., Bellomo, C., Hodara, K., Padula, P.J., Busch, M. (2003). Prevalence of infection with hantavirus in rodent populations of central Argentina. *Memorias do Instituto Oswaldo Cruz* **98**: 727–732.

Sunish, I.P., Reuben, R. (2001). Factors influencing the abundance of Japanese encephalitis vectors in ricefields in India—I. Abiotic. *Medical and Veterinary Entomology* **15**: 381–392.

Svensson, E., Råberg, L., Koch, C., Hasselquist, D. (1998). Energetic stress, immunosuppression and the cost of antibody response. *Functional Ecology* **12**: 912–919.

Tadei, W.P., Thatcher, B.D., Santos, J.M.M., Scarpassa, V.M., Rodrigues, I.B., Rafael, M.S. (1998). Ecologic observations on anopheline vectors of malaria in the Brazilian Amazon. *American Journal of Tropical Medicine and Hygiene* **59**: 325–335.

Tain, L., Perrot-Minnot, M.-J., Cézilly, F. (2006). Altered host behaviour and brain serotonergic activity caused by acanthocephalans: evidence for specificity. *Proceedings of the Royal Society B: Biological Sciences* **273**: 3039–3045.

Tella, J.L. (2002). The evolutionary transition to coloniality promotes higher blood parasitism in birds. *Journal of Evolutionary Biology* **15**: 32–41.

Tella, J.L., Scheuerlein, A., Ricklefs, R.E. (2002). Is cell-mediated immunity related to the evolution of life-history strategies in birds? *Proceedings of the Royal Society B: Biological Sciences* **269**: 1059–1066.

Terenina, N.B., Asatryan, A.M., Movsesyan, S.O. (1997). Content of serotonin in brain and other tissues of rats with experimental trichinellosis. *Doklady Akademii nauk SSSR* **335**: 412–413.

Thomas, F., Mete, K., Helluy, S., Santalla, F., Verneau, O., de Meeüs, Y., Cézilly, F., Renaud, F. (1997). Hitch-hiker parasites or how to benefit from the strategy of another parasite. *Evolution* **51**: 1316–1318.

Thomas, F., Renaud, F., de Meeüs, T., Cézilly, F. (1995). Parasites, age and the Hamilton–Zuk hypothesis: inferential fallacy? *Oikos* **74**: 305–309.

Thomas, F., Renaud, F., Rousset, F., Cézilly, F. (1995). Differential mortality of two closely related host species induced by one parasite. *Proceedings of the Royal Society of London Series B - Biological Sciences* **260**: 349–352.

Thomas, F., Renaud, F., de Meeüs, T., Poulin, R. (1998). Manipulation of host behaviour by parasites : ecosystem engineering in the intertidal zone? *Proceeding of the Royal Society B: Biological Sciences* **265**: 1091–1096.

Thomas, F., Poulin, R., de Meeüs, T., Guégan, J-F., Renaud, F. (1999a). Parasites and ecosystem engineering: what roles could they play? *Oikos* **84**: 167–171.

Thomas, F., Oget, E., Gente, P., Desmots, D., Renaud, F. (1999b). Assortative pairing with respect to parasite load in the beetle *Timarcha maritima* (Chrysomelidae). *Journal of Evolutionary Biology* **12**: 385–390.

Thomas, F., Moore, J., Adamo, S. (2005). Parasitic manipulation: where are we and where should we go? *Behavioural Processes* **68**: 185–199.

Thomas, F., Fauchier, J., Lafferty, K.D. (2002). Conflict of interest between a nematode and a trematode in an amphipod host: test of the 'sabotage' hypothesis. *Behavioral Ecology and Sociobiology* **51**: 296–301.

Thompson, S.N. (1999). Nutrition and culture of entomophagous insects. *Annual Review of Entomology* **44**: 561–592.

Thorne, E.T., Williams, E.S. (1988). Disease and endangered species: the black-footed ferret as a recent example. *Conservation Biology* **2**: 66–74.

Thornhill, J.A., Jones, J.T., Kusel, J.R. (1986). Increased oviposition and growth in immature *Biomphalaria glabrata* after exposure to *Schistosoma mansoni*. *Parasitology* **93**: 443–450.

Tisdell, C.A. (1990). Economic impact of biological control of weeds and insects. In: *Critical issues in biological control* (ed. M. Mackauer, L.E. Ehler, J. Rolland), pp. 301–316. Intercept, Andover.

Toft, C.A., Aeschlimann, A., Bolis, L. (1991). *Parasite–host associations: coexistence or conflict?* Oxford University Press, Oxford.

Tompkins, D.M., Begon, M. (1999). Parasites regulate wildlife populations. *Parasitology Today* **15**: 311–313.

Tompkins, D.M., Poulin, R. (2006). Parasites and biological invasions. In: *Biological invasions in New Zealand* (ed. R.B. Allen, W.G. Lee),pp. 67–84. Springer-Verlag, Berlin.

Tompkins, D.M., Greenman, J.V., Robertson, P.A., Hudson, P.J. (2000). The role of shared parasites in the exclusion of wildlife hosts: *Heterakis gallinarum* in the ring-necked pheasant and the grey partridge. *Journal of Applied Ecology* **37**: 997–1012.

Tompkins, D.M., Dobson, A.P., Arneber, P., Begon, M., Cattadori, I.M., Greenman, J. V., Heesterbeek, J.A.P., Hudson, P.J., Newborn, D., Pugliese, D., Rizzoli, A., Rosa, R., Rosso, F., Wilson, K. (2001). Parasites and host population dynamics. In: *Wildlife diseases* (ed. P.J. Hudson, A. Rizzoli, B.T. Grenfell, H. Heesterbeek, A.P. Dobson), pp. 45–62. Oxford University Press, London.

Tompkins, D.M., Sainsbury, A.W., Nettleton, P., Buxton, D., Gurnell, J. (2002). Parapoxvirus causes a deleterious disease in red squirrels associated with UK population declines. *Proceedings of the Royal Society B: Biological Sciences* **269**: 529–533.

Tompkins, D.M., White, A.R., Boots, M. (2003). Ecological replacement of native red squirrels by invasive greys driven by disease. *Ecology Letters* **6**: 189–196.

Torchin, M.E., Lafferty, K.D., Kuris, A.M. (2002). Parasite and marine invasions. *Parasitology* **124**: S137–S151.

Torchin, M.E., Lafferty, K.D., Dobson, A.P., McKenzie, V.J., Kuris, A.M. (2003). Introduced species and their missing parasites. *Nature* **421**: 628–630.

Tripet, F., Richner, H. (1997). Host responses to ectoparasites: food compensation by parent blue tits. *Oikos* **78**: 557–561.

Tripet, F., Richner, H. (1999). Density-dependent processes in the population dynamics of a bird ectoparasite *Ceratophyllus gallinae*. *Ecology* **80**: 1267–1277.

Trivers, R.L. (1972). Parental investment and sexual selection. In: *Sexual selection* (ed. B. Campbell), pp. 136–207. Aldine Publishing Company, Chicago, IL.

Tunner, H.G. (1974). Die klonale struktur einer wasserfroschpopulation. *Zeitschrift für Zoologische Systematik und Evolutionsforshung* **12**: 309–314.

Tunner, H.G., Nopp, H. (1979). Heterosis in the common European waterfrog. *Naturwissenschaften* **66**: 268–269.

Turchin, P. (2003). *Complex population dynamics: a theoretical/empirical synthesis*. Princeton University Press, Princeton, NJ.

Turlings, T.C.J., Tumlinson, J.H., Lewis, W.J. (1990). Exploitation of herbivore-induced plant odors by host-seeking parasitic wasps. *Science* **250**: 1251–1253.

Valleron, A.J., Boëlle, P.Y., Will, R., Cesbron, J.Y. (2001). Estimation of epidemic size and incubation time based on age characteristics of vCJD in the United Kingdom. *Science* **294**: 1726–1728.

Verhulst, S., Dieleman, S.J., Parmentier, H.K. (1999). A trade off between immunocompetence and sexual ornamentation in domestic fowl. *Proceedings of the National Academy of Sciences USA* **96**: 4478–4481.

Viboud, C., Pakdaman, K., Boëlle, P.-Y., Wilson, M.L., Myers, M.F., Valleron, A.-J., Flahault, A. (2004). Association of influenza epidemics with global climate variability. *European Journal of Epidemiology* **19**: 1055–1059.

Viney, M.E., Riley, E., Buchanan, K.L. (2005). Optimal immune responses: immunocompetence revisited. *Trends in Ecology and Evolution* **20**: 665–669.

Vinson, S.B. (1975). Biochemical coevolution between parasitoids and their hosts. In: *Evolutionary strategies of parasitic insects and mites* (ed. P.W. Price), pp. 14–48. Plenum, New York.

Vinson, S.B. (1976). Host selection by insect parasitoids. *Annual Review of Entomology* **21**: 109–133.

Vinson, S.B. (1981). Habitat location. In: *Semiochemicals: their role in pest control* (ed. D.A. Nordlund, R.L. Jones, W.J. Lewis), pp. 51–77. John Wiley, New York.

Vinson, S.B., Iwantsch, G.F. (1980a). Host suitability for insect parasitoids. *Annual Review of Entomology* **25**: 397–419.

Vinson, S.B., Iwantsch, G.F. (1980b). Host regulation by insect parasitoids. *Quarterly Review of Biology* **55**: 143–165.

Vinson, S.B., Jones, R.L., Sonnet, P.E., Bierl, B.A., Beroza, M. (1975). Isolation, identification and synthesis of host-seeking stimulant for *Cardiochiles nigriceps* a parasitoid of tobacco budworm. *Entomologia Experimentalis et Applicata* **18**: 443–450.

Vitazkova, S.K., Long, E., Paul, A., Glendinning, J.I. (2001). Mice suppress malaria infection by sampling a 'bitter' chemotherapy agent. *Animal Behaviour* **61**: 887–894.

Vitousek, P.M., Dantonio, C.M., Loope, L.L., Rejmanek, M., Westbrooks, R. (1997). Introduced species: a significant component of human-caused global change. *New Zealand Journal of Ecology* **21**: 1–16.

Vittor, A.Y., Gilman, R.H., Tielsch, J., Glass, G., Shields, T., Lozano, W.S., Pinedo-Cancino, V., Patz, J.A. (2006). The effect of deforestation on the human-biting rate of *Anopheles darlingi*, the primary vector of falciparum malaria in the Peruvian Amazon. *The American Society of Tropical Medicine and Hygiene* **74**: 3–11.

Vorburger, C. (2001). Heterozygous fitness effects of clonally transmitted genomes in waterfrogs. *Journal of Evolutionary Biology* **14**: 602–610.

Vorišek, P., Votýpka, J., Zvára, K., Svobodová, M. (1998). Heteroxenous coccidians increase the predation risk of parasitized rodents. *Parasitology* **117**: 521–524.

Waage, J.K. (1990). Ecological theory and the selection of biological control agents. In: *Critical issues in biological control* (ed. M. Mackauer, L.E. Ehler, J. Roland), pp. 135–157. Intercept, Andover.

Wäckers, F.L. (2003). The parasitoids' need for sweets: sugars in mass rearing and biological control. In: *Quality control and production of biological control agents: theory and testing procedures* (ed. J.C. van Lenteren), pp. 59–72. CABI Publishing, Wallingford.

Wajnberg, E. (2004). Measuring genetic variation in natural enemies used for biological control: why and how? In: *Genetics, evolution and biological control* (ed. L.E. Ehler, R. Sforza, T. Mateille), pp. 19–37. CABI Publishing, Wallingford.

Wajnberg, E., Hassan, S.A. (1994). *Biological control with egg parasitoids*. CABI Publishing, Wallingford.

Wajnberg, E., Scott J.K., Quimby, P.C. (2001). *Evaluating indirect ecological effects of biological control*. CAB International, Wallingford.

Waltz, A.M., Whitham, G.T. (1997). Plant development effects arthropod communities: opposing impacts of species removal. *Ecology* **78**: 2133–2144.

Wapshere, A.J. (1974). A strategy for evaluating the safety of organisms for biological weed control. *Annals of Applied Biology* **77**: 201–211.

Ward, P.I. (1988). Sexual dichromatism and parasitism in British and Irish freshwater fish. *Animal Behaviour* **36**: 1210–1215.

Ward, P.I. (1989). Sexual showiness and parasitism in freshwater fish: combined data from several isolated water systems. *Oikos* **55**: 428–439.

Watts, D.J., Strogatz, S.H. (2004). Collective dynamics of 'small-world' networks. *Nature* **393**: 440–442.

Webster, J.P., Gowtage-Sequeira, S., Berdoy, M., Hurd, H. (2000). Predation of beetles (*Tenebrio molitor*) infected with tapeworms (*Hymenolepis diminuta*): a note of caution for the manipulation hypothesis. *Parasitology* **120**: 313–318.

Wedekind, C., Seebeck, T., Bettens, F., Paepke, A.J. (1995). MHC-dependent mate preferences in humans. *Proceedings of the Royal Society B: Biological Sciences* **260**: 245–249.

Wedekind, C., Walker, M., Little, T.J. (2005). The course of malaria in mice: major histocompatibility complex (MHC) effects, but no general MHC heteroygote advantage in single-strain infections. *Genetics* **170**: 1427–1430.

Wegner, K.M., Kalbe, M., Kuntz, J., Reusch, T.B.H., Milinski, M. (2003). Parasite selection for immunogenetic optimality. *Science* **301**: 1343.

Wheelwright, N.T., Freeman-Gallant, C.R., Mauck, R.A. (2006). Asymmetrical incest avoidance in the choice of social and genetic mates. *Animal Behaviour* **71**: 631–639.

White, P.C.L., Newton-Cross, G.A., Gray, M., Ashford, R., White, C., Saunders, G. (2003). Spatial interactions and habitat use of rabbits on pasture and implications for the spread of rabbit haemorrhagic disease in New South Wales. *Wildlife Research* **30**: 49–58.

Whitham, G.T. (1989). Plant hybrid zones as sinks for pests. *Science* **244**: 1490–1493.

Whitham, G.T., Morrow, P.A., Potts, B.M. (1994). Plant hybrid zones as centers of biodiversity: the herbivore community of two endemic Tasmanian eucalypts. *Oecologia* **97**: 481–490.

Whitham, G.T., Martinsen, G.D., Floate, K.D., Dungey, H.S., Potts, B.M., Keim, P. (1999). Plant hybrid zones affect biodiversity: tools for genetic-based understanding of community structure. *Ecology* **80**: 416–428.

Williams, G.C. (1957). Pleiotropy, natural selection, and the evolution of senescence. *Evolution* **11**: 398–411.

Williams, T.D., Christians, J.K., Evanson, M. (1999). Enhanced immune function does not depress reproductive output. *Proceedings of the Royal Society B: Biological Sciences* **266**: 753–757.

Wilson, K., Bjornstad, O.N., Dobson, A.P., Merler, S., Poglayen, G., Randolph, S.E., Read, A.F., Skorping, A. (2001). Heterogeneities in macroparasite infections—patterns and processes. In: *The ecology of wildlife diseases* (ed. P.J. Hudson, A. Rizzoli, B.T. Grenfell, H. Heesterbeek, A.P. Dobson), pp. 6–44. Oxford University Press, Oxford.

Wolday, D., Mayaan, S., Mariam, Z.G., Berhe, N., Seboxa, T., Britton, S., Galai, N., Landay, A., Bentwich, Z. (2002). Treatment of intestinal worms is associated with decreased HIV plasma viral load. *Journal of Acquired Immune Deficiency Syndromes* **31**: 56–62.

Wolinska, J., Keller, B., Bittner, K., Lass, S., Spaak, P. (2004). Do parasites lower *Daphnia* hybrid fitness? *Limnology and Oceanography* **49**: 4.

Woolhouse, M.E.J. (1992). A theoretical framework for the immuno-epidemiology of helminth infection. *Parasite Immunology* **14**: 563–578.

Woolhouse, M.J.E. (1998). Patterns in parasite epidemiology: the peak shift. *Parasitology Today* **14**: 428–434.

Woolhouse, M.E., Dye, C., Etard, J.F., Smith, T., Charlwood, J.D., Garnett, G.P., Hagan, P. Hii, J.L., Ndhlovu, P.D., Quinnell, R.J., Watts, C.H., Chandiwana, S.K., Anderson, R.M. (1997). Heterogeneities in the transmission of infectious agent: implications for design of control programs. *Proceedings of the National Academy of Sciences USA* **94**: 338–342.

Worden, B.D., Parker, P.G. (2005). Females prefer non-infected males as mates in the grain beetle *Tenebrio molitor*: evidence in pre- and postcopulatory behaviours. *Animal Behaviour* **70**: 1047–1053.

Woyke, J., Wilde, J., Reddy, C.C. (2004). Open-air-nesting honey bees *Apis dorsata* and *Apis laboriosa* differ from the cavity-nesting *Apis mellifera* and *Apis cerana* in brood hygiene behaviour. *Journal of Invertebrate Pathology* **86**: 1–6.

Wright, A.N., Gompper, M.E. (2005). Altered parasite assemblages in raccoons in response to manipulated ressource availability. *Oecologia* **144**: 148–156.

Yan, G., Stevens, L., Goodnight, C.J., Schall, J.J. (1998). Effects of a tapeworm parasite on the competition of *Tribolium* beetles. *Ecology* **79**: 1093–1103.

Zahavi, A. (1975). Mate selection—a selection for a handicap. *Journal of Theoretical Biology* **53**: 205–214.

Zakhary, K. (1997). Factors affecting the prevalence of schistosomiasis in the Volta region of Ghana. *McGill Journal of Medicine* **3**: 93–101.

Zimmerman, K. (1949). Zur kenntnis der mitteleuropäischen hausmause. *Zoologisches Jahrbuch, Abteilung für Systematik* **28**: 301–322.

Zohar, A.S., Holmes, J.C. (1998). Pairing success of male *Gammarus lacustris* infected by two acanthocephalans: a comparative study. *Behavioral Ecology* **9**: 206–211.

Zuk M., Johnsen T.S. (1998). Seasonal changes in the relationship between ornamentation and immune response in red jungle fowl. *Proceeding of the Royal Society B: Biological Sciences* **265**: 1631–1635.

Zuk, M., Thornhill, R., Ligon J.D., Johnson, K. (1990). Parasites and mate choice in red jungle fowl. *American Zoologist* **30**: 235–244.

Zuri, I., Gazit, I., Terkel, J. (1997). Effect of scent-marking in delaying territorial invasion in the blind mole-rat *Spalax ehrenbergi*. *Behaviour* **134**: 867–880.

Index

Note: page numbers in *italics* refer to Figures, Tables and Boxes, whilst those in **bold** refer to Glossary entries.